Introduction to Relativity Volume II

$E = mc^2$ is known as the most famous but least understood equation in physics. This two-volume textbook illuminates this equation and much more through clear and detailed explanations, new demonstrations, a more physical approach, and a deep analysis of the concepts and postulates of Relativity.

Volume II progresses into further depth than Volume I, and its scope is more extended than most introductory books on Relativity. It contains:

- complementary explanations,
- alternative demonstrations relying on more advanced means and revealing other physical aspects,
- further topics, notably:
  - In Special Relativity: acceleration in different frames, nuclear reactions, the force in relativity, the use of hyperbolic trigonometry, the Lagrangian approach, Maxwell's equations.
  - In General Relativity: tensors, the affine connection, the covariant derivative, the geodesic equation, the Schwarzschild solution with two of its consequences: black holes and the bending of light, further axiomatic considerations.
  - In Cosmology: the FLRW Metric, the Friedman equation, the cosmological constant, the four ideal cosmological Models.

These subjects are presented in a concrete and incremental manner, and illustrated by many case studies. The emphasis is placed on the theoretical aspects, with rigorous demonstrations based on a minimum set of postulates. The mathematical tools dedicated to Relativity are carefully explained for those without an advanced mathematical background.

Both volumes place an emphasis on the physical aspects of Relativity to aid the reader's understanding and contain numerous questions and problems (147 in total). Solutions are given in a highly detailed manner to provide the maximum benefit to students.

Introduction to Relativity Volume II

In-Depth and Accessible

Paul Bruma

CRC Press
Taylor & Francis Group
Boca Raton London New York

CRC Press is an imprint of the
Taylor & Francis Group, an **informa** business

Cover image: Portrait of Albert Einstein at a writing desk, used with permission of the Albert Einstein Archives, The Hebrew University of Jerusalem. Colorization by Danièle Mémet and Paul Bruma.

First edition published 2022
by CRC Press
6000 Broken Sound Parkway NW, Suite 300, Boca Raton, FL 33487-2742

and by CRC Press
4 Park Square, Milton Park, Abingdon, Oxon, OX14 4RN

CRC Press is an imprint of Taylor & Francis Group, LLC

ISBN: 9781032056760 (hbk)
ISBN: 9781032062440 (pbk)
ISBN: 9781003201359 (ebk)

DOI: 10.1201/9781003201359

Typeset in Palatino
by codeMantra

Contents

Preface by Jean Iliopoulos, Dirac Prize 2007

The beginning of the last century witnessed two major revolutions in the physical sciences which changed profoundly our perception of space and time and our views on the structure of matter. They are known as the theory of Relativity, for the first, and quantum mechanics, for the second. They both have a well-deserved reputation of complexity, and this has scared people away. Only specialized professionals dared to study and understand them. Yet, both are omnipresent in our everyday lives. The tiny chips which form the heart of our smartphones could not be designed without quantum mechanics, and our GPS would be grossly inaccurate without the theory of Relativity. Maybe it is time to try to demystify these theories and make them accessible to those among the younger generation who are interested, well motivated, and have followed a good science course at high school.

This book tries to meet this challenge for the theory of Relativity. By "demystifying" I do not mean a kind of vulgarization which avoids the difficulties by offering more or less convincing plausibility arguments. The ambition of this book is to explain every point by presenting a rigorous and complete derivation. It starts from first principles, sets the axioms and develops the theory step by step in a fully deductive way. Every assumption is solidly anchored in experimental results.

The book contains two volumes. In the first, we find the general principles and the main results, first for the Special theory (Chapters 1 to 5) and then for the General theory of Relativity (Chapters 6 and 7). More advanced points are left for the second volume. The book is not "easy", and the reader should follow every step carefully and repeat the calculations, but the end result is highly rewarding. At the end he, or she, will have a very good working knowledge of this very beautiful theory. A great help is provided by wellchosen problems and exercises, which we find at the end of each chapter.

I strongly advise the reader to try hard to solve them. The solutions can be found at the end of each volume, but it would be a mistake to look there directly for the answers.

There exist many excellent books on the theory of Relativity, but, contrary to most of them, the author of this one does not require the reader to have a background in physics and mathematics beyond what one can reasonably expect from a good high school graduate. The result is amazing. Incredible as it may sound, the author wins his bet. He shows that the theory of Relativity is not "difficult". It is fully accessible to the kind of readership he set to meet.

It is the book I wish I had 65 years ago, when I finished high school.

J. ILIOPOULOS
Director of Research Emeritus
Ecole Normale Superieure
Paris

1

Steps before the Lorentz Transformation

Introduction

Firstly, Stellar Aberrations and the Michelson–Morley experiment will be described since these are the main elements that induced Einstein to reject the existence of ether. The principle of Relativity will then be examined in-depth, all the more as it is the cornerstone of the new theory. Several fundamental laws will be derived from this principle, including the identity of the proper time in all inertial frames and the inertial frame velocity reciprocity. The case of non-inertial frames will also be addressed, particularly the inertial forces. Then, methodological considerations will be presented as Relativity considerably changed the way of thinking in Physics, in particular, the possibility for different observers to say contradictory statements. A special focus will be given on intrinsic notions, as these play a very important role in Relativity. Then, the main properties leading to the Lorentz transformation will be presented in an in-depth manner: the transverse distance invariance, the linearity of the new coordinates transformation law and some important clarifications on the time dilatation.

1.1 Observations and Experiments That Helped Characterize the Ether

Before Relativity, the vast majority of the scientific community believed in the existence of the luminiferous ether, which was supposed to be a very light substance filling the universe and whose vibrations were responsible for the propagation of light, similarly as the air is responsible for the propagation of sounds. Einstein said that Stellar Aberrations were a main reason that induced him to reject the existence of ether; hence, we will first examine this phenomenon.

DOI: 10.1201/9781003201359-1

1.1.1 Stellar Aberrations

The astronomer James Bradley noticed in 1725 that the stars were not seen at exactly the same apparent position during the year, but were moving along very small circles, each one having an angular diameter of ~41 seconds. This strange phenomenon could be explained as follows: a telescope, fixed on Earth, receives the light rays coming from a star, but as our planet is moving around the Sun, it is also moving relative to the star, and the latter is assumed to be fixed relative to the Sun. Consequently, the observer and his telescope are moving relative to the rays coming from the observed star, as shown in Figure 1.1.

Let's now assume the Earth is at position 1, and an astronomer is pointing his telescope toward the star: The left part of Figure 1.2 shows that he will not see the star because the photon coming from the star will hit first the upper lens of his telescope, and then while it continues toward the astronomer's eye, the telescope and the astronomer are moving toward the left, so that when the photon terminates its trajectory, it is on the right of the telescope.

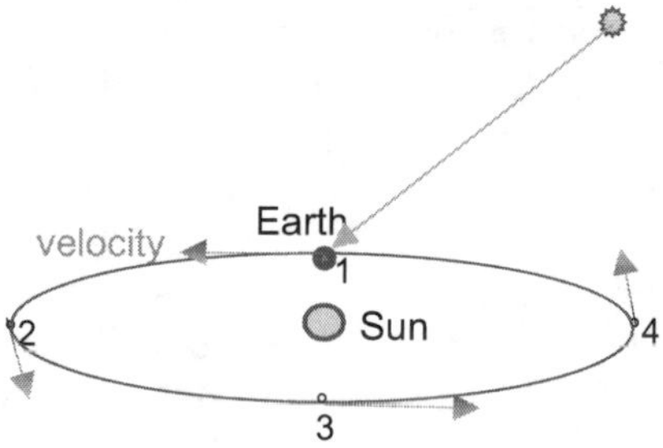

FIGURE 1.1
Photons emitted by a star and received on Earth at different times in a year.

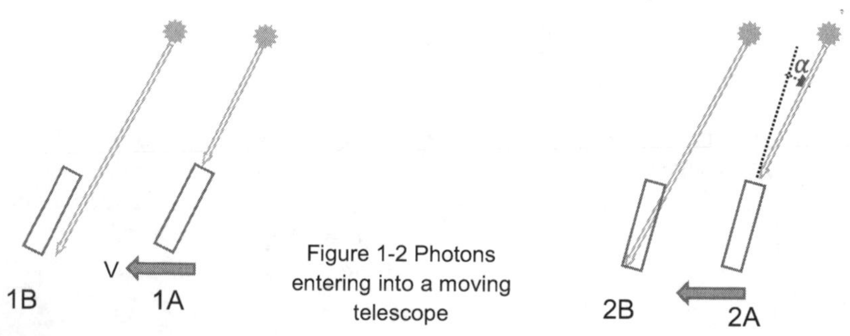

FIGURE 1.2
Photons entering a moving telescope.

To offset this effect, the telescope must point slightly above the star, as shown in the right part of Figure 1.2, making an angle α with the photon trajectory. This angle is such that during the time taken by the photon to continue its descent, the Earth and the telescope have moved the exact distance toward the left, which permits the photon to reach the astronomer's eye.

During the whole year, the Earth makes an elliptic trajectory around the Sun; hence, the direction of the speed of the Earth and the telescope relative to the star changes from 30 km/s in step 1 to –30 km/s in step 3. In step 3, the telescope must point in a lower direction than the actual direction of the star. Consequently, during the whole year, the astronomer sees the star making a small ellipse, and this is the stellar aberration phenomenon.

1.1.1.1 Calculations and Consequences

We will first assess the simpler case where the star is directly above the astronomer, and then make the exact calculation.

Let L be the telescope length and v the telescope velocity (= the Earth velocity) with respect to the star. We previously mentioned that v is about 30,000 m/s. The photon takes the time duration t = L/c to traverse the telescope. During this time t, the telescope has covered the distance vt. Hence, the angle α that the telescope must make with the vertical is such that: hence

$$\tan\alpha = \frac{v\,t}{L} = \frac{v\,t}{L} \approx 10^{-4};\ \text{hence: } \alpha \approx \tan^{-1}(10^{-4}) \approx 10^{-4}\ \text{rad.}$$

During the whole year, the star thus appears making a circle whose diameter is seen with an angle of 2×10^{-4}rad, which represents about 41 seconds of arc and which well matches with the observed results. It is very tiny, like seeing a soccer ball from a distance of 8.5 km.

Note that this effect is identical to the one you encounter when you are inside a moving car and it rains: you see the drops falling not vertically but along oblique lines.

This result shows that the ether cannot be fixed with the Earth; otherwise, the photon would be dragged by the ether (similarly as sound is dragged by the air, which moves at the same speed as the Earth), so that the photon would continue its trajectory inside the telescope and terminate in the eye of the astronomer. The astronomer would then not have to modify the angle of his telescope with respect to the star, meaning that there would be no stellar aberration.

1.1.1.2 Classical and Relativistic Calculations in the General Case

Let K be the frame of the star and K′ the frame of the astronomer on Earth. The O′Z′ axis is a vertical direction for the astronomer and the O′X′ horizontal one (Figure 1.3).

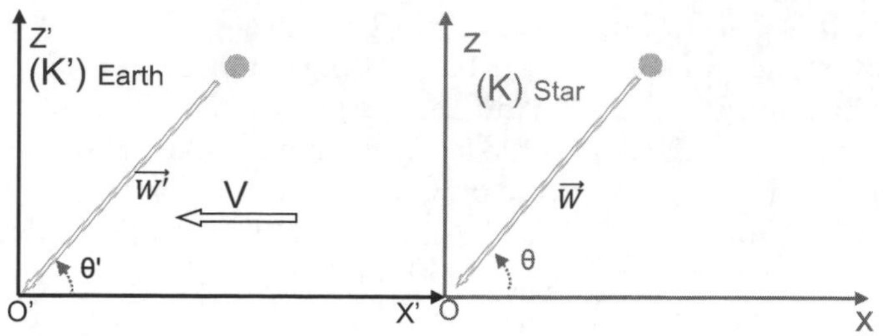

FIGURE 1.3
Photon trajectory seen from the Earth frame and from the Star frame.

The astronomer sees the photon in K′ arriving with a velocity $\overrightarrow{W'}$, which is the result of the composition of the photon velocity $\overrightarrow{W}$ in K with the velocity $\vec{V}$ of K′ relative to K.

In the Star frame K, the photon makes an angle θ with the horizontal, and we want to calculate the angle θ′ with which the astronomer sees the star. The classical velocity composition rule yields:

- The horizontal component of $\overrightarrow{W'}$ is: $W'_h = c\cos\theta - v$.
- The vertical component of $\overrightarrow{W'}$ one is: $W'_t = c\sin\theta$.

Hence: $\tan\theta' = \dfrac{\sin\theta}{\cos\theta - \beta}$, with $\beta = v/c \approx 10^{-4}$. With $\theta = 75°$, we obtain: $\theta' - \theta = 19.9$ seconds of arc. Consequently, the circle (ellipse) seen during the whole year has a diameter of 39.8 seconds of arc, which is very close to the observations' results.

With the relativistic velocity composition law, the theoretical result is extremely close to the classical one, the difference being $5 \times 10^{-7}\%$ as shown in Problem 1.1.

Remark 1.1: Stellar Aberration should not be confused with the effect of parallax, also due to the motion of the Earth around the Sun. The difference is that parallax is due to the displacement of the orbit and decreases as the distance to the star increases, while the stellar aberration is due to the velocity of the orbit and is independent of the distance to the star. The relative size of the two effects can be seen from the fact that the astronomer's unit of distance, parsec, is defined as the distance of a star that has one arc-sec of parallax. For objects within the solar system, parallax is much bigger, while for stars, the aberration is always bigger.

1.1.2 The Michelson–Morley Experiment

One conclusion of the Stellar Aberrations was the impossibility for the ether to be fixed relative to the Earth. Besides, the Fizeau experiment showed that the ether cannot be fully dragged by the atmosphere (cf. Section 1.5.1). The likely hypothesis was then that the ether was fixed relative to the Sun, or very partially dragged the Earth's atmosphere. Consequently, the ether was assumed to be moving relative to the Earth at a velocity in the range of 30 km/s, which continuously changes during the year, reaching 6 months later –30 km/s. This would generate an "ether wind" effect whereby light would propagate faster in the direction of the ether motion than in the opposite direction, similarly as the sound propagates faster in the direction of the wind. Michelson and Morley conducted their famous experiment in 1881–1887 aiming at measuring the ether speed relative to the Earth.

1.1.2.1 Presentation of the Michelson–Morley Experiment

A light beam is emitted by a source on Earth and is then split into two beams by a half-silvered mirror H in two perpendicular directions. One beam follows the same direction as the ether motion, and then reaches the mirror M1 where it is reflected and returns to H. The other beam follows a perpendicular direction relative to the ether motion, then reaches the mirror M2 where it is reflected, and returns to H. Both beams recombine at H and reach an interference detector close to the observer O, as shown in Figure 1.4. The distance d from the half silvered mirror H to the mirror M1 is the same as to the mirror M2. This distance d was 10 m.

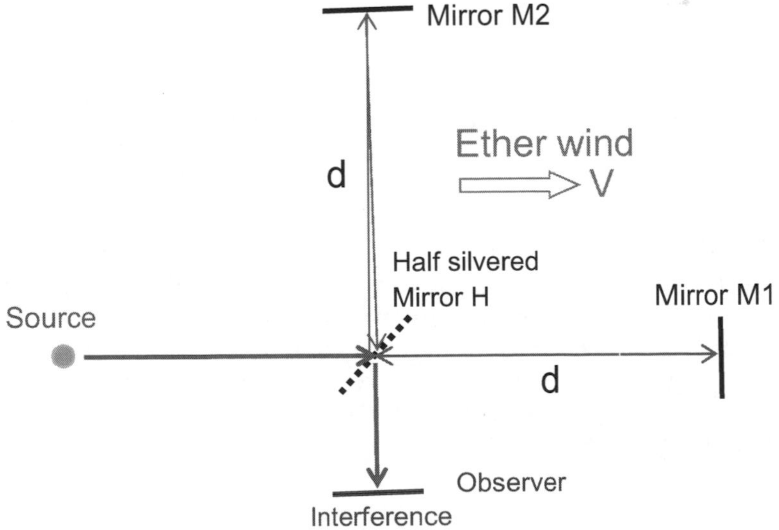

FIGURE 1.4
Michelson–Morley experiment.

If there is an ether wind, these two rays will not go exactly at the same speed; consequently, they will not remain in phase, which will result in a modification of their interference pattern. According to the calculations presented hereafter, this modification is approximately 0.2 fringe, which can be detected.

This experiment was repeated in different directions and different seasons of the year, but no shift in the interference pattern was ever detected. The conclusion is that light propagates at the same speed in all directions, which means that the ether, if it exists, has no influence on light propagation. Since then, many other experiments confirmed that light can propagate in the vacuum without requiring any medium: the ether doesn't exist.

Remark 1.2: It is a common mistake to say that Michelson–Morley's experiment proved that the speed of light is independent from the speed of its source. Both beams are in fact emitted from the same source. It is Maxwell's equations which, when applied in the context of no ether, imply the light speed invariance (independence of its source).

1.1.2.2 Calculations of the Fringe Shift

1.1.2.2.1 Time for the Path to the Mirror M1 (Back and Forth)

The time taken by the beam from the half-silvered mirror to the full-mirror M1 and then from this full-mirror M1 to the half-silvered mirror is: $T_1 = \frac{d}{c+v} + \frac{d}{c-v} = \frac{2d.c}{c^2 - v^2} = \frac{2d}{c\left(1 - v^2/c^2\right)}$. Then, as a first-order approximation, we have:

$$T_1 \approx \frac{2d}{c}\left(1 + v^2/c^2\right) \tag{1.1}$$

where v is the ether speed relative to the Earth, assumed to be: 3×10^4m/s.

1.1.2.2.2 Time for the Path to the Mirror M2 (Back and Forth)

The time taken by the beam from the half-silvered mirror to the full-mirror M2 and then from this full-mirror M2 back to the half-silvered mirror is somewhat more complex to calculate. The beam coming from the source hits the half-silvered mirror in O, and then one part of the beam goes toward the mirror M2. This beam does not go exactly perpendicularly toward M2 since it is dragged by the ether as shown in Figure 1.5, which amplifies this phenomenon for the sake of clarity.

We thus have:

$$T_2 = \frac{OM}{c} + \frac{O'M}{c}. \tag{1.2}$$

Then, as an approximation: $OM = OM'$ and $OO' = vT_2$. (1.3)

Note that this approximation does not seem legitimate when looking at Figure 1.5, but the latter does not reflect the real lengths of the distances

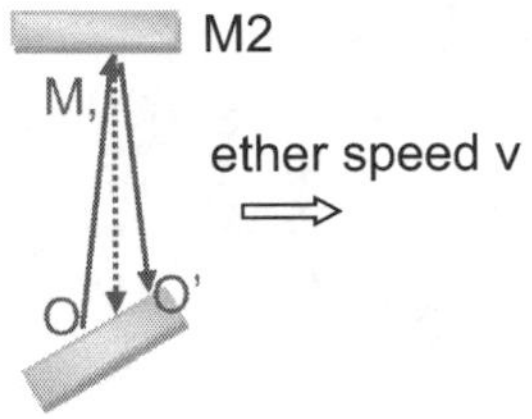

FIGURE 1.5
Beam trajectory in the ether wind.

involved. Indeed, in the experiment, the arm OM was 10 m, whereas the distance OO′ was approximately 0.002 m.

Then, the Pythagorean relation gives:

$$OM^2 = d^2 + \tfrac{1}{2}OO'^2. \tag{1.4}$$

From equations 1.2 to 1.4, we have: $T_2 = \frac{4d^2 + v^2T_2^2}{c^2}$, so finally: $T_2 = \frac{2d}{c\sqrt{1-v^2/c^2}}$.

As a first-order approximation, we have: $T_2 \approx \frac{2d}{c}\left(1 + v^2/2c^2\right)$. (1.5)

1.1.2.2.3 The Path Difference and the Resulting Interference Fringe Shift

We can thus see from equations 1.1 and 1.4 that the beams coming from the mirrors M2 and M1 won't be in phase when recombining in the half-silvered mirror. An approximation of the time difference is:

$$\Delta T = T_2 - T_1 \approx \frac{dv^2}{c^3}.$$

With a distance d of 10 m, we obtain $\Delta T \approx 10^{-15}$ second. This time difference induces a phase difference which generates a fringe shift, noted Δp, in the interference pattern. If the wavelength of the beam is λ, we have: $\Delta p = \frac{c\Delta T}{\lambda}$. With visible light of wavelength $\lambda = 5 \times 10^{-7}$ m, we obtain $\Delta p = 0.2$ fringes, which is detectable.

Remark 1.3: Historically, the Danish scientist L.V. Lorenz, who was one of the very few early opponents to the ether assumption, stated in 1867: "The assumption of an ether would be unreasonable because it is a new nonsubstantial medium which has been thought of only because light was conceived in the same manner as sound and hence, there had to be a medium of exceedingly large elasticity and small density to explain the large velocity of light... It is most unscientific to invent a new substance when its existence is not revealed in a much more definite way."[1]

1.2 Complements to the Principle of Relativity

The principle of Relativity is central to Relativity, and several of its fundamental implications are often given without demonstration because they seem very natural. These fundamental results are:

- The proper time and the proper distance are the same in all inertial frames.
- The speed of light is the same in all inertial frames.
- The speed V of O′ relative to K is the opposite of the speed of O relative to K′.
- Any physical experiment gives the same results in any inertial frame.

The principle of Relativity is commonly defined as follows: "Physical laws take the same form in all inertial frames." We will see how these fundamental results are implied by the principle of Relativity, all the more as Relativity invites us to take nothing for granted.

Our demonstrations will be based on an important result that we will show first: the existence of a permanent center of symmetry between any pair of inertial frames.

1.2.1 The Existence of a Permanent Center of Symmetry between Two Inertial Frames

Our demonstration will be based on universal space-time homogeneity and spatial isotropy.[2] *Mathematically, this corresponds to the invariance of all physical laws under translation in space and in time, and also under space rotation (3D).*

Consider two inertial frames K and K′, with K′ moving at the constant speed V relative to K. We set the origin O of K to coincide with the origin O′ of K′ at $t=t'=0$.

Let's have a third inertial frame K″ going at the constant speed S″ relative to K, with O″ also coinciding with O and O′ at $t=t'=t''=0$. We remind that the origins and the axis directions of a frame can be chosen arbitrarily.

We will first show that there is a certain speed S″ such that an observer in the frame K″ sees at the time $t''=1$ second the two points O′ and O at equal distance and in opposite directions. We will then show that if K″ moves at this speed S″, the observer in O″ always sees O and O′ at equal distance and moving at equal speed.

1.2.1.1 The Existence of a Speed of K″ Such That at $t''=1$, an Observer in O″ States: $O''O=O''O'$

Let an observer in K″, at the time $t''=1$ second, measure the distances O″O and O″O′. He or she can then calculate the ratio: $R=O''O/O''O'$. This ratio R is a

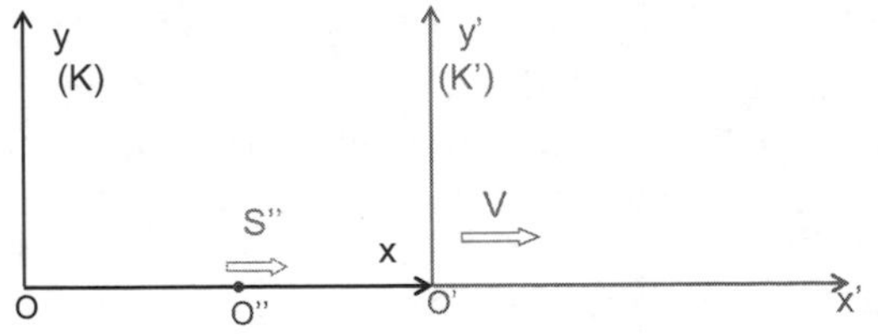

FIGURE 1.6
The points O′ and O″, respectively, go at the velocity V and S″ relative to O.

function of the speed S″ of K″ relative to K; we thus note: R(S″)=O″O/O″O′. Let's calculate this ratio for S″=0 and for S″= V: if S″=0, O″ and O coincide, implying R(0) = 0; and if S″=V, O″ coincide with O′, implying: R(V)=∞ (Figure 1.6).

The function R(S″) is continuous since it is assumed that our space-time universe is a continuum. According to the *Intermediate Value Theorem*, any continuous curve going from the point [0, R(0)=0] to the point [V, R(V)=∞] passes by all points having an ordinate between 0 and ∞; hence, this curve must go by the point having 1 for ordinate. Consequently, there is a value of S″, denoted by S, such that R(S)=1. Thus with this speed S, we have after 1 second in K″: O″O=O″O′ (Figure 1.7).

Remark 1.4: Figure 1.7 shows a random continuous curve going from (0, 0) to (V, ∞). It is unimportant for our demonstration to know if this curve is monotonically increasing or not.

Remark 1.5: An intuitive result is: S=V/2. However, it is wrong, as shown in Problem 1.3.

Remark 1.6: The continuity of the function R(S″) can be further justified by the linearity of the Lorentz transformation and consequently of the velocity composition law (which is a homographic function).

We will now show that with S″ = S, the equality O″O′ = O″O is true in K″ at any time, and not only at t″ = 1. This will prove that O″ is the permanent center of symmetry between O of K and O′ of K′.

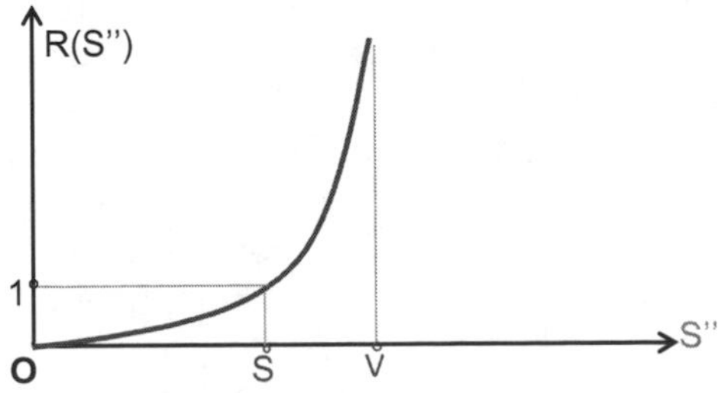

FIGURE 1.7
There exists a velocity S such that R(S)=1.

1.2.1.1.1 The point O″ Is the Permanent Center of Symmetry between K and K′

The three frames K, K′ and K″ being inertial (by definition), the speed of any of the three frames relative to any of the two others is constant. Consequently, after 1 second in K″, the speed of K relative to K″ is equal to the distance O″O, which is equal to O″O′, but this latter segment represents the speed of K′ relative to K″. Thus at t″ = 1, the observer in O″ sees both frames going at the same speed (and in opposite directions). Then, since the speeds of these inertial frames are constant, the observer in O″ always sees both frames K and K′ moving with equal speeds (in absolute value and in the opposite direction), which means that O″ is the permanent center of symmetry between the two inertial frames K and K′.

Remark 1.7: In General Relativity, this property of the existence of a permanent center of symmetry between two inertial frames is not valid in the general case, but it is locally.

1.2.2 Fundamental Implications of the Principle of Relativity

We will demonstrate the following key results that are generally covered by the principle of Relativity:

1. The proper time and the proper distance are the same in all inertial frames.
2. The speed of light is the same in all inertial frames.
3. The speed V of O′ relative to K is the opposite of the speed of O relative to K′.
4. Any physical experiment gives the same results in any inertial frame.

We will use the existence of a permanent center of symmetry between two inertial frames K and K′, as shown previously: an observer in the center of symmetry, noted O″, permanently sees O of K at an equal distance to O′ of K′; moreover, he sees O and O′ going at the same speed and in opposite directions.

This observer in O″ can apply the universal isotropy property (associated with the homogeneity postulate) and state that at any point of time (of his time), all that is on his right-hand side must be identical as its symmetrical part on his left-hand side. We can thus say that he is in a referee position, and hence we will use him in the following scenarios to tell what happens in K knowing what happens in K′:

In the frame K′, we place a mirror in a point A′ on the X′ axis such that a light pulse emitted from O′ toward A′ automatically returns to O′. We choose the distance O′A′ such that the turnaround time of the light pulse is 1 second in K′ (from the emission to the arrival in A′ and then back in O′) (Figure 1.8).

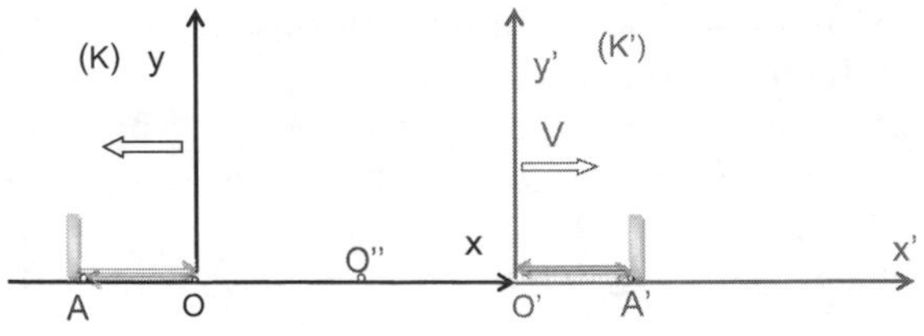

FIGURE 1.8
Two simple and symmetrical experiments seen from the permanent center of symmetry.

Let's now do the same experiment in K: we place a mirror in point A which is symmetrical to A′ relative to O″ (we remind O″ is the permanent center of symmetry). We then simultaneously emit light pulses from O and O′, respectively, toward A and A′, the simultaneity being relative to the referee observer in O″.

Due to the symmetry of this scenario relative to O″, the referee observer in O″ will say that the light pulses simultaneously reach A and A′, and then they simultaneously reach O and O′. This has important consequences as we will see.

Remark 1.8: One may wonder how the person in O can emit his or her pulse simultaneously as the one in O′, the simultaneity being relative to the referee in O″. We can proceed as follows: let the referee in O″ simultaneously emit two pulses, one toward O and the other one toward O′; then at the moment when O and O′ receive these pulses, they emit their pulses toward A and A′.

1.2.2.1 The Proper Time and the Proper Distance Are the Same in All Inertial Frames

1.2.2.1.1 The Proper Time Is the Same in K and in K′

The turnaround time of the pulse in K′ represents 1 second of the proper time in O′. Similarly in K, the turnaround time measured by an observer in O is a proper time, and the referee observer in O″ states that both turnaround times are equal, which means that these proper time durations in O and O′ are equal. Since the one of K′ represents 1 second, the one in K also represents 1 second.

Generalization: What is true for 1 second is also true for any time duration, and the same reasoning indeed applies for any distance O′A′. Besides, the universal homogeneity implies that time flows at the same pace in any point of an inertial frame. Hence, the proper time in O flows at the same pace as in any point of K, and the same for K′. Having shown that the proper time in O flows at the same pace as the proper time in O′, we deduce that the proper time flows at the same pace in all inertial frames. This is the precise meaning of "the proper time is the same in all inertial frames." This important property enables us to have a common unit of time (e.g., the second) in all inertial frames, which represents the same duration.

Remark 1.9: We knew that the proper time was an intrinsic notion, but it did not necessarily imply that 1 second in O represents the same time duration as 1 second in O′.

Remark 1.10: This scenario doesn't prove that the proper time is *universal*, meaning that it is also independent of the acceleration of the observer watching his or her proper time. We indeed need a specific postulate (the clock postulate, cf. more in Volume I Section 3.5.2).

1.2.2.1.2 The Proper Distance Is the Same in K and K′

The referee observer in O″ can say that the distance O′A′ seen by an observer in O′ must be the same as the distance OA seen by an observer in O, because of the full symmetry of the layout. These distances are proper distances since the segments are considered by observers who are fixed relative to them.

The same generalization reasoning as for the proper times applies, leading to the conclusion that the proper distance is the same in all inertial frames: a reference segment of 1 m in K represents exactly the same length as the same segment placed in any other inertial frame K′ (as seen by observers in K′).

1.2.2.2 The Speed of Light Is the Same in K and in K′

Having seen that the distance covered by the pulses is the same in K and K′, and that the turnaround times are the same, the speed of the pulses must be the same in K and in K′.

Remark 1.11: We could have even more directly reached this result: the referee observer in O″ can say that the way an observer in O′ sees the pulse propagating in K′ is the same as the way an observer in O of K sees his pulse propagating in K. However, a speed being a distance divided by a time, we first need to make sure that we can have common units of distance and time in K and K′ before being able to compare speeds in K and K′.

Generalization: The universal homogeneity postulate implies that within the same inertial frame, the speed of light is the same everywhere and in all directions (isotropy). Consequently, the speed of light is the same in all inertial frames. However, this does not mean that the speed of light is independent of the speed of the source which has emitted it. This latter formulation requires an additional postulate, which was stated by Einstein.

1.2.2.3 The Reciprocity of Velocities between Inertial Frames

We will show that the speed of O seen from O′ is the opposite of the Speed of O′ seen from O.

This is again directly implied by the symmetry of this scenario relative to O″: the referee observer in O″ can indeed tell that the speed of O′ seen by an observer in O must be the same as the speed of O seen by an observer in O′.

Generalization: This applies whatever the speed v of K relative to K′, and the universal homogeneity implies that the speed of the same object seen by different observers fixed in the same inertial frame is the same.

Note that this result is not valid in General Relativity since the symmetry argument doesn't apply, except locally (and in regions without chrono-geometry deformations).

Remark 1.12: The reader may wonder why the statement issued by a single observer, even located in the center of symmetry, is enough to establish these results, which are supposed to be valid for anyone in any frame. This issue is addressed in Section 1.3.1. In the present case, the referee in O″ can say that his bias is the same regarding his perception of symmetrical objects in the frames K and K′ (due to the homogeneity-isotropy postulate).

Remark 1.13: Another demonstration is given in Section 1.4.4.

1.2.2.4 Any Physical Experiment Gives the Same Results in K as the Same Experiment in K′

Let's have an experiment in K′ which comprises different object points, for example, the point A′. We can make a second experiment by performing a symmetry of the first experiment relative to the plane O″Z″Y″ of the frame K″. The point A, which is symmetrical to A′, is such that the segment AA′ perpendicularly crosses the plane O″Y″Z″ in its middle M. This symmetry thus operates as if the plane O″Z″Y″ was a mirror.

The universal homogeneity (including isotropy) implies that observers located in the plane O″Y″Z″ must see the evolution of the experiment on their right-hand side identically as the one on their left-hand side (Figure 1.9).

However, since it is a mirror-type symmetry, the distances and the angles between any couple of points are conserved, but not their orientations (e.g., if

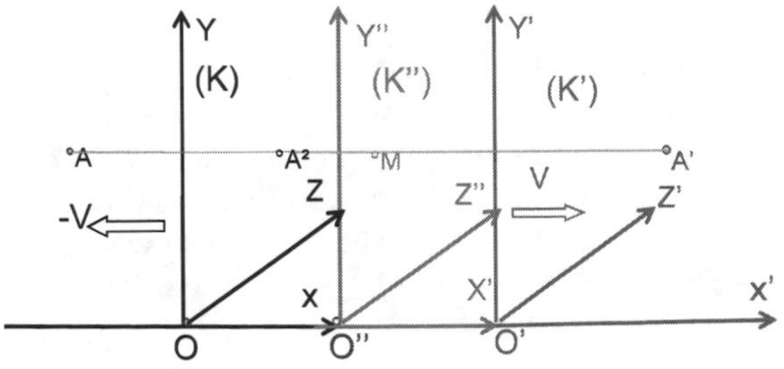

FIGURE 1.9
Two symmetrical experiments seen from the permanent plane of symmetry.

you wear a watch on your left hand, your image in a mirror wears the watch on the right one). In order to return to the initial orientation, we again make a symmetry of the second experiment relative to the plane OYZ of the frame K, and we thus obtain a third experiment.

This third experiment is in the frame K since both the second experiment and the plane of symmetry OYZ are in K, but it now has the orientations as the first experiment. For instance, in the above Figure 1.9, the point corresponding to A relative to the plane OYZ is the point noted A^2, which has the same orientation as A′ in K′.

Observers in the plane OYZ can say that this third experiment evolves symmetrically as the second one, meaning identically to the first one. This third experiment in K is thus identical to the first one in K′.

As the evolution of the experiment in K′ is dictated by the law of Physics in K′, this scenario shows that the laws of Physics must be the same in K and K′; otherwise, at least one of our observers in O and O′ will see a breach in the homogeneity–isotropy postulate.

1.2.2.5 Illustration with Galileo's Famous Sailboat Scenario, and Further Comments

1.2.2.5.1 Illustration with Galileo's Sailboat Scenario

One boat is fixed relative to the ground (frame K), and an identical one (frame K′) is moving with a constant velocity relative to K. In each boat, a sailor is on the top of the mast, holding a heavy bowl. Let's have a third boat (K″), which is permanently at equal distance between the two boats. In the middle O″ of this boat (K″), we place a referee, and when he emits a flash, the two sailors drop their heavy bowls. In the fixed boat (relative to K), it is clear that the bowl arrives at the bottom of its mast. Consequently, the referee must see that the ball of the moving boat K′ also falls at the bottom of its mast; otherwise, there would be a breach in the homogeneity–isotropy postulate.

Before Galileo, people considered that the bowl of the moving boat K′ cannot arrive at the bottom of its mast for the following reason: from the moment the bowl is dropped from the top of the mast, its horizontal motion (relative to the Earth) is stopped because of the absence of any horizontal force to push it horizontally. People indeed believed in Aristotle's statement that "everything that is in motion must be moved by something else.[3]"

Galileo was an astronomer who discovered celestial bodies moving in the cosmos at constant velocity along straight lines, and without being submitted to any force. He deduced that this was also the normal behavior on Earth for any object submitted to no force, but it wasn't ordinarily apparent because we did not consider the forces due to friction with the ground, air, water, etc.

1.2.2.5.2 Further Comments on the Expression of Principle of Relativity

The principle of Relativity often has the following definition: *"Physical laws take the same form in all inertial frames"*. However, the formulation "same form" is subject to interpretation; hence, the following definition is preferable: *"All natural phenomena can be described according to the same physical laws in all inertial frames"*.

The formulation using the words "same form" can be meaningful in the context where we have different systems of coordinates in different inertial frames, for instance, one frame has an orthonormal system, whereas the other frame has a non-orthonormal one (or even a curvilinear one). In such a case, the same physical law has different mathematical expressions, but still with the "same form."

Besides, the above demonstrations show that another definition of the principle of Relativity is possible: "**The homogeneity and isotropy postulate is valid in any inertial frame.**" Indeed, the above four properties which characterize the principle of Relativity only required this content.

1.2.3 Case of Non-Inertial Frames: Inertial Forces

If an object has an acceleration relative to an inertial frame, it incurs a Newtonian force, $\vec{F_n} = m\vec{a}$. Consider the object's own frame K′: this frame is said to be an accelerated frame since it has an acceleration relative to an inertial frame. In K′, the object is at rest, its speed and acceleration are constantly null, and hence its equilibrium is due to the existence of an inertial force, $\vec{F_i}$, which offsets the Newtonian force: $\vec{F_i} + \vec{F_n} = \vec{0}$. We thus have:

$$\vec{F_i} = \vec{F_n} = m\vec{a}. \tag{1.6}$$

The inertial force opposes the acceleration relative to an inertial frame, hence its name "inertial force."

Examples are numerous in our day-to-day life: If you are in a lift starting its descent, you feel the inertial force pushing you upward. If you have a scale for weighing inside this lift cabin, you can see a lower figure than your normal weight.

A special case is a free-falling person: he or she doesn't feel any gravity since the inertial force totally offsets his or her weight. This observation led Einstein to formulate the principle of equivalence, which is the key to General Relativity.

An object that follows a circular trajectory incurs a centrifugal force since its acceleration relative to an inertial frame is centripetal (with the intensity v^2/r). This explains, for instance, the motion of the Moon, where the gravitational force from the Earth is offset by the centrifugal force due to its circular trajectory around the Earth.

1.2.3.1 Generalization

Consider that you are in a frame K′ that has an acceleration $\vec{a}$ relative to an inertial frame K: you not only incur this inertial force, but you notice that any object of mass m′ also incurs an inertial force: $\vec{F}_i' = -m'\vec{a}$. Let's see two special cases:

- The object that you see is fixed relative to you: its acceleration being $\vec{a}$ relative to an inertial frame, it incurs a Newtonian force $\vec{F}_n' = -m'\vec{a}$. Its equilibrium in K′ is due to the inertial force:

$$\vec{F}_i' = -m'\vec{a} = -\vec{F}_n'. \tag{1.7}$$

  - Example: Following another Galileo's thought experiment, two persons are on the top of a tower: one is heavy, while the other is light. They simultaneously jump from this tower, side by side. Galileo stated that if we neglect the friction from the air, both will remain side by side. The heavy person can explain that his light friend remains fixed relative to him because he incurs an inertial force, which offsets gravity. Thus, like gravitation, the inertial force is proportional to the mass. In addition, this scenario shows that acceleration annihilates gravity.
- The object that you see is inertial: it incurs no Newtonian force, but an inertial force $\vec{F}_i' = -m'\vec{a}$. The acceleration of this object relative to you is: $-\vec{a}$; hence, for you, this inertial force has the same expression and magnitude as the Newtonian force.

Note that the weight is different from the gravitational force: what you measure with a weighing scale is the sum of the gravitational force and the inertial force resulting from the fact that the Earth is turning. Hence, inertial centrifugal forces are generated, which offset a small part of the gravitational force. These centrifugal forces also explain why the Earth is larger in the Equator than in the Poles by 22 km. In addition, if an object is moving with constant velocity relative to the Earth, its acceleration relative to a real inertial frame generates the Coriolis force, which in some cases is responsible for turning significant masses of air.

1.2.3.2 Fundamental Explanation of Inertial Forces: The Mach Principle

The inertial force shows that the world has a special class of frames: the inertial frames. The very absence of inertial force characterizes an inertial frame: you leave an object without giving it any impulsion and submitting it to any force: if it stays still, then you are in an inertial frame; if it moves, you are in an accelerated frame.

The fundamental reason behind the existence of inertial force is not straightforward: E. Mach, who was a strong opponent to the Newtonian absolute frame, saw in the inertial frames another privileged frame (or class of frames). For him, this special status must be explained by a physical reality. Hence, he issued the idea, known as Mach's principle, that the existence of this privileged class of inertial frames is due to the presence of distant great masses in the universe, and that the inertial forces are due to these distant masses by a mechanism that he could not explain. This principle helped Einstein to formulate his theory of General Relativity.

Another illustration of this principle: The Earth is turning around itself, hence, someone on the equator incurs a centrifugal force. Before Copernic, the Earth was considered fixed, but the rest of the universe was turning; hence, the person on the equator would not have incurred an inertial force, but he actually does. According to Mach's principle, this dissymmetry between the Earth frame, and the one of the rest of the universe, is due to the presence of great masses in the latter.

These principles underlying the inertial force remain valid in Relativity, but a major adaptation is required in General Relativity: an inertial frame is a free-falling frame (cf. more in Volume I Section 6.1).

1.3 Methodological Considerations

Relativity has introduced a radical change of paradigm: some fundamental postulates and laws proved to be wrong; a new counterintuitive postulate was introduced: light speed invariance. Observations made by an observer can be contradicted by other observers (e.g., the simultaneity of two events). One may then wonder if an observer is entitled to make an observation, which questions the basic methodology by which Physics is elaborated. It thus appears important to make precise some key methodological rules and see their adaptions in the context of Relativity. We will then elaborate on the major distinction introduced by Relativity between notions which are intrinsic and those which aren't.

1.3.1 The Observer's Role in Relativity: Possibilities and Limits of Contradictory Statements

Physics is a hard science, meaning that laws are derived from postulates with the rigor of mathematics. Moreover, in Physics all postulates and laws must be confronted with observations and experiments. It is a long process to establish a postulate: the scientific community must make sure before adopting it that there is enough evidence showing that it is valid under all circumstances and for any observer. Moreover, it must be consistent with all existing postulates and laws. A single clear counterexample is enough to invalidate a postulate and a law.

Besides, observations and experiments are hard facts, as opposed to impressions, beliefs... A camera and other instruments can be placed at the same location as the observer, and these can provide all evidence confirming his or her statements. Then if two observers disagree on the same piece of reality, the only possibility is that one observer is wrong; for example, his or her instrument was misaligned. Let's also note that Physics is fundamentally a shared knowledge: it could not be elaborated if two observers could say contradictory statements about the same piece of reality.

Relativity is part of Physics and makes no exception to all these considerations. Hence, the possibility introduced by Relativity of contradictory statements issued by different observers is particularly disturbing. However, the insight of the French scientist and philosopher René Descartes will guide us: "It is not our senses that deceive us, but the judgment we formulate on the basis of their testimonies."

The observer must indeed take into account the postulates and laws of Relativity when formulating his statements (which actually are judgments) after making an observation. Relativity has indeed introduced important changes which restrict the possibilities of an observer to make general statements from a given observation, for example, the distance contraction law. More generally, Relativity shows that there are two types of concepts:

- Those which have a meaning only in a given inertial frame, such as the time of an inertial frame, the simultaneity of events, then the kinetic energy... These notions are described as "non-intrinsic" (cf. more in Chapter 2).
- Those which have the same meaning for any observer anywhere, such as the event, the proper time, the proper distance... These notions are said to be "intrinsic."

Consequently, if an observer wants to issue a statement that is valid for any observer in any frame, he must use intrinsic concepts. However, the reality directly seen by an observer is often based on non-intrinsic notions (time, length...). Nevertheless, Relativity fortunately provides the tools enabling the observer to offset his bias (the Lorentz transformation in particular).

Then if contradictory statements between observers are encountered, these contradictions can only be apparent in the sense that they can be reformulated in non-contradictory statements, as shown with the following examples:

- An observer on the railway platform looks at a clock inside a moving train and says: "Time passes more slowly inside the train." A passenger inside the train looks at a clock on the platform and says: "Time passes more slowly on the ground." These two statements are contradictory, but they can be transformed into consistent statements by saying: "When I look at a moving clock, I see its space slower than my own clock," and now both observers agree on this statement.
- An observer says: "Two events are simultaneous," whereas another observer disagrees. The first observer should rather have said: "In my inertial frame, these two events are simultaneous."

Take note that we have assumed that the observer has no impact on the observed piece of reality. This is indeed true in the scales that are considered by Relativity. Quantum Mechanics has, however, shown that this cannot be the case at the smallest scales.

In conclusion, a single observer is still entitled to make statements (judgments) which are valid for all provided that he or she applies the physical laws including the relativistic ones.

1.3.1.1 Further Axiomatic Consideration: The Event Coincidence Invariance Postulate

As many phenomena are counterintuitive in Relativity, one can only be sure that a given statement is absurd if it contradicts a postulate or a law.

In Einstein's train scenario, we have stated that it is absurd that observers inside the train say that the pulses meet in the middle M of the wagon, whereas observers on the platform say the opposite. Which postulate is violated?

It is *the event coincidence invariance* postulate, which states that **if two or more events are seen in one frame occurring at the same time and at the same place, then these events occur at the same time and at the same place in every frame**.

According to this postulate, if passengers see the pulses arriving simultaneously at point M, then observers on the ground *must* also see the pulses meeting at point M, which they don't.

The rationale behind this postulate is the following: Consider a collision between two objects seen from a given frame, this collision has a physical reality (e.g., objects can break) and it would be impossible if, in another frame, such a collision would not have occurred (and the objects remain intact). In many instances, collisions can be recorded, and such records can be shown to any observer in any other frame. A collision can thus be considered an event by itself (provided that it is located at one point).

The event coincidence invariance implies that there is a one-to-one relationship (bijection) between the four coordinates of an event in one frame, and those in another frame.

Another illustration of this postulate is presented in the Question 1.7, and concerns the tennis ball scenario of Volume I section 2.1.1.1.

1.3.2 "Intrinsic" Notions: Definition and Properties

In any science, it is fundamental to identify invariants in order to define concepts and laws. Relativity showed that some fundamental concepts which were thought to be invariant (and even absolute such as the time), actually were not. Hence, new important concepts were introduced which are invariant, and this invariance stems from their "intrinsic" character. We will then make precise the meaning of intrinsic and see the properties of intrinsic notions.

1.3.2.1 Definition of an Intrinsic Notion

A notion is intrinsic if the physical reality that it characterizes is fully independent of the frame where it is considered. In other words, the observer's frame (and the observer himself) plays no role in an intrinsic physical reality, but only in the way he sees this reality. A typical example is the event: an event exists independently of the frame where it is considered. It is then an intrinsic notion, and it is characterized in a frame by a set of values. However, each value is not intrinsic because it is specific to this frame (time and distance being relative to a frame).

- *Other examples of intrinsic notions:* the proper time, the proper distance, the space-time interval, the relativistic Momentum (P), the mass (defined as the Lorentzian norm of P), the Four-Force ($dP/d\tau$).
- *Examples of non-intrinsic notions:* the time of an inertial frame, the simultaneity of events, the classical distance, the inertia of an object, the energy (due to its kinetic part), the Newtonian-like Force (dP/dt).

Remark 1.14: In mathematics, which is a discipline where each concept has a precise meaning, the word "intrinsic" means a notion or a function which is independent of any frame. There are many examples, from simple ones such as the distance between two points, to more complex ones such as the sum of the diagonal of a square matrix (trace), the divergence or the covariant derivative...

Other words than intrinsic are frequently used; however, the word "intrinsic" appears to be more appropriate.

Absolute: The Newtonian space and time were said to be absolute. However, the word "absolute" has several other meanings such as perfect, independent from arbitrary systems of measurements, and unmitigated. The word "absolute" thus appears too general. For instance, when saying that time is not absolute, it does not convey the idea that time is specific to an inertial frame. In contrast, the word "non-intrinsic" does.

Proper: The word "proper" is often used to qualify intrinsic notions such as the proper time, the proper distance… One may then wonder if "proper" and "intrinsic" are synonyms. The answer is negative because the word "proper" also has other meanings, such as true, real, appropriate, and correct. Then regarding, for example, the notion of time, one cannot say that the time that we use every day is not true, or incorrect, nor even inappropriate. Again, saying that time is not intrinsic doesn't have these meanings.

Invariant: The word "invariant" has several meanings, in particular, invariant over the time or invariant whatever the frame where it is considered. In Relativity, "invariant" always refers to the latter and is thus similar to "intrinsic." However, "invariant" can be misleading when applied to notions having several dimensions, such as the event or the Momentum, because the coordinates are not invariant (they are different in different frames). In contrast, "intrinsic" also applies to such notions: the event, the Momentum, the Four-Force…

Not-Relative: The terms "not-relative to a frame" and "intrinsic" are equivalent. However, "intrinsic" is preferable because it is not logical for a key notion to be defined by a negation, and "intrinsic" conveys the idea that an object exists by itself, independently of the observer and his or her frame.

Remark 1.15: The term "intrinsic" doesn't mean that an intrinsic element has no relation with anything in its exterior.

Remark 1.16: Not all physical concepts are intrinsic: e.g., the energy is composed of an intrinsic part plus a non-intrinsic one (the kinetic energy). The Newtonian-like Force (dP/dt) also is not intrinsic. However, these physical notions are derived from physical entities that are intrinsic (The Momentum for the Energy; The Four-Force for the Newtonian-like). These intrinsic physical entities are fundamental and hence follow Einstein's general covariance postulate, as further explained.

1.3.2.2 Properties of Intrinsic Notions

1.3.2.2.1 Intrinsic Notions Characterized by a Single Scalar Must Be Invariant

If an intrinsic notion is characterized by a single value (e.g., the proper time, the mass), this value must be invariant; otherwise, this notion would be frame-dependent, meaning not intrinsic. This invariant value is classically called a "scalar."

1.3.2.2.2 Intrinsic Notions Characterized by a Set of Values

An event is seen in an inertial frame by a set of four values, each representing a notion that is not intrinsic (the time and the distance within an inertial frame), but this set of four values altogether constitutes a vector that fully specifies this event. Then, the event being an intrinsic notion, there must be a one-to-one relationship between this vector in one frame and the vector in another frame representing the same event (and it is the Lorentz transformation). This applies to other intrinsic notions having several dimensions.

1.3.2.2.3 Intrinsic Expressions Are Simpler

In many instances, it is possible to express physical laws with intrinsic terms only, meaning without referring to any particular system of coordinates, and this leads to simpler expressions.

For example, the relativistic Momentum expressed with intrinsic terms is $\vec{P} = m\frac{\overrightarrow{dM}}{d\tau}$: it is simpler than its expression in an inertial frame: $\vec{P} = m\,\gamma_v\left(c, \frac{dx}{dt}, \frac{dy}{dt}, \frac{dz}{dt}\right)$. This simplicity stems from the fact that intrinsic expressions don't carry the terms that are necessary to make the translation between the intrinsic notions and their expressions in a given frame.

Moreover, when expressing the fundamental system Momentum conservation law, the simplicity of its intrinsic formulations is useful: $\vec{P}_A + \vec{P}_B = \overrightarrow{Cst} = m_a\frac{\overrightarrow{dM_a}}{d\tau_a} + m_b\frac{\overrightarrow{dM_b}}{d\tau_b}$. This intrinsic expression indeed shows that the system Momentum conservation law respects the inertial frame equivalence postulate, its translation in any frame being:

$$m_a\gamma_{v_a}\left(c, \vec{v_a}\right) + m_b\gamma_{v_b}\left(c, \vec{v_b}\right) = \overrightarrow{Cst}.$$

1.3.2.2.4 Intrinsic Notions and Einstein's General Covariance Postulate

Einstein's General Covariance[4] Postulate states that **fundamental physical laws take the same form in any frame.** The system of coordinates chosen plays no role in the formulation of fundamental physical laws. This is an extension of the Inertial Frames Equivalence Principles to any frame, hence the name General Relativity.

Mathematical expressions made of intrinsic terms and using intrinsic operators are covariant since they can be translated in any frame by translating each term according to its own frame changing rule (e.g., the proper time follows the rule $dt = \gamma d\tau$).

1.3.3 "Primitive" Notions: Definition and Properties

Gilbert de B. Robinson explained in his theory of knowledge: "To a non-mathematician it often comes as a surprise that it is impossible to define explicitly all the terms which are used. This is not a superficial problem but lies at the root of all knowledge; it is necessary to begin somewhere, and to make progress one must clearly state those elements and relations which are undefined and those properties which are taken for granted."[5]

These concepts that are at the root of any knowledge are called primitive concepts.

Another way to realize the necessity of primitive concepts is the following: the definition of any concept requires other concepts, but the latter also require definitions, a sequence that cannot be iterated indefinitely without returning to the initial concept. Hence, the need for primitive concepts to avoid such vicious circles.

1.3.3.1 Primitive Notions Reflect the State of Knowledge at a Given Time

A primitive notion relies on an axiom or a postulate, otherwise a primitive notion could be characterized by non-primitive notions, meaning that it would not be primitive.

Remark 1.17: The words axioms, postulates, assumptions, and even some principles and laws are equivalent in the sense that they cannot be demonstrated. For the sake of simplicity, all these will be called postulates, even if there are some differences in these terms.

- The difference between a postulate and an axiom is slim: an axiom generally refers to a single element, whereas a postulate often concerns a more general reality. A whole theory is based on some postulates and axioms, and hence these must be "rock-solid."
- An assumption is a statement that is assumed to be true, but with a lower level of confidence than an axiom or a postulate. An assumption and hypothesis are almost synonyms.
- A principle is a guideline, which can eventually suffer exceptions. A postulate doesn't.
- Some laws actually were postulates, in particular the second famous Newtonian law.

Primitive notions and their corresponding postulates reflect the state of scientific knowledge at a given time. A typical example is the concept of time: Relativity showed that the time of an inertial frame cannot be a primitive notion because it relies on a synchronization method starting with the time at one point, i.e., the proper time (cf. Volume 1, Section 1.3.2.2). The proper time is then a more primitive notion than the time of an inertial frame, and it can be considered a primitive or quasi-primitive notion as further discussed.

A set of primitive notions and postulates constitutes a theory explaining a certain domain of reality. One theory is preferable to another one if its set of primitive concepts and postulates is reduced, and if it explains a wider diversity of phenomena; this was the case of Relativity versus previously.

However, the state of scientific knowledge is often not completely well defined: some theories are issued before being confirmed by experiments and observations; hence, they may take time to convince the majority of the scientific community (this was the case of Relativity). Some may be contradicted by new observations, even if they were based on solid facts (e.g., Newtonian laws).

E. Mach's views is worth mentioning: "Simple concepts are created to describe reality for economic reasons (so as to make simple reasoning out of complex situations), but as science progresses, these concepts should be challenged and evolve."

1.3.3.2 Quasi-Primitive Notions

Different sets of primitive notions are possible within the same theory: in Classical Physics, for instance, we could have chosen either the Newtonian force or the inertial mass as the primitive notion, the postulate being in both cases the second famous Newtonian law: $\vec{f} = m\vec{a}$. If the inertial mass were chosen as a primitive notion, then the force would not have been primitive (since it would have been defined by: $m\vec{a}$). Alternatively, if we had chosen the force as primitive, the inertial mass would not have been primitive. Hence, if we call "quasi-primitive" a notion that is a valid candidate to be primitive, then both notions (force, mass) can be considered quasi-primitive. It is clear that all not concepts are quasi-primitive: for example, the time of an inertial frame.

In Relativity, we will further see that the event, the proper time and even the Momentum are quasi-primitive notions (cf. Section 1.7).

1.3.3.3 Primitive and Quasi-Primitive Notions Are Intrinsic

The inertial frame is a complex and ideal construction of men; hence, notions that depend on an inertial frame cannot be primitive. We saw an example with the time of an inertial frame. This means that non-intrinsic notions cannot be primitive, which also means that primitive (or quasi-primitive) notions must be intrinsic. This is in particular the case of the proper time, the proper distance, the event, and the Momentum. Conversely, notions such as the energy and the Newtonian force cannot be primitive.

1.4 Characterization of the New Coordinate Transformation Function

1.4.1 The New Transformation Must Be Linear

This demonstration is for readers having a higher mathematical level, as it involves the partial derivatives of functions having four inputs and four outputs. In addition, this demonstration gives a confirmation that the image in K′ of the speed of an object in K, must be independent of the location of this object.

We saw[6] that if two inertial frames K and K′ share a common origin-event, the function Φ giving the components of a space-time separation vector in K′ knowing those in K, is the same function that gives the four coordinates of an event in K′ knowing those in K: (M′)=Φ (M).

We will now consider very small space-time separation vectors, which will enable us to use the differential calculation. Let's then have an event N* very close to an event M*; the very small space-time separation vector M*N* will be denoted by dM*.

The coordinates of M* are denoted in K by (X, Y, Z, T) and in K′ by (X′, Y′, Z′, T′). The function Φ can be expressed with four equations having the following general form:

$$X' = \Phi_X(X, Y, Z, T);\ Y' = \Phi_Y(X, Y, Z, T);\ Z' = \Phi_Z(X, Y, Z, T);\ T' = \Phi_t(X, Y, Z, T). \tag{1.8}$$

The coordinates of dM* are in K: dM=(dX, dY, dZ, dT) and in K′: dM′=(dX′, dY′, dZ′, dT′).

We assume that the function Φ is differentiable, which means that the four Φi are differentiable. This corresponds to the postulate that the universe is locally smooth. Let's then express Φ(dM) using the relation (1.8) and the partial derivatives of the functions Φx, Φy, Φz and Φt at the event M*. We obtain the four following equations expressing dX′, dY′, dZ′ and dT′:

$$dX' = \frac{\partial \Phi_x(X,Y,Z,T)}{\partial x} dX + \frac{\partial \Phi_x(X,Y,Z,T)}{\partial y} dY + \frac{\partial \Phi_x(X,Y,Z,T)}{\partial z} dZ + \frac{\partial \Phi_x(X,Y,Z,T)}{\partial t} dT \tag{1.9}$$

and the same for the three other equations giving dY′, dZ′ and dT′.

The invariance under translation of the function Φ implies that each of the above four partial partial derivatives does not depend on the coordinates of M*. Consequently, the above four partial derivatives are not functions of (X, Y, Z, T), meaning that they are constant: they are four parameters which are functions of the speed V of the frame K′ relative to K.

We can thus write: $\frac{\partial \Phi_x(X,Y,Z,T)}{\partial x} = a(v)$ and so forth for the three other partial derivatives. We then have:

$$dX' = a(v)dX + b(v)dY + c(v)dZ + d(v)dT. \tag{1.10}$$

Then, by integrating equation 1.10, we obtain an affine function:

$$X' = a(v)X + b(v)Y + c(v)Z + d(v)T + k.$$

The constant k must be null because the image of O* is O′* since we have set that K and K′ share a common origin-event.

This reasoning can be repeated for dY′, dZ′ and dT′, leading to the result that Φ is a linear function ■

**

A direct consequence is worth mentioning: the image in K′ of the speed S of an object in K is independent of the location of this object; indeed, the speed S′ of the object in K′ is:

$$S' = \left(dX'/dT',\ dY'/dT',\ dZ'/dT'\right).$$

Let's consider dX′/dT′, knowing that the following reasoning will apply to the other components of S′: the relation (1.10) shows that neither dX′ nor dT′ is a function of the coordinates (X, Y, Z, T) of the event M* since the coefficients (a, b, c, d) are independent of (X, Y, Z, T), and the same for the coefficients relative to dT′. Hence, dX′/dT′ is not a function of (X, Y, Z, T).

This complies with the universal homogeneity postulate: the image of the speed of an object cannot be a function of the location of this object, nor the time at which we consider its speed, but only of the speed of this object in K and the speed of K′ relative to K.

Remark 1.18: Alternatively, the linearity of function Φ can be shown from the fact that the image of the speed of an object is independent of the location of this object, in accordance with the universal homogeneity.

1.4.2 Invariance of Transverse Distances: Complete Demonstration

The demonstration of Volume I Section 2.1.1 leaves some subjects open, therefore a complete demonstration is given hereafter. It also shows that there is no time relativity along a transverse direction.

For the sake of simplicity, we choose the inertial frames K and K′ to share the same origin-events. We also choose the OX and O′X′ axes parallel to the velocity $\vec{V}$ of K′ relative to K.

We will first show that the image in K′ of the OX and OY axes in K are the axes O′X′ and O′Y′.

1.4.2.1 The Image of the OX and OY Axes of K Must Be the O′X′ and O′Y′ Axes of K′

1.4.2.1.1 The Image of OX Is O′X′

The transformation being linear, the image of OX is a straight line passing by O (since K and K′ share a common origin). Then, OX being parallel to $\vec{V}$, its image O′X′ in K′ must also be parallel to $\vec{V}$ since V is the only privileged direction in this scenario; hence, it is O′X′. Else, there would be a new privileged direction, which would violate the isotropy postulate.

1.4.2.1.2 The Image of OY Is O′Y′

Firstly, the previous reasoning says that the image of OY is a straight line passing by O. Then, consider the following scenario: we place two lights along the OY axis of K, at the points P and Q which are symmetric relative to O. At t=0, they simultaneously emit a light pulse, which defines two events, denoted by P* and Q* (Figure 1.10).

In K, all four coordinates of Q* are the opposite of those of P*, and we thus have: Q=− P. In K′, these events P* and Q* are seen at points P′ and Q′. The transformation being linear, we have: P′=− Q′. Then, if P′ and Q′ were not in a transverse direction, it would mean that the abscissa of one event is positive (P′ in our figure), whereas the abscissa of the other one is negative (Q′ in our figure). This means that the upward direction of OY in K is privileged, since all events having positive ordinates are seen in K′ with positive abscissas (again due to the linearity of the new transformation), whereas all those having negative ordinates are seen with negative abscissas. This privileged upward direction of OY contradicts the isotropy postulate since there is only one privileged direction: the motion of K′ relative to K. Consequently, the image of OY is O′Y′.

Then, the image of a parallel line to OY is also parallel to O′Y′ since the origin-events of K and K′ are arbitrary, and we can choose their common origin-event to be the intersection between the considered parallel line to OY and OX. The same reasoning applies to all transverse directions, such as OZ.

1.4.2.2 There Is No Relativity of Simultaneity along the Transverse Directions

Consider the following scenario: In K, two points P and Q are on the OY axis such that: OP=OQ. These points simultaneously emit a flash, defining two

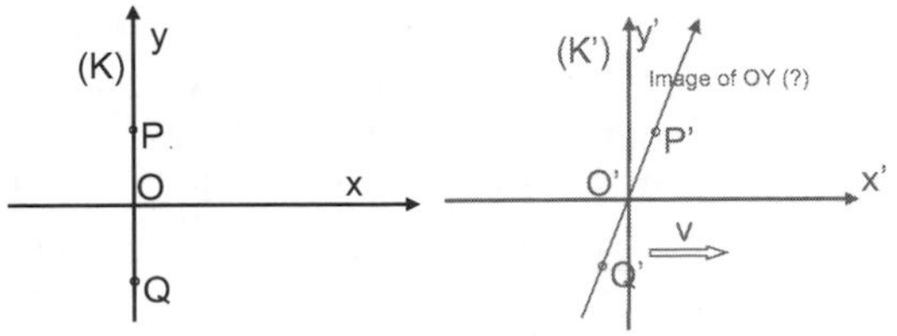

FIGURE 1.10
The image of a transverse segment.

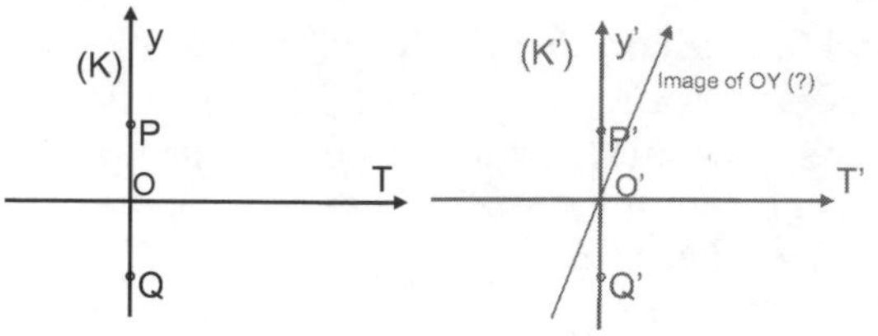

FIGURE 1.11
The image of two simultaneous events.

events: P* and Q*. We replace the OX axis by the time axis in Figure 1.11, and we want to show that in K′, the images of P and Q, denoted by P′ and Q′, are simultaneous.

The linearity of the function Φ implies that P′ and Q′ are symmetrical relative to O′. Then, if they were not along the O′Y′ axis, it would imply that one of the two events occurs after the other one in K′. This would also mean that all events on the OY axis having a positive ordinate are seen occurring after those having a negative ordinate. This contradicts the isotropy postulate since there is no privileged direction along the OY axis.

This shows that simultaneous events in a transverse direction (OY or OZ) are also simultaneous in K′: there is no time relativity along the transverse directions.

These results enable us to show the invariance of transverse distances.

1.4.2.3 Distance Invariance along the Transverse Directions

Consider the following scenario: a segment OA is fixed in K along the OY axis. Both ends of this segment have a lamp, and at t=0, O and A simultaneously emit a light pulse. These events are denoted by O* and A*. In K′, these events are seen in O′ and A′. The previous demonstration shows that the events O* and A* are also simultaneous in K′ (Figure 1.12).

We denote by α the distance OA in K. The coordinates of A* in K are thus: A(0, α, 0, 0).

In K′, we denote by α′ the distance O′A′. Thus, the coordinates of A* in K′ are: A′(0, α′, 0, 0).

We have: A′= Φ(A) with Φ being the new transformation. We know that Φ is linear; hence, $\alpha' = k_v \alpha$ with k_v being a constant which is only a function of the speed v of the frame K′ relative to K.

We will show that $k_v = 1$. To this purpose, we will enact the same scenario, but inverting the roles of K and K′; we will use the property that the inverse of the new transformation, Φ^{-1}, is the same function but with –v instead of v.

We thus have: $A = \Phi^{-1}(A')$, so: $A = \Phi^{-1}\Phi(A)$, meaning that: $\alpha = k_{-v}k_v\alpha$.

We will now show that $k_{-v} = k_v = 1$. Let's place a point B′ on the O′Y′ axis, such that O′B′=OA=α. The observer in O sees the segment O′B′ with the length$=k_{-v}\,\alpha$.

We now place a referee observer in the permanent center of symmetry O″ between O and O′ (cf. Section 1.2.1). This referee observer will say that the way O′ sees OA is the same as the way O sees O′B′, meaning that: $O'A' = k_v\, \alpha = k_{-v}\, \alpha$. Hence, $k_{-v} = k_v$ ■

This means that the length of a segment along a transverse direction has the same measure in K and K′.

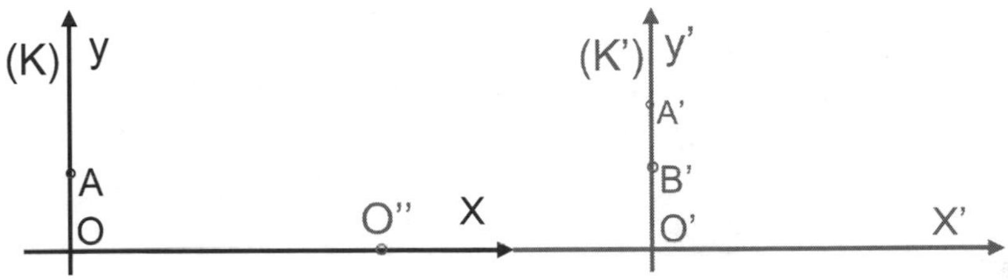

FIGURE 1.12
The referee in O″.

1.4.2.4 Comment

From a theoretical perspective, the invariance of transverse distances provides another way to have a reference unit of distance which is the same in all inertial frames: if both K and K′ adopt a standard segment (e.g., a ruler made of platinum and measuring 1 m), then by placing this segment perpendicularly to the frame motion, they will have a common unit of distance. Subsequently, the universal homogeneity–isotropy will assure that a segment in a given frame has the same length whatever its orientation. Then from this common distance unit, it is possible to define a common unit of time by applying the law $t = d/c$, taking advantage of the invariance of c.

1.4.3 Length Contraction Law Demonstration Using Basic Relativistic Schemes

The following scenario shows that the time dilatation law implies the distance contraction law. From an axiomatic perspective, this demonstration has the interest of not requiring the result that the speed of K′ relative to K is the opposite of the speed of K relative to K′.

A rectangular box is fixed in a frame K′ that is moving at the constant velocity v relative to a frame K along the OX and O′X′ axis. The longest side of the box, which is parallel to the motion, measures L′ in K′. Thus, L′ is the proper length of the box, denoted by L_P. In K, the length of the box is denoted by L, and we want to know the relationship between L and L_P.

A perfect clock is built inside this box by the following means: we place two mirrors inside the box, one on its left side and the other one on its right side. We define three events E1*, E2* and E3* as follows:

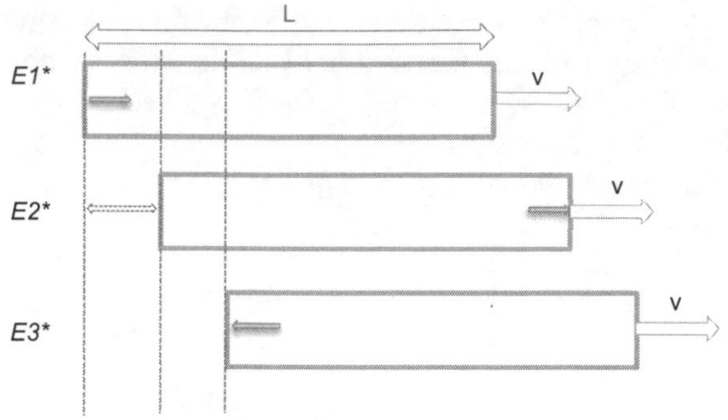

FIGURE 1.13
Light pulse emitted inside a moving box.

- E1*: A light pulse is emitted at the left side in the direction of the right side.
- E2*: This light pulse reaches the mirror on the right side, where it is instantaneously reflected.
- E3*: This pulse reaches the left side, where it initially started (at E1*).

In K′, the total time, T′, taken by the pulse from E1* to E3*, is a proper time since E1* and E3* are co-located (Figure 1.13). Thus, T′ is the proper period of this perfect clock, and hence T′ will be denoted by τ and we have:

$$\tau = \frac{2L_P}{c}. \tag{1.11}$$

In K, the total time, T, taken by the pulse from E1* to E2*, can be calculated in two steps: from E1* to E2*, then from E2* to E3*:

From E1* to E2*: The time taken by the pulse in K is denoted by t_1. During this time, the right side of the box has covered the distance v t_1. Consequently, the pulse has covered the distance: L+v t_1. The pulse moving at light speed, we have: c t_1=L+v t_1, and hence:

$$t_1 = L/(c - v). \tag{1.12}$$

From E2* to E3*: The pulse takes t_2, and during this time, the left side of the box has covered the distance v.t_2. Consequently, the pulse has covered the distance: L−v t_2. Then: ct_2=L−v t_2, so that: t_2=L/(c+v).

In K, the total turnaround time of the pulse is then:

$$T = t_1 + t_2 = L/(c - v) + L/(c + v) = \frac{2c\,L}{c^2 - v^2} = \frac{2\gamma_v^2\,L}{c}. \tag{1.13}$$

This time T in K corresponds to the time between the same events E1* and E3* as the time T′ in K′, and we saw that T′ is a proper time: $\tau = \frac{2L_P}{c}$. Consequently, $T = \gamma_v \tau$.

Then, from equations 1.11 and 1.13: $\frac{2\gamma_v^2 L}{c} = \frac{2\gamma_v L_P}{c}$, so finally: $\mathbf{L = L_P/\gamma}$ ■

This means that in the frame K, **the length of the moving box is seen contracted compared to its proper length by the γ factor.**

1.4.4 The Inertial Frames Velocity Reciprocity Law

We will show that the velocity of O in the frame K′ is the opposite of the velocity of O′ in the frame K. This property was already shown from an axiomatic perspective (cf. Section 1.2.2.3), but we will have another demonstration with the following scenario that is based on the previous length contraction demonstration, which did not require knowing the Inertial Frames Velocity Reciprocity.

The inertial frame K′ is moving at the velocity v relative to K. We denote by w the velocity of K relative to K′, and we will assume that we do not know the law: "$w = -v$", but our objective is to demonstrate it (Figure 1.14).

Let K and K′ have the same origin-event O*. At the time t′=1 second in K′, the point O′ coincides with the fixed point B of K. The measure of OB in K′ is denoted by (OB)′. We have:

$$(OB)' = w.1 . \tag{1.14}$$

Let's then calculate OB in K.

The point O′ is at the point B after 1 second of the proper time of O′, and hence the time t in K at which O′ is at the point B is: $\gamma_v \cdot 1$. Thus in K, the segment OB measures:

$$OB = v.\gamma_v . \tag{1.15}$$

Note that this length is a *proper length* since both O and B are fixed in K. Consequently, the length contraction law says that the proper length OB is seen contracted in K′: $(OB)' = OB/\gamma_v$.

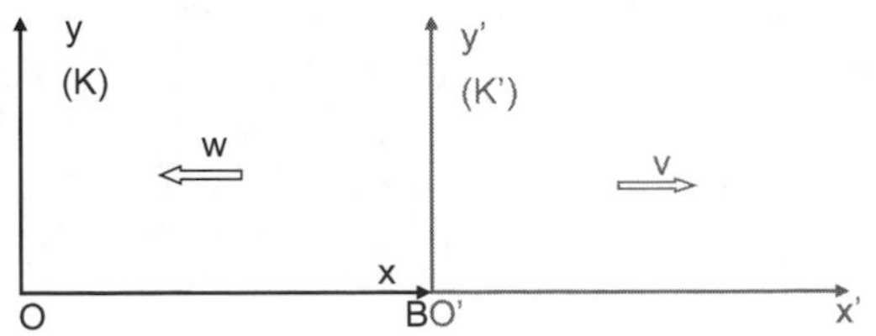

FIGURE 1.14
Is w equal to –v?

With equations 1.14 and 1.15, we finally obtain: $w = -v\,\gamma_v/\gamma_v = -v$ ■

1.4.5 Time Dilatation: Beware of Possible Misinterpretations

It is a common error to deduce from the time dilatation law that the proper period of a moving clock flows more slowly compared to when the clock was at rest. The following case study illustrates this point.

You are in a train going at constant speed v=200 km/h. The train frame is denoted by K′, and the ground frame by K. There is a perfect clock in the window of the train, which can be seen by observers on the ground. The train passes in front of a point A on the ground, and there is a perfect clock fixed in A. We take a photo from the ground that displays both clocks, and this constitutes the event A*. The time origin of a frame being arbitrary, let's assume that this photo shows that both clocks display the same time: 00:00.

The train continues at the same speed v=200 km/h and passes in front of a second point B on the ground, the distance AB being 200 km in K. At point B, there is another perfect clock synchronized with the ground time of K, hence with the clock in A. When the train passes in front of B, a second photo is taken from the ground, and this constitutes the event B*.

Question 1: What time does the fixed clock in B display at the event B*?

Answer: Let's denote by ΔT the duration in K taken by the train to cover 200 km at the speed of 200 km/h. All these data refer to the ground frame K, and hence we can apply the classical laws in K: $\Delta T = AB/v = 200/200 = 1$ hour. Thus, the clock in B displays 1:00.

Question 2: What time does the clock of the train display at the event B*?

Answer: Let's denote by $\Delta\tau$ the duration seen by the train's clock between A* and B*. This proper time duration is seen dilated from the ground: $\Delta\tau = \Delta T/\gamma = 1h/\gamma$. Thus, at the event B*, the train's clock marks $1/\gamma$ hour, meaning that the train's clock is behind the ground time.

Question 3: Does it mean that between A* and B*, the number of seconds that the train's clock has beaten is less than the ground clock? If yes, does it mean the duration of 1 second in the train is longer than 1 second on the ground, so that the number of seconds multiplied by the second duration is the same in K and K′?

Answer: First part: yes (obviously). Second part: no, because the proper time of any inertial clock is the same, independently of the speed of one clock relative to the other, as shown in Section 1.2.2.1.1. There is no reason for the number of seconds multiplied by the second duration to be the same in K and K′. Both clocks beat at the same pace, but the distance covered between A* and B* is not considered to be the same from the train and the ground (distance contraction).

1.4.5.1 Further Comments

Anyone's proper time is also his or her biological time. Hence, in the above scenario, let's take an extreme example where the duration from A* to B* is 80 years of the ground time, and 40 years seen inside the extremely fast vehicle. Let's imagine at the event A*, one baby is born in the train and another one on the ground. At the event B*, the baby who stayed fixed on the ground is 80 years old, but the one in the train is only 40; and this is real!

Besides, we place a perfect clock at each window of the train, all these clocks being synchronized with the train frame K′. On the ground, we can take a photo at the event A*, and we notice that all these clocks show different times. Thus in K, they are seen all desynchronized, even if they are all seen beating at the same dilated pace.

**

The following simple scenario confirms that the proper time is the same in all inertial frames: consider the time taken by a photon to cover 1 m. This time duration is 1/c seconds. The transverse distances being invariant, let's emit another photon in K′ along a transverse direction relative to the velocity of K′ relative to K. In K′. Light speed being also invariant, this photon covers in K′ the same distance as in K, and at the same speed; hence, it will take also 1/c seconds to cover 1 m. One can then build perfect clocks in K and K′ based on this mechanism (adding a mirror as in Einstein's perfect clock, cf. Volume I Section 2.1.2) and the proper period of each clock will be the same, whatever their relative velocities.

1.5 Supplements

1.5.1 The Fizeau Experiment: Light in a Moving Medium

The French scientist H. Fizeau made an experiment in 1851 aiming at characterizing the degree to which the ether is dragged by the motion of the medium where it is (a similar motivation as Michelson and Morley). He thus made a light beam passing inside a transparent tube, where water was circulating at the speed $v = 7.07$ m/s.

The speed of light is c in the vacuum, but inside a medium (a gas, a liquid or a solid), light speed is lower due to interactions between photons and atoms (in particular absorptions, re-emissions). It is then: c/n with n being the refraction index of the medium. For water, $n = 1.33$. For air, it is $n = 1.000293$.

Fizeau designed an apparatus where a light beam passes through a first half-silvered mirror, which acts as a beam splitter (BS1). Then, one of the two

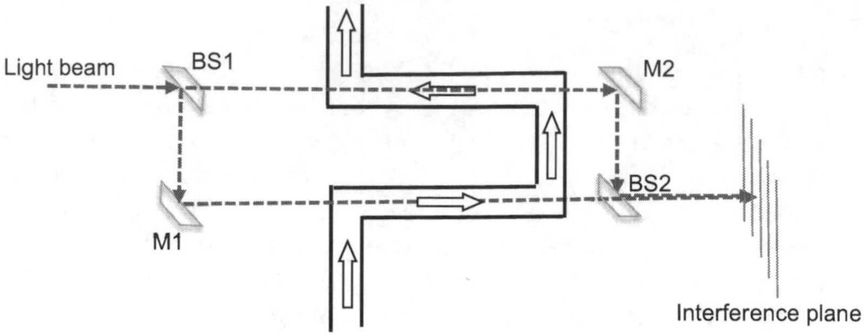

FIGURE 1.15
Simplified schematic of Fizeau experiment.

beams passes directly in the water, against the water flow, and the other beam meets the mirror M1 which orientates it in the water, and in the same direction as the water flow. These two beams then recombine at the half-silvered mirror BS2, and the combined beam goes to an interference plane. Thus, if there is a difference in travel times of these two beams, it will generate interferences that can be measured by this apparatus as shown in Figure 1.15.

Fizeau expected that the ether was fully dragged by the medium motion. Consequently, the beam in the direction of the water motion should have had the following speed relative to the ground: $s=c/n+v$. Symmetrically, the beam in the opposite direction of the water should have had the speed: $s=c/n-v$. He then expected to see an interference movement of 0.46 fringes, but only 0.23 were observed. This was in favor of the ether being partially dragged by the matter.

Besides, this result matched with a formula that was issued by A. Fresnel, but unexplained and even mysterious: $s=c/n+v-v/n^2$ (for the beam in the direction of the water motion; otherwise, $c/n-v+v/n^2$).

Einstein's view was that there is no ether, and that light propagates at the speed c in the vacuum, independently of the speed of its source. However in the medium, photons incur collisions with atoms which result in a reduction of light speed: the speed of light inside the tube containing water is: c/n relative to this medium. Consequently, if this medium goes at the speed v relative to the ground, the speed of the beam relative to the ground is $c/n \oplus v$ relative to the ground, with $\oplus$ meaning the relativistic velocity composition law. This translates to:

$$S=\frac{c/n+v}{1+\frac{vc/n}{c^2}}=\frac{c/n+v}{1+\frac{v}{cn}}\approx(c/n+v)\cdot(1-v/cn)\approx c/n+v-v/n^2.$$

This relativistic result[7] perfectly matched Fresnel's formula; hence, it was for Einstein an important confirmation of his recent theory, and in particular the new velocity composition law.

1.5.2 The Gamma Factor: Approximation with the Taylor Polynomial Method

1.5.2.1 The Taylor Polynomial Method

Taylor and Young showed that any differentiable function f can be approximated in the neighborhood of any point of abscissa a by the following polynomial function which, by construction, has the same derivatives for all orders as the function f at the point of abscissa a:

$$f(x) = f(a) + \frac{f'(a)}{1!}(x-a) + \frac{f''(a)}{2!}(x-a)^2 + \cdots \frac{f^n(a)}{n!}(x-a)^n + R_n(x).$$

where $R_n(x)$ is negligible compared to $(x-a)^n$ for x close to a.

1.5.2.2 Gamma Factor Approximation with the Taylor Polynomial Method

Let's apply the above Taylor polynomial approximation method to the gamma factor. We will use the convention to denote the ratio v/c by β. We thus have:

$$\gamma_v = \sqrt[2]{\frac{1}{1 - v^2/c^2}}, \quad \text{and} \quad \gamma_\beta = \sqrt[2]{\frac{1}{1-\beta^2}}.$$

For usual speeds, β is extremely small: for instance, if v=36,000 km/h, β=0.0000334. Hence, it is possible to apply the Taylor polynomial approximation to the function γ(β) around the point of abscissa a=0. We thus have:

$$\gamma(\beta) = \gamma(0) + \frac{\gamma'(0)}{1!}\beta + \frac{\gamma''(0)}{2!}\beta^2 + \frac{\gamma'''(0)}{3!}\beta^3 + \frac{\gamma''''(0)}{4!}\beta^4 + R_4.$$

The derivatives of the γ function are presented in Section 1.5.3, and the results are as follows: γ′(0)=0; γ″(0)=1; γ‴(0)=0; γ⁗(0)=9. Note, all derivatives of an odd order are null since the gamma function is a function of β^2.

These derivatives give the following Taylor approximation for γ(β), knowing that γ(0)=1:

$$\gamma(\beta) = 1 + \frac{1}{2}\beta^2 + \frac{3}{8}\beta^4 + 0.\beta^5 + R_5 \qquad (1.16)$$

with R_5 being negligible compared to β^5.

In the case where v=100,000 km/h, the third term of this relation is in the range of 10^{-16}, and R_5 is negligible compared to 10^{-20}. In most usual cases, the relation (1.16) can be approximated with the first two terms:

$$\gamma(\beta) \sim 1 + \beta^2/2, \qquad (1.17)$$

while having a precision better than β^3. For example, if v = 10,000 km/h, the approximation for using equation 1.17 is better than 7×10^{-13}.

1.5.3 The Derivatives of the Gamma Factor

We recall that: $\gamma(\beta) = \left(1-\beta^2\right)^{-1/2}$. The derivative of $\gamma(\beta)$ is:

$$\gamma'(\beta) = (-1/2)(-2\beta)\left(1-\beta^2\right)^{-3/2} = \beta\left(1-\beta^2\right)^{-3/2} = \beta\gamma^3. \tag{1.18}$$

When applying to $\beta=0$, we have: $\gamma'(0)=0$.

- Let's calculate the second-order derivative:

$$\gamma'' = \left(\beta\gamma^3\right)' = \gamma^3 + 3\beta\gamma^2\gamma' = \gamma^3 + 3\beta\gamma^2\beta\gamma^3 = \gamma^3 + 3\beta^2\gamma^5.$$

When applying to $\beta=0$, we have: $\gamma''(0)=1$.

- Let's calculate the third-order derivative, and we have:

$$\gamma''' = \left(\gamma^3 + 3\beta^2\gamma^5\right)' = 3\gamma'\gamma^2 + 3\left(2\beta\gamma^5 + \beta^2\,5\gamma'\gamma^4\right)$$

$$= 3\beta\,\gamma^3\gamma^2 + 6\beta\,\gamma^5 + 15\beta^2\gamma^4\beta\,\gamma^3 = 9\beta\,\gamma^5 + 15\beta^3\gamma^7.$$

When applying to $\beta=0$, we have: $\gamma''(0)=0$.

- The fourth-order derivative:

$$\gamma'''' = \left(9\beta\,\gamma^5 + 15\beta^3\gamma^7\right)' = 9\gamma^5 + 45\beta\,\gamma^4\,\gamma' + 45\beta^2\gamma^7 + 315\beta^7\gamma^6\gamma'.$$

When applying to $\beta=0$, we have: $\gamma''''(0)=9$.

- The fifth-order derivative is null, being an odd rank.

1.5.3.1 The Derivative of the Gamma Factor when β Is a Function of the Time

When the object velocity is a function of the time, its acceleration may not be parallel to the velocity (e.g., circular trajectories). Hence, we must consider the velocity vector, $\vec{v}$, or $\vec{\beta}=\frac{\vec{v}}{c}$, and its acceleration: $\vec{a}=\frac{\overrightarrow{dv}}{dt}$ or $\vec{\beta}'=\frac{\overrightarrow{d\beta}}{dt}$.

The gamma function is: $\gamma(t)=\sqrt[2]{\frac{1}{1-\vec{\beta}^2(t)}}$ then $\frac{1}{\gamma^2(t)}=1-\vec{\beta}^2(t)$.

Let's differentiate with respect to the time t: $\frac{-2}{\gamma^3}\cdot\frac{d\gamma}{dt}=-2\vec{\beta}\cdot\vec{\beta}'$. We can see that the derivative of $\vec{\beta}^2(t)$ depends on the orientation of the acceleration relative to the velocity. So finally,

$$\frac{d\gamma}{dt}=\gamma^3\left(\vec{\beta}\cdot\vec{\beta}'\right)\ \blacksquare \tag{1.19}$$

1.6 Questions and Problems

Question 1.1) Michelson–Morley. Does the Michelson–Morley experiment prove that light speed is independent of the source that has emitted it?

Question 1.2) Principle of Relativity. Imagine that you have a circular trajectory, as seen from an inertial frame. Can you consider that according to the principle of Relativity, it is not your frame that is turning, but it is the rest of the world which is turning around you?

Question 1.3) The Observer in Relativity. Can a single observer rely on his senses (eyes) and his instruments, and make valid statements in Relativity? Why is an observer wrong when he or she states that two events are simultaneous?

Question 1.4) Intrinsic notion – Part 1. The proper time is an intrinsic notion. Can we deduce from this intrinsic property that the proper time flows at the same pace in all inertial frames?

Question 1.5) Intrinsic notion – Part 2. Can an intrinsic notion which is characterized by a set of numbers in a frame, have different numbers in different frames? What kind of relationship is there between the set of numbers in K and in K′?

Question 1.6) Intrinsic and primitive notion. Can a primitive, or quasi-primitive notion, be not intrinsic?

Question 1.7) Axiomatic. In the scenario of the tennis ball of Volume I section 2.1.1.1, why is it absurd that one observer states that the ball traverses whereas another one states the contrary? Which postulate is violated?

*

Problem 1.1) Stellar Aberrations: Calculation of the stellar aberration angle θ' specified in Section 1.1.1.2.

Q1) Applying the rules of classical physics, give the general formula and make the calculation for $\theta=75°$.

Q2) Do the same, but apply the rules of Relativity. Then, comment on the difference. We give the value of the gamma factor for $V=30$ km/s: $\gamma=1.000000005$.

The reader should have previously read the chapter on the relativistic velocity composition law.

Problem 1.2) Michelson and Morley's Experiment

Q1) If the ether wind is along the bisector between the mirrors M1 and M2, how will the interference pattern be affected?

Q2) Same question, but if the ether wind is parallel to the line between the mirror M2 and the half-silvered mirror?

Q3) In our calculation of the time for the Path to the Mirror M1 (back and forth), are the laws of Relativity applied correctly? *(The reader should have previously read the chapter on the relativistic velocity composition law.)*

Q4) Make the calculation with the relativistic rules, assuming that there is an ether but that the relativistic velocity composition law applies. Make the calculation with d=10m and the ether wind speed: 30 km/s.

Problem 1.3) Calculation of the speed S of the Center of Symmetry between K and K′

Q1) We saw in Section 1.2.1 that between two inertial frames, K and K′, there always exists an inertial frame K″ where O″ is the permanent center of symmetry between O and O′. This means that an observer in O′ sees K and K′ moving at the same speed but in opposite directions. Calculate the speed S of K″ relative to K, knowing that K′ is moving at the speed V relative to K.

Q2) Make an approximation of S for usual speeds that are far from c. Then, compare this approximation with the exact value in the case of $\beta = c/100$.

Problem 1.4) How to Solve an Apparent Contradiction: Einstein's Train Scenario

We will again consider Einstein's famous scenario of the moving train observed from the railway platform. The reader should have read the Length Contraction Law (Section 1.4.3).

For observers on the platform, the moving train, denoted by TX, is shorter than when it was at rest because of the length contraction law. Consider a second train TY identical to the first one TX when both are at rest. The train TY is now at rest in front of the platform. When the moving train TX reaches the level of the fixed train TY, passengers in TX see the train TY moving backward, hence they see TY as shorter than their own train TX.

This contradicts the above picture, which is taken by an observer on the platform.

Q1) How can we solve this contradiction?

Q2) Why can't Figure 1.16 be a piece of evidence for passengers in the train TX that their statement is false?

Q3) Is there a frame where both trains are seen moving at equal speeds in absolute value? If yes, what is the speed of this frame relative to the ground frame?

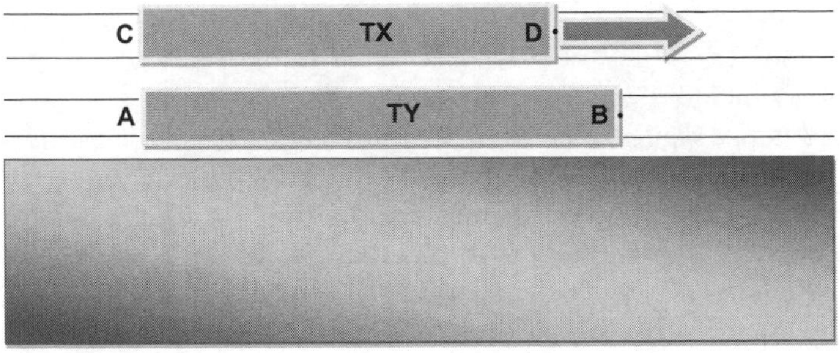

FIGURE 1.16
Identical trains seen from a platform.

Problem 1.5) Time Dilatation and Distance Contraction

You undertake a trip in a fast rocket between two very distant points A and B of an inertial frame K. When you leave the point A, it is the first of January 2000 in K. It takes you 20 years of your time inside the rocket to reach B.

Q1) When you reach B, is it sooner or later in K than the year 2020?

Q2) You see a calendar in B which displays the year 2040 of the frame K. What was your speed, assuming it was constant? You may use Table 2.4 of Volume I, and give a result with an approximation better than 3%.

Q3) What distance did you cover? Seen by you, then, seen in the frame K.

Problem 1.6) Gamma Factor Approximation Using Gamma = 1 + β²/2

Q1) What percentage represents the estimate of the approximation error that is made when using the relation $\gamma \approx 1+\beta^2/2$, if the object speed is: 1,000 km/h; 10,000 km/s; c/10; c/3?

Q2) What is the maximum speed for which the estimate of the approximation error is less than: 0.1%; less than: 0.5%? Less than 1%?

Q3) Is the exact value of γ greater or less than: $1+\beta^2/2$?

Problem 1.7) The Gamma Factor Approximation with Gamma = 1 + β²/2 + 3/8 β⁴

Q1) What percentage represents the estimate of the approximation error that is obtained when using the relation $\gamma \approx 1+\beta^2/2+3/8\,\beta^4$ if the object speed is: 10,000 km/s; c/10; c/3; 0.7c?

Q2) What is the maximum speed so that the estimate of the approximation error is less than 0.5%? Less than 1%?

Problem 1.8) The Fizeau Experiment

The reader should know the relativistic velocity composition law

Calculate the speed of the beam which propagates against the water flow. Then, indicate the formula giving the time difference of the two beams when arriving at the interference plane, assuming that each beam covers 10 m inside its tube.

Notes

1 Quoted in J.D. Jackson and L.B. Okun, *Historical roots of gauge invariance.*
2 In theory, a space can be homogeneous but not isotropic, but the likely hypothesis is that our universe is both.
3 Cf. https://www.britannica.com/biography/Aristotle/The-unmoved-mover.
4 The term "covariant" used here has nothing to do with the term "covariant" used with tensors.
5 Gilbert de B. Robinson (1959) *Foundations of Geometry*, 4th ed., p. 8, University of Toronto Press.
6 cf. Volume I Section 2.2.1.
7 This result was found by M. von Laue in 1907.

2

Lorentz Transformation, New Metric and Accelerated Objects

Introduction

The Lorentzian norm invariance will be first demonstrated from the light speed invariance in a mathematical manner. From this result, two demonstrations of the Lorentz transformation will be presented, one using the time–distance equivalence, the other one hyperbolic trigonometry. The very useful and versatile Minkowski diagrams will be carefully presented, starting with the demonstration why they function so well. Then Minkowski space with complex numbers will be presented together with hyperbolic trigonometric tools that facilitate calculations. Then, the general case of accelerated objects will be addressed, starting with the surprising triangle inequality of the Lorentzian distance, its amazing consequences with in particular Langevin's twin paradox, and then the relation between the acceleration of an object seen from different inertial frames.

2.1 The Lorentzian Norm Invariance

The invariance of the Lorentzian norm (ds^2) plays a very important role in Relativity, and hence several demonstrations will be proposed revealing different aspects. From an axiomatic perspective, it is interesting to know that this invariance is not a consequence of the Lorentz transformation, but that both are implied by the light speed invariance postulate.

2.1.1 Mathematical Demonstration of the Lorentzian Norm Invariance (ds^2)

In an inertial frame K, a light pulse is emitted at the origin point O and at $t = 0$. This light pulse then propagates in all directions at speed c, so that at time t, an event N* meets this pulse if its coordinates satisfy the relation:

DOI: 10.1201/9781003201359-2

$$x_n^2 + y_n^2 + z_n^2 = c^2t_n^2, \text{ or: } c^2t_n^2 - x_n^2 - y_n^2 - z_n^2 = 0\,.$$

Consider another inertial frame K′ sharing the same origin-event as K, and moving along the OX axis at the velocity v. The same pulse is seen moving in K′ at the same speed c; hence, the coordinates of the event N* in K′ satisfy the relation: $c^2t'^2 - x^2 - y'^2 - z'^2 = 0$.

We will show the fact that all events which are on the pulse trajectory satisfy these two relations, implies that all events, M*, not necessarily on the photon trajectory, satisfy the following relation:

$$c^2t^2 - x^2 - y^2 - z^2 = c^2t'^2 - x'^2 - y'^2 - z'^2.$$

The transverse directions are invariant, and hence for the sake of clarity and simplicity, we will only consider the OX and OT dimensions. The coordinates of a random event M* are denoted by M (x, t) in K, and by M′(x′, t′) in K′. We saw that the event coordinate transformation function, denoted by Φ, is linear: M′=Φ(M).

Let's now consider the function F which for any couple of numbers (x, t) gives the number: $F(x, t) = c^2t^2 - x^2$. We will show that for any event M*, we have: F(M)=F(M′), which means: $c^2t^2 - x^2 = c^2t'^2 - x'^2$.

We saw that the function Φ is linear (cf. Section 1.4.1); hence, x′ and t′ are linear combinations of x and t. Consequently, the expression $(c^2t'^2 - x'^2)$ is a polynomial of rank 2 of the variables x and ct, also called a quadratic form:

$$(ct')^2 - x'^2 = a\,(ct)^2 + bxct + dx^2.$$

For any event N* on the light pulse trajectory, we have: F(N)=0=F(N′), so:

$$0 = (ct')^2 - x'^2 = a\,c^2t^2 + bxct + dx^2. \tag{2.1}$$

Mathematically, the fact that the two quadratic forms $(c^2t^2 - x^2)$ and $(a\,c^2t^2 + b\,xct + d\,x^2)$ have an infinity of common roots (all events N* on the pulse trajectory) implies that they are proportional; hence, b=0 and d=−a.

For readers who don't know this mathematical result, the following demonstration is given:

Applying the relation (2.1) to: x=ct, yields: $ac^2t^2 + bc^2t^2 + dc^2t^2 = 0$; hence:

$$a + b + d = 0. \tag{2.2}$$

Applying the same relation to: x=−ct, yields: $0 = a\,c^2t^2 - b\,c^2t^2 + d\,c^2t^2 = 0$; hence:

$$a - b + d = 0\,. \tag{2.2a}$$

Adding equations 2.2 and 2.2a gives: a=−d, and so: $c^2t^2 - x^2 = a\,c^2t^2 + bxct - a\,x^2 = 0$.

If we apply this relation to (ct=1, x=1), we obtain: b=0. Consequently: F(M′)=a F(M).

*

We will now show that a=1: let's apply the function F to a simple case: a flash is emitted at the point O of K at time 1 second in K. This defines the event E* having coordinates in K: E(0, 1). We have: $F(E)=c^2$. Then in K′, this flash is seen at the time γ due to the time dilation law; its abscissa is: $-v\gamma$; and so: E′ $(-v\gamma, \gamma)$.

Then, $F(E')=c^2\gamma^2-v^2\gamma^2=\gamma^2c^2\left(1-\beta^2\right)=c^2=F(E)$. Hence: a=1, and finally: F(M′)=F(M) ■

**

This result also applies to any space–time separation vector, since the coordinates of the event M* are the same as the components of the 4D space–time separation vector $\overrightarrow{OM}$. In the context of space–time separation vectors, the function F is usually denoted by $ds^2()$; thus, we have shown that:

$$ds^2\left(\overrightarrow{OM}\right)=ds^2\left(\overrightarrow{O'M'}\right) \; ■$$

Moreover, this demonstration has the interest of showing that the function $ds^2()$ is the **only invariant** quadratic form made of the components of space–time separation vectors. Hence, mathematically, it is the norm of the 4D space–time universe, called the Lorentzian norm; its consequences are very important, as we will see.

2.1.2 Important Consequences of the Lorentzian Norm Invariance

The Lorentzian norm is the new metric of our 4D space–time universe, and this has fundamental implications: it first reveals the time–distance equivalence distance equivalence, as shown in Volume I Section 3.4, with the important consequence that time and distance can be naturally expressed with the same unit ($t \rightarrow t_d=ct$).

Furthermore, the full interest of this new metric will arise in General Relativity: we will see that the space will no longer be Euclidean and that the fundamental law of inertia stating that inertial objects follow straight lines will no longer be valid. Hence, the Lorentzian norm will be precious as it will enable us to determine inertial trajectories, called geodesics, since these extremize the Lorentzian distance covered by inertial objects, as we will further see.

Other implications are also very important:

- The Lorentz transformation can be demonstrated from the Lorentzian norm invariance, as we will see in the next chapter.

- The relation $(X^2-c^2T^2)$ invites us to express the time with complex numbers: indeed, if the time t is expressed by: $\theta=icT$, then the relation $(X^2-c^2T^2)$ becomes $(X^2+\theta^2)$, which is the classical norm. Subsequently, calculations are facilitated, as we will see in Section 2.4.
- The linear functions which leave the quantity $(c^2T^2-X^2)$ invariant are hyperbolic rotations. This aspect will be further developed, and we will again yield calculation facilities.
- Mathematically, the norm of a vector is the scalar product of this vector with itself. However, it is a different scalar product than the classical one, and this point will be further developed in the chapter on tensors.

Remark 2.1: The notation $ds^2()$ refers to small or even infinitesimal quantities, which are commonly used in differential calculation. This will indeed be necessary in General Relativity, where the laws of Special Relativity only apply locally. However, in Special Relativity, the laws also apply on long distances; hence, the notation $\Delta s^2()$ is also used to express the Lorentzian norm of a space–time interval which is not infinitesimal.

Remark 2.2: The Lorentzian norm could as well have been defined as: X^2-cT^2. It is only a matter of convention, and this latter possibility is indeed used in some contexts.

2.2 Lorentz Transformation: Two Demonstrations

Two demonstrations of the Lorentz transformations will be given, both using the invariance of the Lorentzian norm that has just been established. This first one has the advantage of being relatively rapid and is based on the observation that the coordinates x and ct play a symmetrical role in the fundamental relation expressing the Lorentzian norm invariance. The second demonstration uses hyperbolic trigonometry.

2.2.1 Lorentz Transformation Demonstration with the Symmetry Argument

We will again assume that two inertial frames K and K′ share the same origin-event O*, and that K′ moves along the OX axis at the speed V relative to K. The axes of K′ are parallel to those of K.

For the sake of simplicity, we will use natural units. An event M* has its coordinates in K: (x, t_d), and in K′: (x', t'_d). We seek the function Φ such that: $(x', t'_d)=\Phi(x, t_d)$, and we know that Φ is linear. We also know that the Lorentzian norm of O*M* is invariant, meaning:

$$ds^2(O^*M^*) = (t_d^2 - x^2) = (t_d'^2 - x'^2).$$

We notice that x^2 and t_d^2 play symmetrical roles in this fundamental relation (time-distance equivalence). Hence, it is logical to deduce that the function Φ respects the symmetrical roles of the coordinates of M^*. Consequently, the matrix performing the function Φ has one of the following forms:

$$\begin{pmatrix} x' \\ t_d' \end{pmatrix} = \begin{pmatrix} a & b \\ b & a \end{pmatrix} \begin{pmatrix} x \\ t_d \end{pmatrix} \text{ or } \begin{pmatrix} x' \\ t_d' \end{pmatrix} = \begin{pmatrix} a & -b \\ b & a \end{pmatrix} \begin{pmatrix} x \\ t_d \end{pmatrix}.$$

The terms on the diagonal cannot be a and $-a$ since we know that for speeds that are very low relative to c, we must have: $a \approx 1$ and $b \approx 0$, the classical Galilean transformation being a very good approximation in this context.

The matrix of the second form can be invalidated because it does not give correct results for events that are on the trajectory of a photon: consider a photon leaving O at $t = t' = 0$ and moving along the OX and O′X′ axis. The events characterizing this photon always satisfy: $t_d = x$ in K, and $t_d' = x'$ in K′. Let's then consider the event whose coordinates in K are: $t_d = x = 1$; we can see that the symmetric matrix satisfies the condition $t_d' = x'$ as both terms are equal to: $1+1=2$. Conversely, the matrix of the second form gives: $x' = 1-1=0$ while $t_d' = 1+1=2$.

We can then find the parameters a and b with a simple time dilatation scenario:

At the time $t=1$ second in K, a light pulse is emitted at the origin point O of K. This defines the event E*, with its coordinates in K being in natural units: $E'(0, c.1)$.

In K′, the time dilatation law says that E* is seen at the time $\gamma \cdot 1$ second, meaning $t_d' = \gamma\, c$ (with natural units). Then $x' = -\gamma\, V$. Thus, the coordinates of E* in K′ with natural units are: $(-V\gamma, \gamma c)$.

We then have: $\begin{pmatrix} -V\gamma \\ c\gamma \end{pmatrix} = \begin{pmatrix} a & b \\ b & a \end{pmatrix} \begin{pmatrix} 0 \\ c \end{pmatrix}.$

The first line yields: $-V\gamma = bc$, so: $b = -\beta\gamma$, with $\beta = \frac{V}{c}$; the second line: $c\gamma = ac$, so: $a = \gamma$.

Hence, we finally obtain:

$$\begin{pmatrix} x' \\ t_d' \end{pmatrix} = \gamma \begin{pmatrix} 1 & -\beta \\ -\beta & 1 \end{pmatrix} \begin{pmatrix} x \\ t_d \end{pmatrix} \tag{2.3}$$

which is the Lorentz transformation in natural units. It is the same matrix as in Volume I Section 3.4.2.3, which was obtained by adapting the classical

Lorentz transformation to the natural system of units. We will see in the next chapter that it corresponds to a hyperbolic rotation.

The Reverse Lorentz Transformation

We saw that the reverse Lorentz transform is obtained by replacing v with –v. Hence, we have:

$$\begin{pmatrix} x' \\ t'_d \end{pmatrix} = \gamma_v \begin{pmatrix} 1 & \beta \\ \beta & 1 \end{pmatrix} \begin{pmatrix} x' \\ t'_d \end{pmatrix}. \tag{2.4}$$

2.2.2 Lorentz Transformation Demonstration Using Hyperbolic Trigonometry

This demonstration is for readers who know some hyperbolic trigonometry. A reminder on hyperbolic trigonometry is in section 2.8.3. We will again use natural (homogeneous) units, meaning $t \rightarrow ct$, which brings more simplicity in physical laws:

The new transformation must leave invariant the Lorentzian norm:

$$c^2T'^2 - X'^2 = c^2T^2 - X^2.$$

We will use the following property of hyperbolic trigonometry: For any time-like interval, meaning $c^2T^2 > X^2$, there exists a couple (k, α) such that: $X = k.\sinh(\alpha)$ and $cT = k.\cosh(\alpha)$.

We indeed have: $\mathbf{k^2 = c^2T^2 - X^2}$ since whatever α: $\cosh^2(\alpha) - \sinh^2(\alpha) = 1$.

Then, $X/cT = k.\sinh(\alpha)/k.\cosh(\alpha) = \tanh(\alpha)$; so: $\boldsymbol{\alpha = \text{artanh}(X/cT)}$.

Applying this property to (X′, cT′) yields: $X' = k'.\sinh(\alpha')$ and $cT' = k'.\cosh(\alpha')$.

The relation $c^2T'^2 - X'^2 = c^2T^2 - X^2$ implies: $k^2 = k'^2$. Consequently, the couple (k, α′) is obtained from (k, α) by the **hyperbolic rotation** of angle: $\theta = \alpha' - \alpha$. This hyperbolic rotation indeed reads:

$$\begin{pmatrix} k\sinh(\theta+\alpha) \\ k\cosh(\theta+\alpha) \end{pmatrix} = \begin{pmatrix} \cosh(\theta) & \sinh(\theta) \\ \sinh(\theta) & \cosh(\theta) \end{pmatrix} \begin{pmatrix} k\sinh(\alpha) \\ k\cosh(\alpha) \end{pmatrix}$$

Then, in order to find θ, we can apply the same simple time dilatation scenario as in the previous Section 2.2.1, which shows that the event E* having coordinates in K: (X=0, T=1) has its coordinates with natural units as follows: in K: (X=0, cT= c) and in K′: ($X' = -s\gamma_s$, $cT' = c\gamma_s$).

We thus have for the event E* in K: $k.\sinh(\alpha) = 0$, so: $\alpha = 0$, then: $\cosh(\alpha) = 1$; then: $k.\cosh(\alpha) = c$, so: $k = c$.

Then, for the event E* in K′: $\alpha + \theta = \theta = \text{artanh}(-s\gamma_s/c\gamma_s) = \text{artanh}(-\beta)$.

We are then induced to replace θ with $\theta' = -\theta$, so that: $\boldsymbol{\theta' = \text{artanh}(\beta)}$.

The rotation becomes: $\begin{pmatrix} x' \\ ct' \end{pmatrix} = \begin{pmatrix} \cosh(\theta') & -\sinh(\theta') \\ -\sinh(\theta') & \cosh(\theta') \end{pmatrix} \begin{pmatrix} x \\ ct \end{pmatrix}$

with **θ′ = artanh(β).**

The angle θ′ is called the "**rapidity**": **θ′ = artanh(β).**

Applying this relation to E*, we deduce: $X' = -s\gamma_s = -\sinh(\theta')\cdot c$; hence: $\mathbf{\sinh(\theta') = \beta\ \gamma_s}$.

And: $cT' = c\gamma_s = c.\cosh(\theta')$; hence: $\mathbf{\cosh(\theta') = \gamma_s}$.

We thus again find the important relation (2.4): $\begin{pmatrix} x' \\ ct' \end{pmatrix} = \gamma \begin{pmatrix} 1 & -\beta \\ -\beta & 1 \end{pmatrix} \begin{pmatrix} x \\ ct \end{pmatrix}$.

Note that the angle θ′ can be denoted by any letter.

We will further see in section 2.5 simplifications brought by the fact that the Lorentz transformation is a hyperbolic rotation.

2.3 Minkowski Diagrams

Minkowski diagrams are very useful and powerful tools. Hence, we will first see why they function so well, and then some applications with in particular the Loedl variant which has additional avantages.

2.3.1 Why Do the Combined Minkowski Diagrams Work Like Magic?

The fundamental reason why the Minkowski diagram works like magic is that the Lorentz transformation is a linear function, and a linear function can be used to implement the coordinate changes from one frame to another.

Let's consider a random event P in the 4D universe. It exists independently of the frame where it is considered, meaning that it is an intrinsic geometrical object. It can be represented in any frame, and we will see how its representation changes when changing frames.

For the sake of clarity and simplicity, we will consider only two dimensions, OX and OT, knowing that the transformation leaves invariant the transverse dimensions; we will also consider frames that share a common origin-event O.

Our first frame K has its basis vectors $\vec{i}$ and $\vec{j}$. A second frame K′ has its basis vectors $\vec{u}$ and $\vec{v}$.

The random point P has coordinates (x, t) in K, meaning that: $\overrightarrow{OP} = x\vec{i} + t\vec{j}$.

Similarly, P has coordinates (x′, t′) in K′, meaning that: $\overrightarrow{OP} = x'\,\vec{u} + t'\,\vec{v}$.

The vector $\overrightarrow{OP}$ is the same in both relations since it is an intrinsic geometrical object (Figure 2.1).

The vectors $\vec{u}$ and $\vec{v}$ can be expressed in the frame K: we have: $\vec{u} = a\,\vec{i} + b\vec{j}$ and $\vec{v} = e\vec{i} + f\vec{j}$.

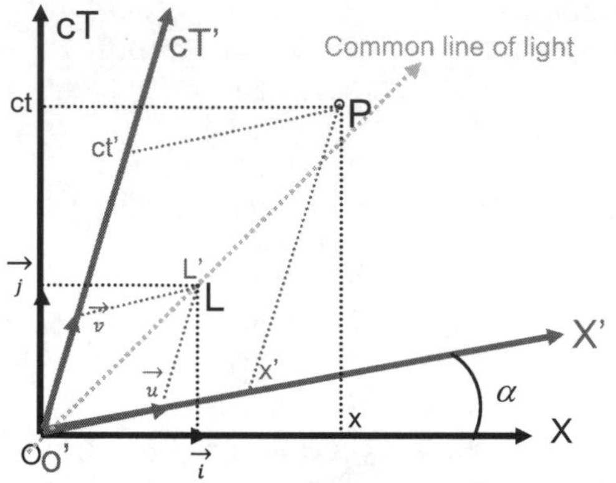

FIGURE 2.1
The Minkowski diagram.

Let's denote by M the matrix: $\begin{pmatrix} a & b \\ e & f \end{pmatrix}$.

It is a well-known mathematical result that the coordinates (x′, t′) in K′ of a vector $\overline{\text{OP}}$ are obtained from its coordinates (x, t) in K by applying the matrix M^{-1}, which is the inverse of the matrix M.[1] This is the reason why the coordinates of the vector $\overline{\text{OP}}$ are said to be *contravariant*. Applying this inverse matrix defines a linear function.

Note that the Lorentz transformation is also a linear function, which can be characterized by a matrix. Subsequently, if we choose the above matrix M to be the reverse Lorentz transformation, then the coordinates (x′, t′) obtained from (x, t) by the frame changing rule will match with the coordinates (x′, t′) obtained by applying the Lorentz transformation to (x, t).

We recall that the reverse Lorentz transformation is with natural units:

$$\gamma\begin{pmatrix} 1 & \beta \\ \beta & 1 \end{pmatrix}.$$

Consequently, if we choose the basis vectors of K′ to be: $\vec{u} = \gamma\left(\vec{i} + \beta\,\vec{j}\right)$ and $\vec{v} = \gamma\left(\beta\,\vec{i} + \vec{j}\right)$, then the above diagram is a Minkowski diagram, since the same point P has coordinates (x, ct) in K and (x′, ct′) in K′, with (x′, ct′) being obtained from (x, ct) by applying the inverse of the inverse of the Lorentz transformation, meaning the Lorentz transformation.

We notice the symmetry in the coordinates (resulting from the time–distance equivalence with ct as the time unit). Consequently, if $\vec{i}$ and $\vec{j}$ have equal lengths, then it is also the case for $\vec{u}$ and $\vec{v}$. Besides, the line of light is common because it is the set of points invariant under the Lorentz transformation.

In this reasoning, we never used the fact that the frame K is orthonormal; hence, this result is valid for non-orthonormal frames such as the Minkowski–Loedel diagrams that we will see now.

2.3.2 The Minkowski–Loedel Diagram

The Minkowski–Loedel diagram is a variant of the Minkowski diagram, which presents an important advantage that the basis vectors have the same lengths in both frames: $\vec{i}$, $\vec{u}$, $\vec{j}$ and $\vec{v}$ all have the same length.

This enables us to directly visualize the effects of time dilatation and distance contraction laws, with the same proportions between K and K′.

We indeed saw in Volume I Section 2.4.2.1 that with the Minkowski diagram, the length of a segment appears to be longer than its proper length, which is not consistent with the length contraction law, but it is explained by the different scales used by the basis vectors in K and K′.

Thus, with the Minkowski–Loedel Diagram, this drawback disappears. The Minkowski–Loedel Diagram has another interesting aspect as well: it does not privilege any frame, consistent with the spirit of Relativity.

A Minkowski–Loedel Diagram consists of placing the O′cT′ axis of the second frame K′, perpendicular to the OX axis of the first frame K. No frame is orthonormal now, no frame is privileged. All basis vectors of K and K′ have the same length.

We previously saw that K and K′ have the same bisector (common line of light), which implies that the axis OcT is perpendicular to OX′ (Figure 2.2).

2.3.2.1 The Angle α between the Two Axes: OX and OX′ (or OcT and OcT′)

We saw in Section 2.3.1 the relations between the unit basis vectors: $\vec{u} = \gamma\left(\vec{i} + \beta\,\vec{j}\right)$ and $\vec{v} = \gamma\left(\beta\vec{i} + \vec{j}\right)$. Let A be the point of the OX axis such that $\overrightarrow{OA} = \vec{i}$; and A′ the point of O′C′ such that $\overrightarrow{O'A'} = \vec{u}$. We saw that the lengths of OA and O′A′ are equal.

From the relation $\vec{u} = \gamma\left(\vec{i} + \beta\vec{j}\right)$, we deduce that: $\vec{i} + \beta\vec{j} = \overrightarrow{O'H}$ with H being the orthogonal projection of A on O′X′. Consequently, $\overrightarrow{AH} = \beta\vec{j}$, and so finally $\sin\alpha = \dfrac{AH}{OA} = \beta$ and $\cos\alpha = \sqrt[2]{1-\alpha^2} = \dfrac{1}{\gamma}$ ■

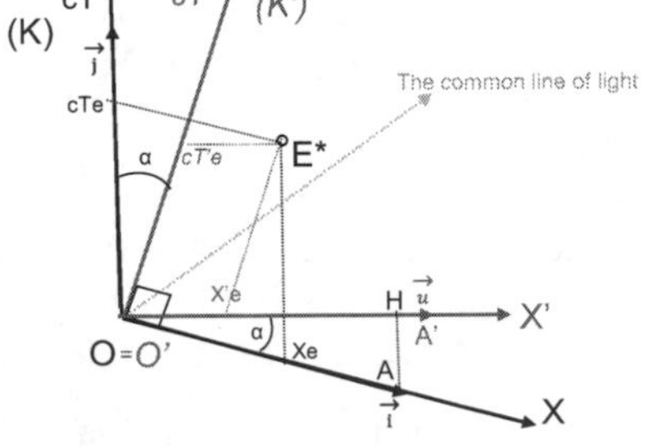

FIGURE 2.2
Generic Minkowski–Loedel diagram.

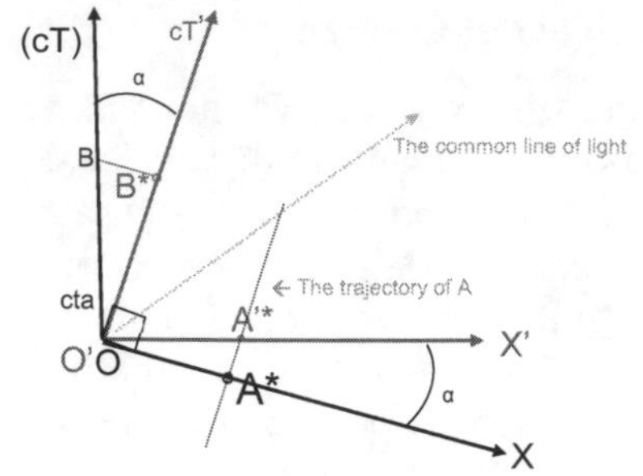

FIGURE 2.3
Minkowski–Loedel diagram – time dilatation.

2.3.2.2 Examples of Application: The Time Dilatation Law

Let the point O′ emit a flash at the time $t'=1$ second; this defines the event B*, which is on the OcT′ axis since its abscissa is null in K′ (B* takes place in O′) (Figure 2.3).

In K, the event B* has the time coordinate marked by the point B, which is such that BB* is parallel to OX. We can see that OB is longer than OB*, which is consistent with the time dilatation law; and their length ratio is in accordance with this law: $OB^*/OB=\cos\alpha=1/\gamma$.

2.3.2.3 Second Example of Application: The Distance Contraction Law

Consider a fixed segment OA in K′. We want to measure this segment in K, and for this, let both ends of the segment OA simultaneously emit flashes, the simultaneity being relative to K. This defines two events O* and A* which are not simultaneous in K′.

The extremity A of the segment is moving at the same velocity as O′; hence, its trajectory (worldline) is parallel to O′cT′. This trajectory crosses O′cX′ at the event A′*. Since the segment is fixed in K′, the abscissa of A′* is the proper length of this segment.

We can see in the diagram that: $OA^*/OA'^*=\cos\alpha=1/\gamma$, which is in accordance with this length contraction law.

Other examples of Minkowski–Loedel diagram: The Bob and Alice space travel, and the Train and tunnel scenario (cf. Problems 2.1, 2.2 and 2.4).

2.4 The Minkowski Space with Complex Numbers

The invariance of the Lorentzian norm of any space–time interval, $ds^2(V)=c^2t^2-x^2-y^2-z^2$, induced Minkowski to associate to any vector V(x, y, z, t) the vector V°(x, y, z, ict), with i being the complex number such that $i^2=-1$, and c the speed of light. The interest of this association is that the Lorentzian norm of V is equal to the opposite of the classical norm of the vector V°, denoted by $\left|\overrightarrow{V^{\circ}}\right|^2$. We indeed have: $\mathbf{ds^2(V)} = -\left|\overrightarrow{\mathbf{V^0}}\right|^2$.

The relation t → ict is bijective; hence, the Minkowski complex space is isomorphic to the classical 4D space–time universe. We will see that the Lorentz transformation is a classical rotation in the Minkowski space with complex numbers, which will enable us to benefit from the mathematical properties of rotations.

2.4.1 The Lorentz Transformation Is a Rotation in Minkowski's Complex Space

Two demonstrations will be presented:

- the first one, hereafter, assumes that the Lorentzian norm invariance is known,
- the second one, in Section 2.7.5 , assumes that the classical Lorentz transformation is known.

We saw that the Lorentz transformation must leave the Lorentzian norm invariant. This means that in the Minkowski complex space, the Lorentz transformation leaves the classical norm invariant, and such functions are well known mathematically: these are isometries, comprising rotations, translations and reflections; or even a combination of these functions.

The Lorentz transformation is not a translation (the image of the vector (0, 0, 0, 0) being (0, 0, 0, 0); it is not a reflection (it does not change the left-right orientation); hence, it is a rotation. This means that **a frame boost in the classical space is a classical rotation in the Minkowski space with complex numbers.** Besides, the transverse directions being invariant, we will focus only on X and T.

Remark 2.3: The isotropy property of the real 3D space can thus be extended to 4D in the Minkowski complex space, but with the reservation that it is impossible to travel in the past (a 180° rotation of the time axis is thus impossible).

Any rotation is defined by an angle α, and its expression is facilitated with the use of trigonometric functions. We indeed have:

$$\begin{pmatrix} X' \\ icT' \end{pmatrix} = \begin{pmatrix} \cos(\alpha) & \sin(\alpha) \\ -\sin(\alpha) & \cos(\alpha) \end{pmatrix} \begin{pmatrix} X \\ icT \end{pmatrix} \tag{2.5}$$

In order to find the angle α, we can apply the same time dilatation scenario as in Section 2.2. At the time 1 second in K, the point O of K emits a flash, which defines the event E*. The time t=1 second in K is seen dilated in K′ by the gamma factor, γ_s, with s being the speed of K′ relative to K. Thus, the event E* has coordinates in K(0, ic) and in K′ $K'(-\gamma_s, ic \cdot \gamma_s)$. Applying equation 2.5

we have: $\begin{pmatrix} \gamma_s s \\ ic\gamma_s \end{pmatrix} = \begin{pmatrix} \cos(\alpha) & \sin(\alpha) \\ -\sin(\alpha) & \cos(\alpha) \end{pmatrix} \begin{pmatrix} 0 \\ ic \end{pmatrix}$ so that: $-\gamma_s s = \sin\alpha \cdot ic$; hence:

$\mathbf{\sin\alpha = i\gamma_s\beta}$ with $\beta = \frac{s}{c}$. Then: $ic\gamma_s = \cos\alpha \cdot ic$, so: $\mathbf{\cos\alpha = \gamma_s}$. The relation (2.5) thus becomes:

$$\begin{pmatrix} X' \\ icT' \end{pmatrix} = \gamma_s \begin{pmatrix} 1 & i\beta \\ -i\beta & 1 \end{pmatrix} \begin{pmatrix} X \\ icT \end{pmatrix} \tag{2.6}$$

From this relation, we can retrieve the Lorentz transformation in the 4D space with real numbers: Let's indeed expand (2.6): $X' = \gamma X + i\gamma\beta icT = \gamma_s(X - \beta cT)$, and: $icT' = \gamma_s(-\beta iX + icT)$, so: $cT' = \gamma_s(-\beta X + cT)$, which matches with the Lorentz transformation expressed with natural units (equation 2.3) ■

We notice that γ_s is greater than 1; hence, the angle α is a complex number. We will see in Section 2.5 that it means that the rotation is a hyperbolic rotation.

2.4.2 Consequences

2.4.2.1 The Reverse Minkowski–Lorentz Matrix Is Its Transpose

The inverse of a rotation is a rotation having the opposite angle. Hence, the inverse Lorentz transformation is:

$$\begin{pmatrix} X' \\ icT' \end{pmatrix} = \begin{pmatrix} \cos(\alpha) & -\sin(\alpha) \\ \sin(\alpha) & \cos(\alpha) \end{pmatrix} \begin{pmatrix} X' \\ icT' \end{pmatrix}$$

We can see that this is equivalent to replacing β with $-\beta$, meaning s with $-$s. We can also see that the inverse of a rotation matrix is the transpose of the initial rotation matrix.

2.4.2.2 The Velocity Composition Law

The composition of two boosts in the real 4D space is the composition of two rotations in Minkowski's complex space. In case these boosts are along the same axis, the composition of two rotations leads to a rotation whose angle is the sum of the angles of each rotation.

We will see in Section 5.2.1 that the hyperbolic trigonometric tools facilitate calculations in this context.

2.4.2.3 After a Boost, the Basis Vectors Are Complex

Let the basis vectors of the real frame K be e_1 along the OX axis and e_2 along the OT (Time Axis). In the Minkowski complex space, e_1 is unchanged and e_2 is along an imaginary axis. After performing a boost, the coordinates of a vector change with the reverse matrix than the one giving the new basis vectors from the initial ones (contravariance). This means that we have:

$$\begin{pmatrix} e_1' \\ e_2' \end{pmatrix} = \gamma \begin{pmatrix} 1 & -i\beta \\ i\beta & 1 \end{pmatrix} \begin{pmatrix} e_1 \\ e_2 \end{pmatrix}$$

This shows that the two new reference frame vectors are complex. However, time-like intervals remain time-like, and space-like ones remain space-like (the norm being invariant). Thus, e_1' remains space-like, and e_2' remains time-like.

2.5 From Complex Numbers to Hyperbolic Trigonometry

We just saw that Minkowski–Lorentz transformation is a rotation in the Minkowski complex space. We will again show that in our world with real numbers, it is a hyperbolic rotation.

We previously saw that:

$$\begin{pmatrix} X' \\ icT' \end{pmatrix} = \begin{pmatrix} \cos(\alpha) & \sin(\alpha) \\ -\sin(\alpha) & \cos(\alpha) \end{pmatrix} \begin{pmatrix} X \\ icT \end{pmatrix} = \begin{pmatrix} \gamma & i\gamma\beta \\ -i\gamma\beta & \gamma \end{pmatrix} \begin{pmatrix} X \\ icT \end{pmatrix}.$$

We also saw that: $\cos(\alpha) = \gamma_s$, which is greater than 1, implying that the angle α is a complex number. Nevertheless, we still have:

$$\cos\alpha = \frac{1}{2}\left(e^{i\alpha} + e^{-i\alpha}\right) \quad \text{and} \quad \sin\alpha = \frac{1}{2i}\left(e^{i\alpha} - e^{-i\alpha}\right).$$

We will first show that α must be a purely imaginary number.

Let's call $z = \mathbf{e}^{i.\alpha}$. The relation $\cos(\alpha) = \gamma_s$ becomes: $(z + 1/z)/2 = \gamma_s$; then: $z^2 - 2\gamma_s z + 1 = 0$. We notice that the discriminant $(4\gamma_s^2 - 4)$ is always positive (since $\gamma_s > 1$). Consequently, the two solutions for z are real; then, having $z = \mathbf{e}^{i.\alpha}$, α is a purely imaginary number.

Let's call θ *the real number such that:* $\alpha = i.\theta$, *and we will introduce the hyperbolic sine and cosine functions which are adapted to this context:*

By definition, $\cosh(\theta) = \frac{1}{2}\left(e^{\theta} + e^{-\theta}\right)$ and $\sinh(\theta) = \frac{1}{2}\left(e^{\theta} - e^{-\theta}\right)$ *with* θ being a real number.

We then have: $\cosh(\theta) = \cos(i.\theta)$ and $\sinh(i\theta) = -i.\sin(\theta)$.

Thus: $\cos(\alpha)=\cos(i\,\theta)=\cosh(\theta)$, and as: $\cos(\alpha)=\gamma_s$, we have: $\cosh\theta=\gamma_s$.
Similarly, $\sin(\alpha)=\sin(i\,\theta)=-i.\text{sinsh}(\theta)$, and as: $\sin(\alpha)=i.\beta.\gamma_s$,
we have: $\text{sinsh}(\theta)=\beta.\gamma_s$.
The Minkowski–Lorentz matrix then becomes:

$$\begin{pmatrix} x' \\ ict' \end{pmatrix} = \begin{pmatrix} \cosh(\theta) & i.\sinh(\theta) \\ -i.\sinh(\theta) & \cosh(\theta) \end{pmatrix} \begin{pmatrix} x \\ ict \end{pmatrix}$$

Consequently, we have: $\tanh(\theta)=\beta$, so finally: $\boldsymbol{\theta} = \mathbf{artanh}(\beta)$ (equation 2.6).

By expanding this matrix relation, we can return to the space with real numbers as we have: $x' = \cosh(\theta)x - \sinh(\theta)ct$, and: $ct' = -\sinh(\theta)\,x + \cosh(\theta)\,ct$ (we divided both sides by i).

So we finally obtain: $\begin{pmatrix} x' \\ ct' \end{pmatrix} = \begin{pmatrix} \cosh(\theta) & -\sinh(\theta) \\ -\sinh(\theta) & \cosh(\theta) \end{pmatrix} \begin{pmatrix} x \\ ct \end{pmatrix}$

This matrix relation shows that the Lorentz transformation is a hyperbolic rotation.

Then, when replacing sinh(θ) and cosh(θ) with their values, we obtain the same important relation (2.3):

$$\begin{pmatrix} x' \\ ct' \end{pmatrix} = \gamma \begin{pmatrix} 1 & -\beta \\ -\beta & 1 \end{pmatrix} \begin{pmatrix} x \\ ct \end{pmatrix} \blacksquare$$

Remark 2.4: The name hyperbolic stems from the fact that events that are equidistant to a given event are along the hyperbola: $c^2T^2 - X^2 = \text{cst}$.

2.5.2 Examples of Application

2.5.2.1 The Velocity Composition Law Obtained with Hyperbolic Trigonometry

We saw that the Lorentz transformation is a hyperbolic rotation. Consequently, composing two velocities, v and w, both along the OX axis is equivalent to successively composing two rotations. The result is another rotation with its angle being the sum of the angles of the two initial rotations. We recall that the Minkowski–Lorentz transformation matrix is:

$$\begin{pmatrix} \cosh(\theta) & -\sinh(\theta) \\ -\sinh(\theta) & \cosh(\theta) \end{pmatrix} \text{with } \theta=\text{artanh}(\beta).$$

We want to calculate the composition of the speeds v and w, denoted by: $s = v \oplus w$. We have:

$$\text{artanh}(s/c) = \text{artanh}(v/c) + \text{artanh}(w/c). \tag{2.7}$$

Let's compose tanh() on both sides of 2.7:

$$\tanh\left[\text{artanh}(s/c)\right] = s/c = \tanh\left[\text{artanh}(v/c) + \text{artanh}(w/c)\right].$$

Using the general formula: $\tanh(m+n) = \dfrac{\tanh(m) + \tanh(n)}{1 + \tanh(m) \cdot \tanh(n)}$, we obtain:

$$s/c = \frac{\tanh\left[\text{artanh}(v/c)\right] + \tanh\left[\text{artanh}(w/c)\right]}{1 + \tanh\left[\text{artanh}(v/c)\right] \cdot \tanh\left[\text{artanh}(w/c)\right]} = \frac{\mathbf{v/c + w/c}}{1 + \mathbf{vw/c^2}}$$

which matches the classical result ■

2.5.2.2 The Invariance of the Lorentzian Norm: A Property of Hyperbolic Rotations

We saw that the Lorentz transformation can be determined without knowing the Lorentzian norm invariance. Let's place ourselves in the situation where we don't know the Lorentzian norm invariance, and we will demonstrate this law using the fact that the Lorentz transformation is a hyperbolic rotation:

We saw that the Lorentz transformation can be expressed as:

$$\begin{pmatrix} x' \\ ct' \end{pmatrix} = \gamma \begin{pmatrix} 1 & -\beta \\ -\beta & 1 \end{pmatrix} \begin{pmatrix} x \\ ct \end{pmatrix}. \tag{2.8}$$

It is then possible to identify an angle θ such that: $\cosh(\theta) = \gamma$ since $\gamma > 1$, and $\sinh(\theta) = \gamma \cdot \beta$ since: $\gamma^2 - \gamma^2 \cdot \beta^2 = 1$.

In addition, we saw that to any time-like pair (x, ct), we can associate a pair (k, α) such that: $x = k.\sinh(\alpha)$ and $ct = k.\cosh(\alpha)$. The relation (2.8) then becomes a rotation:

$$\begin{pmatrix} x' = k\sinh(\theta + \alpha) \\ ct' = k\cosh(\theta + \alpha) \end{pmatrix} = \begin{pmatrix} \cosh(\theta) & \sinh(\theta) \\ \sinh(\theta) & \cosh(\theta) \end{pmatrix} \begin{pmatrix} k\sinh(\alpha) \\ k\cosh(\alpha) \end{pmatrix}$$

We can see that the Lorentzian norm of (x′, ct′) is equal to k^2, which is the Lorentzian norm of (x, ct) ■

2.6 Special Relativity in the Real World

We will first demonstrate the surprising ds^2 triangle inequality, as well as the solution of Langevin's twin paradox in a relatively simple way using mixed-mode graphs. We will then address mathematically the correspondence of the acceleration between different frames.

2.6.1 Demonstration of the ds^2 Triangle Inequality with the Mixed-Mode Graph

An object M is moving along the OX axis of an inertial frame K. At each time τ of the object's proper time corresponds the event $M(\tau)$, which is seen at the abscissa $x(\tau)$ in the frame K, its other spatial coordinates (y and z) being always null.

In a mixed-mode graph, the event $M(\tau)$ is represented by a point in a frame, whose abscissa is $x(\tau)$, and its ordinate: $c.\tau$. The times are indeed expressed with natural units: $\tau \to c\tau$.

Let's then represent Langevin's twins scenario with such a graph: two twins are at a point A on the ground. The twin F (fixed) stays at the point A, whereas the traveling twin M (mobile) leaves A, goes with constant velocity along a straight line until the point B, and then immediately returns to A, again with a constant velocity. He then meets with his twin F who remained still. The different events encountered are denoted as follows:

A*=The twin M leaves the twin F located on the ground at the point A $(x_A, c.\tau_A)$.

B*=The twin M arrives at the point B $(x_B, c.\tau_B)$.

C*=The twin M arrives at A (and meets again his twin F). The event C* is represented by the point C $(x_A, c.\tau_C)$.

These events are respectively represented in the diagram below by the points A, B and C (Figure 2.4).

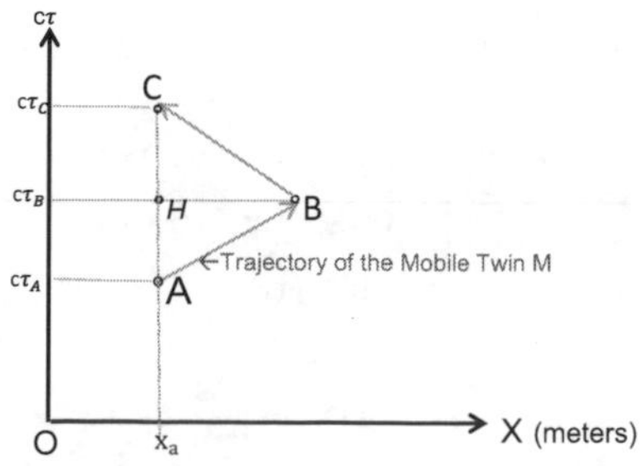

FIGURE 2.4
Langevin's twins in a mixed-mode graph.

In the right triangle ABH, we have: $AB^2=AH^2+HB^2$. Besides, $AH^2 = c^2.\tau_{AB}^2$, and: $HB^2 = (X_B - X_A)^2$.

In the frame K, the events A*, B* and C* are seen at the times T_A, T_B and T_C, respectively.

Let's express the Lorentzian norm of A*B*:

$d\ s^2(A^*B^*) = c^2.\tau_{AB}^2 = c^2(T_B - T_A)^2 - (X_B - X_A)^2$.

We thus have: $AH^2 = c^2(T_B - T_A)^2 - HB^2$. Consequently, $c^2(T_B - T_A)^2 - AB^2$.

This shows that the **classical length of the segment AB represents the time duration considered from the ground (K) for the trip from A* to B*,** and denoted by T_{AB} (The time is expressed with natural units: ct).

The same reasoning applies to the rectangle triangle BCH, showing that the length BC represents the time duration considered from K for the trip from B* to C*.

Thus, the total travel time (from A* to B*, and then back to A*) considered from K is: $cT_{AC} = AB + BC$.

On the other hand, the proper time incurred by M for the same trip is equal to: $c\tau_{AC} = AH + HC = AC$. We always have $AC < AB + BC$, which means that: $c\tau_{AC} < cT_{AC}$.

Regarding T_{AC}, it is the time measured by the fixed twin between the events A* and C*, which are co-located with him. The proper time of the fixed twin, denoted by τ^F, flows at the same pace as the time of the frame K since he remained fixed in K; hence, we have: $cT_{AC} = c\tau_{AC}^F$. Consequently, $c\tau_{AC} < c\tau_{AC}^F$.

This shows that the trip duration measured by the moving twin with his proper time, is shorter than that measured by the fixed twin. The latter is then older than the former when they meet again after the trip. ■

Let's see the impact of this surprising inequality on the Lorentzian norm:

The fixed twin can state: $c^2\left(\tau_{AC}^F\right)^2 = \Delta s^2(A^*C^*)$, while the mobile twin can state: $c\tau_{AC} = c\tau_{AB} + c\tau_{BC}$.

So finally, the inequality $c\tau_{AC} < c\tau_{AC}^F$ implies:

$$\sqrt{\Delta s^2(A^*B^*)} + \sqrt{\Delta s^2(B^*C^*)} < \sqrt{\Delta s^2(A^*C^*)}$$

which is the opposite of the classical distance triangle inequality ■

Remark 2.6: This general rule was demonstrated in the context where A*B* and B*C* are time-like intervals.

2.6.2 Accelerated Trajectories, and Problems for Time Synchronization

2.6.2.1 Langevin's Twins Paradox: General Case

In the general case, there are accelerations and the mobile trajectory is a differentiable curve. We will again use the same mixed-mode graph:

The traveling twin's accelerated trajectory is represented by the curve starting in A and ending in C, with A and C having the same abscissas since the journey ends at the same place as it starts (Figure 2.5).

The curvilinear length of the curve from A* to C* represents the time taken for the journey considered from the Earth frame. Indeed, the inertial tangent frame law states that from an inertial frame, we can consider any object trajectory as a succession of very small inertial sections, apply the laws of relativity with each inertial tangent frame, and then integrate these results along the trajectory. This curvilinear length can then be considered as the sum of a succession of infinitesimal *linear* sections, and in each section the previous reasoning applies, which shows that the length of each small section represents the time on the ground of the considered section of the trip. Consequently, the **curvilinear length of the curve from A* to C* is equal to the time duration seen in the ground frame K for the trip from A* to C***.

The duration incurred by the traveler, measured with his own proper time, is given by the length of the segment A*C* as previously. Hence, this diagram shows that this proper duration is always shorter than the curvilinear line between A* and C*, which represents the time duration seen by the fixed twin on the ground for the same trip.

Remark 2.7: The mixed-mode graph is consistent with principles of Relativity, as the relativistic Momentum, which is a key quantity in Relativity, has a mixed-mode expression: $P = m \cdot \frac{dM}{d\tau}$, which gives for objects moving along the OX axis: $m\,\frac{dx}{d\tau}$. We can then see that in a mixed-mode diagram, the Momentum of an object is proportional to the inverse of the slope of its trajectory.[3]

Other applications are presented in Section 2.7.4, showing in particular the problem for synchronizing time.

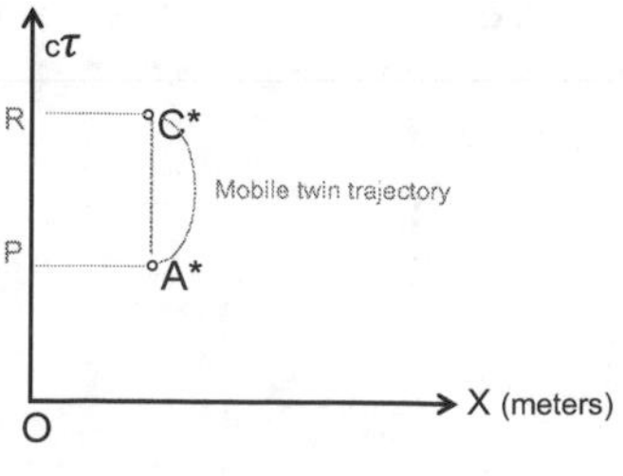

FIGURE 2.5
Twins paradox – general accelerated trajectory.

2.6.3 Mathematical Relations between the Acceleration in K and in K′

In an inertial frame K, an object M has a classical speed V(t) which varies in time.

We thus have: $V(t)=\left[V_x(t)=\frac{dx(t)}{dt},\ V_y(t)=\frac{dy(t)}{dt},\ V_z(t)=\frac{dz(t)}{dt}\right].$

Its classical acceleration in K is:

$$A(t)=\left[A_x(t)=\frac{dV_x(t)}{dt},\ A_y(t)=\frac{dV_y(t)}{dt},\ A_z(t)=\frac{dV_z(t)}{dt}\right].$$

Let's calculate the acceleration of M considered in another inertial frame K′ moving at the speed W relative to K.

The object is characterized in K at the time t by the event M(t); this same event is seen in K′ at the time t′. The velocity composition law yields:

$$V_x'(t')=\frac{V_x(t)-W}{1-V_x(t)W/c^2} \quad \text{and} \quad V_y'(t')=\frac{V_y(t)(t)}{\gamma_W\cdot\left(1-\frac{V_x(t)W}{c^2}\right)}.$$

The acceleration in K′ is then:

$$A_x'(t')=\frac{dV_x'(t')}{dt'}=\frac{dV_x'(t')}{dt}\frac{dt}{dt'}=\frac{dV_x'(t')}{dV(t)}\frac{dV(t)}{dt}\frac{dt}{dt'}=A_x(t)\frac{dV_x'(t')}{dV(t)}\frac{dt}{dt'}.$$

The term $\frac{dt}{dt'}$ is obtained by the Lorentz transformation, which gives: $dt'=\gamma_W\left(1-V_x w/c^2\right)dt.$

From the above relation, we have:

$$\frac{dV_x'(t')}{dV(t)}=\frac{1-\frac{V_x(t)W}{c^2}+\frac{W}{c^2}\cdot\left[V_x(t)-W\right]}{\left[1-\frac{V_x(t)W}{c^2}\right]^2}$$

$$=\frac{1-\frac{W^2}{c^2}}{\left[1-\frac{V_x(t)W}{c^2}\right]^2}=\frac{1}{\gamma_W^2\left[1-\frac{V_x(t)W}{c^2}\right]^2}.$$

We finally obtain:

$$\mathbf{A_x'(t')=\frac{A_x(t)}{\gamma_W^3\left(1-V_x(t)W/c^2\right)^3}}\ \blacksquare \qquad (2.9)$$

Regarding the transverse direction, the calculation is somewhat more complicated and gives:

$$A'_y(t') = \frac{dV'_y(t')}{dt}\frac{dt}{dt'} = \frac{A_y(t)}{\gamma_W^2\left(1 - V_x W/c^2\right)^2} + \frac{A_x(t)\cdot V_y W}{c^2\gamma_W^2\left(1 - V_x W/c^2\right)^3} \quad (2.10)$$

The derivative of the gamma factor is given in Section 1.5.2.1.

These relations show that, unlike in classical physics, the acceleration is not the same in all inertial frames. This implies in particular that the classical Newtonian force is not the same in all inertial frames.[4] However, there is an element of absoluteness: an object having an acceleration null in one inertial frame, has its acceleration null in all inertial frames; and this confirms the special role of inertial frames.

A case is of special interest: when the frame K′ is the inertial tangent frame (ITF) of the moving object.

2.6.3.1 Case Where the Acceleration Is Measured in the Inertial Tangent Frame of the Moving Object

Let's now consider that the frame K′ is the inertial tangent frame (ITF) at the time τ of the moving object, which is still going along the OX and O′X′ axis of K and K′ respectively. We place an accelerometer inside the moving object, which enables us to know its acceleration relative to K′. Hence, at each moment of its proper time τ, we can know: $A'(\tau) = \frac{dV'}{dt'} = \frac{dV'}{d\tau}$. *(We indeed saw that during a small time lapse, dt′=dτ since the object's speed is null at the time τ in its ITF.)*

In the frame K, let's denote by t the time when the object's proper time is τ. Its acceleration in K is given by equation 2.9:

$$A(t) = \frac{dV(t)}{dt} = \frac{dV'(t)}{dt'}\cdot\gamma_V^3\left(1 - V^2(t)/c^2\right)^3 = A'(\tau)\cdot\gamma_V^{-3}\cdot\gamma_V^{-2\times 3} = A'(\tau)\cdot\gamma_V^{-3}. \quad (2.11)$$

Let's express the object's speed V in K as a function of τ, knowing that: $dt = \gamma_V \cdot d\tau$.

We thus have: $\frac{dV}{d\tau} = \frac{dV}{dt}\cdot\frac{dt}{d\tau} = A'(\tau)\cdot\gamma_V^{-3}\cdot\gamma_V = A'(\tau)\cdot\gamma_V^{-2}$.

Or, if we use the notation with $\beta(\tau) = V(\tau)/c$, we have:

$$\frac{c\cdot d\beta(\tau)}{d\tau} = A'(\tau)\cdot\left(1 - \beta^2(\tau)\right). \quad (2.12)$$

**

To solve this equation, let's make the variable change: $\beta(\tau)=\tanh(f(\tau))$.

We indeed have: $\frac{d\beta(\tau)}{d\tau}=f'(\tau)\cdot\left[1-\tanh^2(f(\tau))\right]=f'(\tau)\cdot\left[1-\beta^2(\tau)\right]$.

Then with equation 2.12, we have: $A'(\tau)=c\cdot f'(\tau)$; hence: $f(\tau)=\int_0^\tau \frac{A'(u)}{c}du$.

We can then express the coordinates x and ct of the object as a function of τ:

From $dt=\gamma_v d\tau$, we deduce: $\mathbf{dt}=\left(1-\beta^2\right)^{-1/2}d\tau=\cosh\left(f(\tau)\right)d\tau$.

Then, from $dx=V\cdot dt$, we have:

$\mathbf{dx}=c\cdot\tanh(f(\tau))\cdot\cosh(f(\tau))d\tau=\mathbf{c\cdot\sinh(f(\tau))\cdot d\tau}$.

A Case is of Special Interest: when A′ is Constant:

2.6.3.1.1 *Case of Special Interest: A′ Is Constant*

Let's denote by a′ this constant, and we have: $f(\tau)=a\,\tau/c$.

Then, the relation $\beta(\tau)=\tanh(f(\tau))$ yields: $V(\tau)=c\cdot\tanh(a'\,\tau/c)$.

Then, $dt=\cosh(a'\,\tau/c)\,d\tau$, and so finally:

$$\mathbf{t=\frac{c}{a'}\cdot\sinh\left(a'\,\tau/c\right)+t_0}. \tag{2.13}$$

Similarly, $dx=c\cdot\sinh(a'\,\tau/c)\cdot d\tau$, and so finally:

$$\mathbf{x=\frac{c^2}{a'}\left(\cosh\left(a'\,\tau/c\right)-1\right)+x_0}. \tag{2.14}$$

Remark 2.8: Regarding the order of magnitude, if we consider natural units with the time expressed in years and the distance in light-years,[5] then for an acceleration of $a'=9.5\ m/s^2$, meaning close to the gravity on Earth, the ratio a′/c is: 0.9993, meaning very close to 1 per annum.

The above relations open the possibility of another surprising effect in Relativity, the event horizon that we will see now.

2.6.4 Another Surprising Effect: The Event Horizon

Bob is inside a rocket, which will soon take off. His twin sister Alice stays on Earth. The rocket takes off and then moves along a straight line; its speed always increases. Bob chooses a trajectory that has an asymptote parallel to the line of light, as shown in Figure 2.6. Such trajectory makes possible an amazing effect: the event horizon that we will show now:

In the above Minkowski diagram, K is the Earth frame (assumed to be inertial), and the origin-event O is when Bob takes off. The OX axis is along

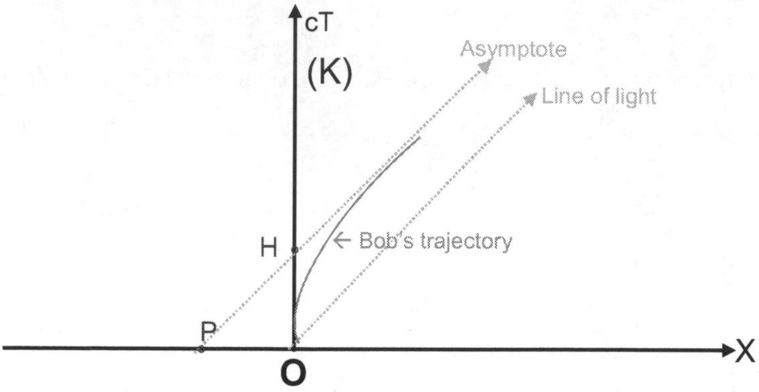

FIGURE 2.6
The event horizon.

the rocket trajectory. Bob's worldline has an ever-decreasing slope, with an asymptote that is parallel to the line of light: his speed is ever increasing and tends to c.

This asymptote line meets the OcT axis at the event H. Alice is staying at the same place; hence, her worldline is the OcT axis. Alice will then encounter the event H, and from this event on, she won't be able to communicate with Bob because he won't be able to receive any signal from her. Indeed, the fastest transmission means use photons, but photons emitted by Alice after the event H will go beyond this asymptote; hence, they will never meet Bob's trajectory.

This is all the more surprising as Alice can make the reasoning that the photons emitted toward Bob move faster than his rocket; hence, they must reach him. This is another example of the counterintuitive effect of Relativity.

Similarly, Bob won't be able to receive messages sent by persons that are on the OX axis and with an abscissa, which is less than that of the point P, this point P being at the intersection between the OX axis and the asymptote.

Examples of such rocket trajectories are given with Problems 2.9–2.10.

2.6.5 The Proper Acceleration

We saw that unlike in classical physics, the acceleration is not the same in all inertial frames, meaning that the acceleration is not an intrinsic notion. However, an intrinsic notion of acceleration can be defined, and to this end, we should express the mobile velocity and the time in intrinsic terms. We saw in Volume 1 Section 4.2.1.2 that the intrinsic velocity notion is the proper velocity, defined as: $\mathbf{S}(\tau) = \dfrac{\mathbf{dM}(\tau)}{d\tau}$, with τ being the object's proper time.

Subsequently, the proper acceleration was defined in an intrinsic manner as being: $\mathbf{A} = \frac{\mathbf{dS}(\tau)}{d\tau}$. This 4-vector **A** is is also called Four-Acceleration, and we notice that its spatial part matches with the classical acceleration in the inertial tangent frame.

In an inertial frame K where the object moves at the 3D velocity $\vec{v}$ and with the acceleration $\vec{a} = \frac{d\vec{v}}{dt}$, the components of the proper acceleration are:

$$\mathbf{A} = \frac{\mathbf{dS}(\tau)}{d\tau} = \left(\frac{dc\gamma}{d\tau}, \frac{d\vec{v}\gamma}{d\tau}\right) = \left[\gamma^4\left(\vec{a}\cdot\vec{\beta}\right), \gamma^4\left(\vec{a}\cdot\vec{\beta}\right)\vec{\beta} + \gamma^2\vec{a}\right] \tag{2.15}$$

where $\vec{\beta} = \frac{\vec{v}}{c}$. The corresponding demonstration is given with Problem 2.12.

2.7 Supplements

2.7.1 Geometrical Demonstration of the Lorentzian Distance Triangle Inequality

The following notations will be used: consider two events A* and B*, the classical distance of the spatial part of A*B* is denoted by AB, with $AB^2 = (x_b - x_a)^2 + (y_b - y_a)^2 + (z_b - z_a)^2$.

The time part of A*B* is denoted by $T_{AB} = c(t_b - t_a)$, natural units being used.

We then have: $\Delta s^2(A^*B^*) = T_{AB}^2 - AB^2$.

In the Figure 2.7, these terms of the Δs^2 expressions are represented the lengths of sides of right triangles, enabling us to take advantage of the Pythagorean relation, as we will see (Figure 2.7).

Consider the events A* and C* forming a time-like interval A*C*: We represent the points A, C and E such that AC is the classical 3D distance of the spatial part of A*C*.

Then, we set the point E to be the point in the perpendicular line to AC, and which is at the distance to A such that: $AE = T_{AC}$, knowing that AE>AC since A*C* is time-like.

Consequently, in the right triangle AEC, the side CE measures:

$$EC^2 = AE^2 - AC^2 = -AC^2 = \Delta s^2(A*C*).$$

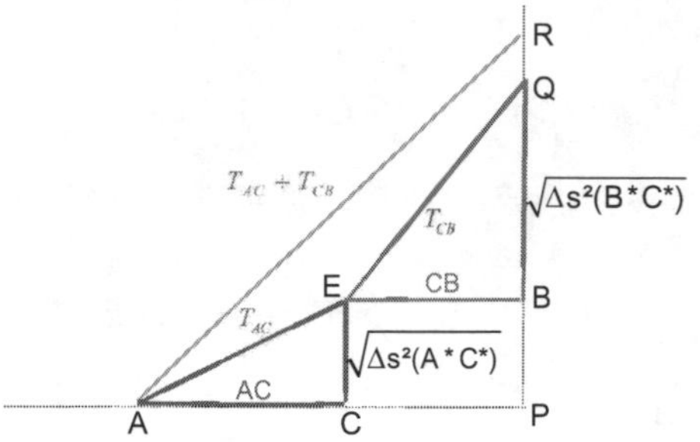

FIGURE 2.7
Lorentzian distance triangle inequality.

Let's now consider another event B* and the same event C*. We locate the point B such that EB is parallel to AC, and such that: $EB^2=(x_c-x_b)^2+(y_c-y_b)^2-(z_c-z_b)^2$.

We then locate the point Q on the perpendicular line to EB and such that: $EQ = T_{BC}$. As before, in the right triangle EBQ, the side QB measures: $QB^2 = EQ^2 - EB^2 = T_{BC}^2 - BC^2 = \Delta s^2(C^*B^*)$.

Then, let's consider the point R, which is on the perpendicular line to AP and such that $AR=T_{BC}+T_{AC}$. The distance PR is greater than PQ because the segment AQ is smaller than AE+EQ. Thus: PR>PQ.

The meaning of PR: we have:

$PR^2=AR^2-AP^2=(T_{BC}+T_{AC})^2-(AC+CB)^2=\Delta s^2\,(A^*C^*+C^*B^*)$.

So finally, PR>PQ means: $\sqrt{\Delta s^2(A^*C^*+C^*B^*)} > \sqrt{\Delta s^2(A^*C^*)} + \sqrt{\Delta s^2(C^*B^*)}$ ■

2.7.2 Twins Scenario Using the Minkowski Diagram: The Line of Simultaneity

Alice and her twin Bob are at the point O on Earth, assumed to be inertial. Bob leaves Alice, goes along a straight line with acceleration, then decelerates, stops and comes back to see Alice.

Let's draw a Minkowski diagram: Alice is at the point O of the frame K, and Bob's trajectory is represented as the curved line from O to C*, which is the event corresponding to Bob's return at the point O. For the sake of simplicity, we set the time origin to be the age of Alice (multiplied by c) and Bob when Bob departs.

Being inertial, Alice can consider Bob's trajectory as a succession of very small inertial sections.[6] Let's then consider an event A* belonging to Bob's trajectory. Bob's trajectory can be considered linear in the vicinity of A*; hence, the time axis of the frame K'a of the Minkowski diagram, denoted by A*cT'a, is tangent to Bob's trajectory in A*. Subsequently, the A*X'a axis of this frame K'a at the event A* can be found thanks to the relation X'a=cT'a, which

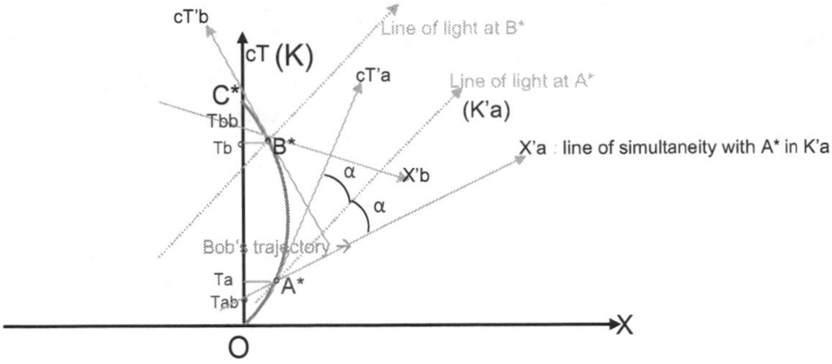

FIGURE 2.8
The lines of simultaneity.

implies that the line of light in A* is the bisector between A*X′a and A*cT′a. Moreover, this line of light can be drawn since it makes an angle of 45° with the OX axis. Consequently, we can draw the line A*X′s, which corresponds to all events that are considered by Bob to be simultaneous with the event A*. This coincides with the axis A*X′a and is labeled as such in Figure 2.8.

This line A*X′s crosses the OcT axis of Alice's frame K at the ordinate Tab. This value corresponds to the age of Alice considered by Bob at the event A* (the simultaneity being relative to Bob). We can see that this time is less than the time of the event A* considered by Alice. However, we cannot easily compare Tab with Bob's age between O and A*, the scales of K and K′ being different (and changing with Bob's speed). Nevertheless, during each small inertial section, we have $dt_a = \gamma . d\tau_b$ (with dt_a being the small time duration incurred by Alice during the small time duration $d\tau_b$ incurred by Bob). Then, since dt_a is always greater than $d\tau_b$, Alice is older than Bob at the event A*. Subsequently, the fact that Tab<Ta does not imply that Bob considers that Alice is younger than him at A*.

Let's now consider an event B*, which occurs in the return phase of Bob's trip: we can construct the similar Minkowski frame K′b with the axis B*cT′ along the small inertial section in the vicinity of B*. We can see that the resulting line X′b, which is the line of simultaneity relative to Bob, crosses the OcT axis at the point Tbb. We notice that unlike in the initial phase, Tbb is greater than Tb, which is the time of the event B* seen by Alice. This time, we can conclude that Bob considers that Alice is older than him, because Tb is greater than Bob's proper time at the event B* due to the same reasoning that in all small sections of Bob's trajectory, we have: $dt_a = \gamma . d\tau_b$. Bob can then conclude that Alice is aging more during his return phase than in his outward journey.

Apparently, if we swap the roles of Alice and Bob in the above diagram, the new diagram will be similar, leading to the conclusion that Bob is older than Alice at the event C*. However, this swap is not allowed because the inertial tangent frame law[7] cannot be applied from Bob's frame since he is not inertial. This means that Bob cannot consider Alice's trajectory as a sum of inertial sections where he could apply the laws of Special Relativity between inertial frames, and then sum up (integrate) the results obtained.

2.7.2.1 Interpretation of This Scenario

This scenario can also help to avoid a frequent misinterpretation, which consists of saying that Bob's proper time has flowed more slowly due to the fact that he is younger than Alice at his arrival. According to this interpretation, Bob's acceleration is responsible for modification in the pace of his proper time. To show that this is not the case, let's imagine a second similar scenario with Bob having a constant proper acceleration: Bob's acceleration is now in the opposite direction to his initial speed, which is not null at the event O*. During his journey, Bob compares his aging rate with Alice's using the same method of the simultaneity lines, as we did in the above scenario. He is then surprised by the results of this method, which say that Alice's proper time does not evolve at a constant pace, despite the constancy of his proper acceleration. This contradicts the assumption that Bob's acceleration affects the pace of his proper time. Besides, such proper pace change would contradict the clock postulate[8] and the proper time universality. Consequently, the explanation for the aging difference is not a pace change in Bob's proper time, but a continuous desynchronization of Bob's and Alice's proper times due to their evolutions along different worldlines (as we saw in Section 1.4.5). Note that the second scenario is more or less the one drawn in the figure, where the trajectory begins and ends with nonzero velocity.

Remark 2.9: It is difficult to imagine such desynchronization effects as we never experience these in real life. Moreover, the wording "time dilatation" is somewhat misleading as it does not convey the idea of desynchronization, but a change in the duration of a standard period of time, which is wrong. We first encountered the desynchronization effect with Einstein's train scenario, where each clock placed at each window appeared from the ground to be desynchronized, but still beating at the same pace. This showed that the clock location has an impact on its synchronization. A parallel illustration with the distance may help realizing the desynchronization effect on the twins: imagine two persons walking at the same pace, and with each step measuring the same length. They make the same number of steps; hence, you expect them to cover the same distance. However, if one of them is walking on a treadmill, it won't be the case.

2.7.3 The Gamma Factor Composition Law

We will show the relation between the Gamma factor in K′ and in K:

$$\gamma_{w'} = \gamma_v^2\left(1 - vw/c^2\right).$$

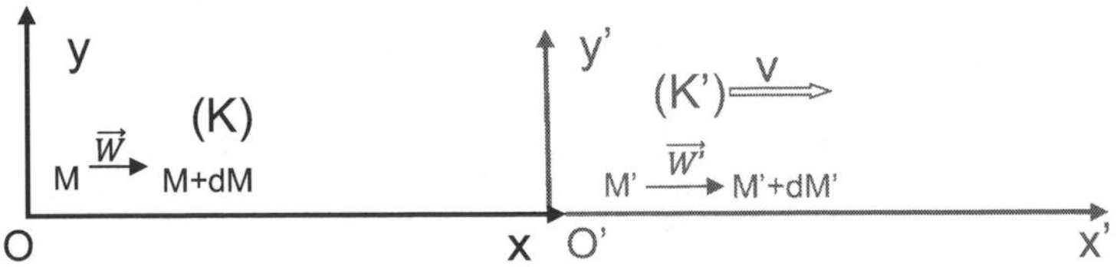

FIGURE 2.9
The object at the event M* has the velocity $\vec{w}$ in K, and $\overrightarrow{w'}$ in K′.

Consider an object that is moving along the OX and O′X′ of K and K′. Its trajectory is a succession of events; hence, we will consider two close events: at the time τ of the object's proper time, the object is characterized by the event $M^*(\tau)$, and then at the time $(\tau + d\tau)$ by the event $M^*(\tau + d\tau)$. The space–time separation vector $[M^*(\tau)\ M^*(\tau + d\tau)]$ is denoted by $\overrightarrow{dM^*}$ (Figure 2.9).

This same vector $\overrightarrow{dM^*}$ is seen in K as $\overrightarrow{dM}$ (dx, dy, dz, dt), and in K′ as $\overrightarrow{dM'}$ (dx′, dy′, dz′, dt′). The velocity of this object in K′ is denoted by $\overrightarrow{W'}$, with its magnitude denoted by w′. We have: $w' = dx'/dt'$. In K, $w = dx/dt$.

The gamma factor in K′ appears in the relation: $dt' = \gamma_{w'} d\tau$. Besides, the Lorentz transformation gives dt′ as a function of dx and dt; indeed:

$dt' = \gamma_v \cdot (dt - v/c^2 \cdot dx) = \gamma_v \cdot dt\ (1 - v/c^2 \cdot w)$. Then: $dt' = \gamma_{w'} d\tau = \gamma_v\ dt\ (1 - vw/c^2)$

Besides, we have: $dt/d\tau = \gamma_v$, so finally: $\gamma_{w'} = \gamma_v^2 (1 - vw/c^2)$ ■

2.7.4 Langevin's Travelers Paradox, and Problems for Synchronizing the Time

In an inertial frame K, we have two twins going from the fixed point A to the fixed point B. Both leave A at the same time, and both arrive at B after having passed the same time as seen on their own perfect watches. One twin, named F, is inertial; the other one, named M, has an accelerated trajectory. We want to know who arrives first at the point B, and their possible age difference after their arrivals.

In the mixed-mode graph (Figure 2.10), the time considered from the ground (K) for the trip duration is:

- For the twin F: the length of the segment AB.
- For the twin M: the length of the curvilinear curve from A to B.

The curvilinear length from A to B being always longer than the segment AB, the twin M is seen from the ground having passed a longer time than the twin F, hence F arrives first at B. This is also true seen from any frame thanks to the invariance of the chronological order of events occurring at one point. Then M arrives at B, and as he has the same age as F when F arrived at B (according to the problem statement), he is younger than F (we assume that F stays at the point B after his arrival).

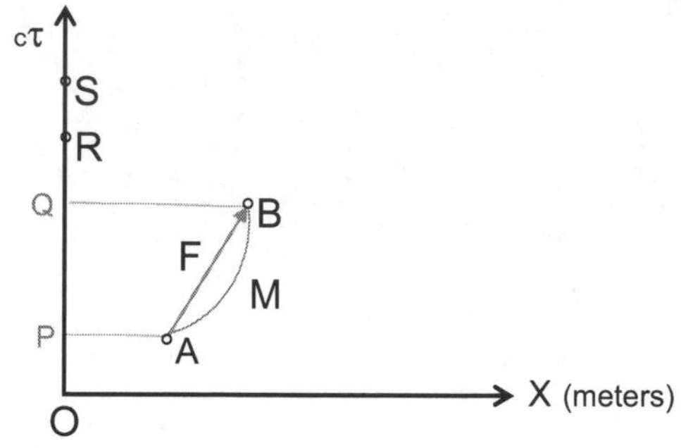

FIGURE 2.10
Langevin's travelers paradox.

Remark 2.10: In the inertial frame where F is fixed, A and B are at the same point, F always has zero velocity and we get the classic version of the paradox.

2.7.4.1 Problem with Time Synchronization

In an inertial frame K, we want to synchronize the times in two points A and B, both fixed in K. To synchronize the time of B with the one of A, we can think of bringing a perfect clock from A to B, this clock being identical to the one of A. We will show that this method does not work, because the motion of the clock will desynchronize it.

We recall that two clocks in A and B are synchronized if an observer located in the middle O of AB constantly sees the times of these clocks displaying the same value.

Let's consider two cases: an ideal case where the moving clock has an inertial trajectory, and a realistic one where it has an accelerated trajectory (Figure 2.11).

Actually, this is the same scenario as previously, with F now being the inertial clock, and M the accelerated one. In the previous schematic, let's mark on the **O**c**τ** axis the points R and S such that: PR=AB and PS is equal to the curvilinear length for M's trajectory between A and B. Then, seen from the ground, the inertial clock arrives at the time given by the length of PR, and the accelerated one, M, at the time given by the length of PS. Also, the length PQ corresponds to the time taken by both clocks measured with their proper times.

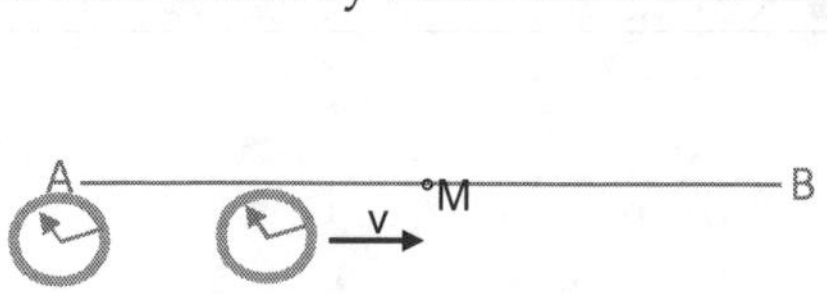

FIGURE 2.11
Time synchronization by moving a perfect clock.

Let's then take an example that is consistent with the considered scenario: PQ=10 ; PR=12; PS=14. These values mean that when the inertial clock beats 10 seconds to go from A to B, the ground time has progressed by 12 seconds. And when the accelerated clock beats 10 seconds to go from A to B, 14 seconds passes in ground time.

This scenario thus shows that an inertial clock will display at the point B a time in behind of the time of the frame K, and that an accelerated clock will display a time, which is even more behind.

Consequently, if we want to synchronize the time in different points of an inertial frame, we cannot do it by moving perfect clocks, but by the method described in Volume I Section 1.3.2.2, which consists of broadcasting signals from a central perfect clock. However, we will further see in the General Relativity part, that the space–time distortions of our universe due to great masses make such a method impossible in the general case.

2.7.5 Minkowski's Complex Space: Lorentz Transformation is a Rotation, Lorentzian Norm is Invariant

We saw in Volume I Section 2.2.2 that the Lorentz transformation can be demonstrated without knowing the Lorentzian norm invariance. Let's now assume that we don't know the law stating the Lorentzian norm invariance, and we will demonstrate it by showing that the Lorentz transformation is a rotation in the Minkowski complex space. Then, it is a mathematical result that any rotation leaves all distances invariant.[9]

Let's then show that the Lorentz transformation is a rotation in the Minkowski complex space: We will adapt the classical Lorentz transformation to Minkowski's complex space, and then show that it is a rotation. We saw that the classical Lorentz transformation is:

$$X' = \gamma(X - vT) \text{ and } T' = \gamma(T - v/c^2\, X)\,.$$

Let's make appear icT and introduce $\beta = v/c$:

$$X' = \gamma(X - v.icT/ic) = \gamma(X + \beta.icT),$$

$$icT' = \gamma(icT - ic.v/c^2 X) = \gamma(icT - i\beta X)\,.$$

Let's put these expressions in the matrix form, and we obtain the Lorentz matrix in Minkowski's complex space:

$$\begin{pmatrix} X' \\ icT' \end{pmatrix} = \gamma \cdot \begin{pmatrix} 1 & -i\beta \\ -i\beta & 1 \end{pmatrix} \begin{pmatrix} X \\ icT \end{pmatrix}$$

This matrix represents a classical rotation as we will see. We will indeed show that the basis vectors of K are transformed into vectors having their norms equal to 1, and that they are all perpendicular one with the other.

Lets' show that the images of the basis vectors have their norms equal to 1:

The image of vector (1, 0) is the vector: $(\gamma, -i\gamma\beta)$; its (classical) norm is: $\gamma^2(1-\beta^2)=1$.

The image of the vector (0, 1) is the vector: $(i\gamma\beta, \gamma)$; its (classical) norm is: $\gamma^2(-\beta^2+1)=1$.

The images of the basis vectors are perpendicular the one with the other: Let's show that the classical scalar product of the images of basis vectors is equal to zero.

This scalar product is: $(\gamma, -i\gamma\beta)\cdot(i\gamma\beta, \gamma) = i\gamma^2\beta + (-i\gamma^2\beta) = 0$.

Consequently, the Minkowski–Lorentz transformation is a rotation, and any rotation preserves the (classical) distances; hence: $X^2 + (icT)^2 = X'^2 + (icT')^2$; so finally:

$$X^2 - c^2T^2 = X'^2 - c^2T'^2 \ \blacksquare$$

Which means that the Lorentzian norm is invariant.

2.8 Mathematical Supplements

2.8.1 A Classical Rotation Conserves the Classical Norm of Any Vector

Any vector $\vec{V}$ (X, cT) can be represented with the point M(X, cT) as we have $\overrightarrow{OM} = \vec{V}$, where O is the origin of the frame. Indeed, given any couple of values (X, cT), there exists a couple (k, α) such that:
$X = k.\sin(\alpha)$ and $cT = k.\cos(\alpha)$. We then have:

- $k^2 = c^2T^2 + X^2$ since for any value of α, we have: $\cos^2(\alpha) + \sin^2(\alpha) = 1$. Thus, k is the classical norm of $\overrightarrow{OM}$.
- $X/cT = k\sin(\alpha)/k\cos(\alpha) = \tan(\alpha)$, then: $\alpha = \text{artan}(X/cT)$.

Any rotation can be expressed as follows:

$$\begin{pmatrix} k\sin(\theta+\alpha) \\ k\cos(\theta+\alpha) \end{pmatrix} = \begin{pmatrix} \cos(\theta) & \sin(\theta) \\ -\sin(\theta) & \cos(\theta) \end{pmatrix} \begin{pmatrix} k\sin(\alpha) \\ k\cos(\alpha) \end{pmatrix}. \tag{2.16}$$

The image of M (X, cT) by the rotation is denoted by M′ (X′, cT′). The relation (2.16) shows that: $X' = k.\sin(\alpha+\theta)$ and $cT' = k.\cos(\alpha+\theta)$. Hence, the classical norm of $\overrightarrow{OM'}$ is k^2, which is equal to the norm of $\overrightarrow{OM}$. ■

2.8.2 A Classical Rotation Conserves the Scalar Product

To show that a classical rotation conserves the scalar product between any couple of vectors, we will first show that the scalar product between any couple of vectors, $\vec{u}$ and $\vec{v}$, verifies the relation:

$$2\vec{u}\cdot\vec{v} = \left|\vec{u}+\vec{v}\right|^2 - \left|\vec{u}\right|^2 - \left|\vec{v}\right|^2. \qquad (2.17)$$

Then, the invariance of the norm of any vector by a rotation will imply the invariance of the scalar product.

Note that the invariance of the scalar product implies the invariance of the angles.

Demonstration of the relation (2.17):

The norm of a vector, $\left|\vec{u}\right|^2$, can be written as the scalar product of this vector with itself: $\left|\vec{u}\right|^2 = \vec{u}\cdot\vec{u}$.

Let's then calculate the norm of the vector $\left(\vec{u}+\vec{v}\right)$:

$\left|\vec{u}+\vec{v}\right|^2 = \left(\vec{u}+\vec{v}\right)\cdot\left(\vec{u}+\vec{v}\right) = \left|\vec{u}\right|^2 + \left|\vec{v}\right|^2 + 2\vec{u}\cdot\vec{v}$

This shows that: $2\vec{u}\cdot\vec{v} = \left|\vec{u}+\vec{v}\right|^2 - \left|\vec{u}\right|^2 - \left|\vec{v}\right|^2$ ■

In addition, this result shows that given a norm, it is possible with this relation to define the corresponding scalar product between any pair of vectors.

Note that this demonstration is general. In particular, this method enables us to define the Lorentzian scalar product knowing the Lorentzian norm, and this will be further developed with the tensors (cf. Section 4.1).

2.8.3 Reminder on the Main Hyperbolic Functions

2.8.3.1 Sinh, Cosh and Tanh Curves

See Figure 2.12.

2.8.3.2 Artanh Curve

See Figure 2.13.

2.8.3.3 Useful Hyperbolic Trigonometric Relations

Definitions: $\cosh(x) = \frac{1}{2}\left(e^x + e^{-x}\right)$ and $\sinh(x) = \frac{1}{2}\left(e^x - e^{-x}\right)$.

Relations with sine and cosine: $\cosh(ix) = \cos(x)$ and $\sinh(ix) = -\, i.\sin(x)$

- $\cosh(x) + \sinh(x) = e^x$; $\cosh(x) - \sinh(x) = e^{-x}$
- $\sinh(\alpha+\beta) = \sinh(\alpha)\cosh(\beta) + \cosh(\alpha)\sinh(\beta)$

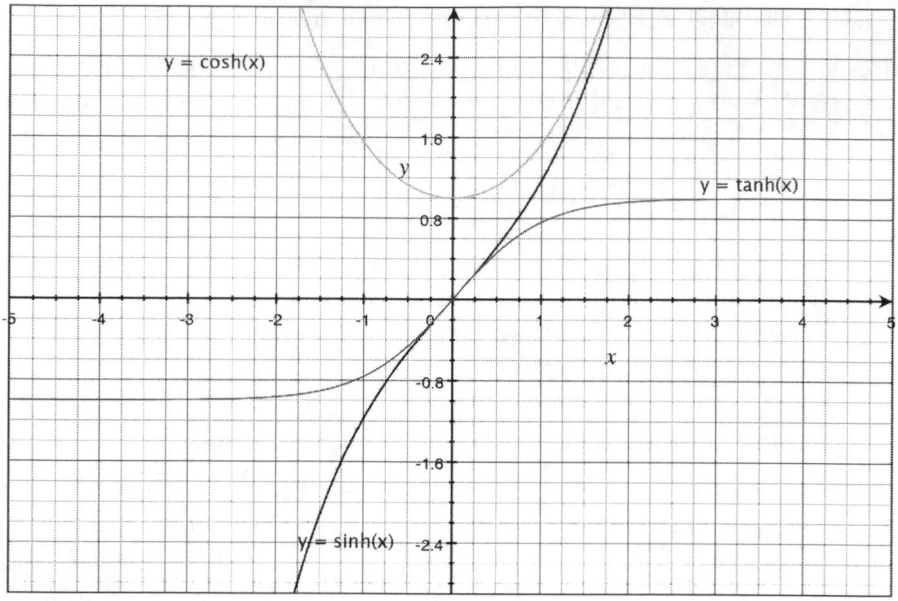

FIGURE 2.12
Sinh, Cosh and Tanh curves. (Robert Harr. Used with Permission.)

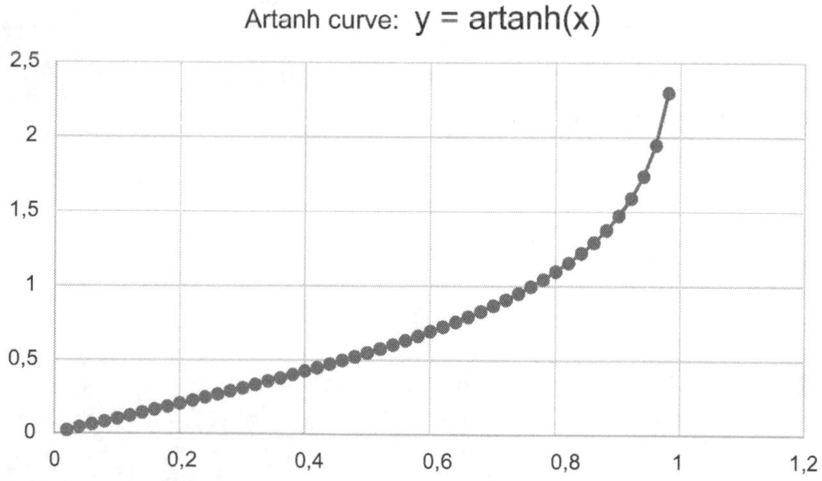

FIGURE 2.13
Artanh curve: y=artanh(x).

- $\cosh(\alpha+\beta)=\cosh(\alpha)\cosh(\beta)+\sinh(\alpha)\sinh(\beta)$
- $\cosh^2(\alpha)-\sinh^2(\alpha)=1$
- $\text{artanh}(\alpha)+\text{artanh}(\beta)=\text{artanh}\left(\frac{\alpha+\beta}{1+\alpha\beta}\right)$

2.8.3.3.1 Useful Hyperbolic Derivatives

- $\sinh'(x)=\cosh(x)$; $\cosh'(x)=\sinh(x)$
- $\tanh'(x)=1-\tanh^2(x)=\frac{1}{\cosh^2(x)}$
- $\text{arsinh}'(x)=\frac{1}{\sqrt[2]{x^2+1}}$; $\text{arcosh}'(x)=\frac{1}{\sqrt[2]{x^2-1}}$
- $\text{artanh}'(x)=\frac{1}{1-x^2}$.

2.9 Questions and Problems

Problem 2.1) The Train and Tunnel Scenario

A train is moving and will soon enter a tunnel. This tunnel has the same length as the train when the train is at rest. A bomb is placed at the front of the train, and a person located at the very back of the train has a button, which triggers the bomb explosion at the front. This person presses the button when he is at the entrance of the tunnel. *Question: Does the bomb explode inside or outside the tunnel?*

Q1) Seen from the ground, is the train shorter or larger than the tunnel?

Q2) Seen from the train, is the train shorter or larger than the tunnel?

Q3) *We will first assume that the signal from the button in the back of the train takes no time to arrive at the front of the train.*

Q3A) For an observer on the ground: does the bomb explode inside or outside the tunnel?

Q3B) Same question, but for an observer inside the train?

We will now assume that the signal from the button in the back of the train moves at the speed c along its wires until it reaches the front of the train. We will denote by K the tunnel frame and K′ the train frame; both share the common origin-event O*, which is: "the back of the train is at the entrance of the tunnel". The OX and O′X′ axes are along the tracks. We denote by L the proper length of the train. We want to know if the explosion will occur inside or outside the tunnel, and to this end, we will successively apply two methods:

Q4) Method 1: Calculate in both frames the time at which the explosion will occur. Subsequently, calculate in both frames the distance at which this explosion will occur.

Q4A) For an observer inside the train, does the bomb explode inside or outside the tunnel?

Q4B) The same question, but for an observer outside the train?

Q5) Method 2: Calculate in the train frame the time at which the explosion will occur. Subsequently, apply the Lorentz transformation to know the distance to the explosion seen on the ground, and conclude.

Q6) Answers 3A) and 3B) led to a contradiction. Is it a real or an apparent contradiction, and why?

Problem 2.2) Solve the Previous Train and Tunnel Problem with a Minkowski–Loedel Diagram

Q1) Make a Minkowski–Loedel Diagram representing the previous scenario, with K being the ground frame, and K′ the train frame. Their common origin-event is: "the back end of the train is in front of the point O, which is at the entrance of the tunnel". Then answer the previous question 1).

Q2) Answer the previous question 2)

Q3) Answer the question: Will the bomb explode inside or outside the tunnel?

Problem 2.3) Bob and Alice Case Study, Part 2: Surprising Effects When Traveling in the Cosmos

The first part of this case study is presented in Volume I, Problem 2.3; we will now address further questions:

Bob and Alice have the same age, 30, when Bob is undertaking a journey into the cosmos while Alice stays on Earth. Bob and Alice Bob set their clocks at zero when Bob leaves Alice. Bob's rocket goes at the very fast speed of 0.943 c along a straight line; this speed corresponds to the gamma value of 3.00. Bob's clock shows 10 years when he arrives at the distance of the first planet, named Proxima. We will consider both Alice's and Bob's frames to be inertial, despite the circular movements of the Earth and the acceleration of the rocket.

Note that the next problem consists of representing in a Minkowski–Loedel diagram the different points addressed in this problem. The reader is invited to do both problems at the same time because the representations in a diagram also help in understanding the phenomena.

Q1) We previously saw that at the event "Bob reaches Proxima", Alice's clock marks 30 years. Then, at the year 30 of Alice's clock, she initiates a phone call in order to congratulate Bob for arriving at the distance of Proxima. Let's now place ourselves in Bob's rocket frame: what time is it when Alice initiates her call?

Q2) Alice wants her phone call to reach Bob exactly when he arrives at Proxima. At what time of her clock must she initiate her call?

Q3) At the year 30 for Alice's clock, she knows Bob arrives at the distance of Proxima, so she takes her very precise telescope, points it in the direction of Bob's rocket, which has a clock on its surface: what hour does she see?

Problem 2.4) Minkowski–Loedel Diagram of Bob and Alice Case Study, Part 2

Q1) Represent the previous scenario in a Minkowski–Loedel diagram. Answer the previous questions by showing the corresponding points on the diagram.

Q2) Show the distance Earth-Proxima considered from the Earth.

Problem 2.5) Time Dilatation and Desynchronization Effect

Represent the scenario of Problem 2.14) of Volume I with a Minkowski–Loedel diagram and answer questions 1) and 2).

Problem 2.6) Hyperbolic Rotation Angle Calculation

Q1) The frame K′ is moving at the speed 0.75 c relative to the frame K. What is the angle of the hyperbolic rotation that corresponds to the Lorentz transformation from K to K′? We give the value of the gamma factor for 0.75 c=1.512. You may either give an approximate value by looking at the hyperbolic curves (cf. Section 2.8.3) or use a calculator and give the exact value.

Q2) We give the angle $\theta=1.76$ for the hyperbolic rotation corresponding to an inertial frame boost. What is the speed of K′ relative to K?

Problem 2.7) Velocity Composition

Using the artanh curve (cf. Section 2.8.3.2), give an approximation of the following velocities composition, given that all velocities are parallel to the same direction.

Case 1: $V1=0.4c$; $V2=0.6c$. What is the value of $W=V1 \oplus V2$ =?

Case 2: $V1=0.6c$; $V2=0.8c$. What is the value of $W=V1 \oplus V2$ =?

Problem 2.8) An Accelerated Rocket with Constant Power

A rocket M moves due to an engine having a constant relativistic power, meaning that its Four-Force is constant. Its relativistic Momentum is denoted by $P(\tau)$, with τ being the rocket's proper time. We thus have: $\frac{dP}{d\tau}=k=m\cdot\frac{d^2M}{d\tau^2}$. We will assume that the rocket mass remains constant (which is questionable in reality). We want to represent the rocket trajectory in a mixed-mode graph, where the OX axis is along the rocket trajectory, and the rocket's proper time τ is along the time OcT axis. We set the origin-event of the frame so that the rocket is in front of O at $\tau=0$.

Q1A) Express the rocket trajectory, meaning the relation between c τ and its abscissa x.

Q1B) What curve is it? Represent this curve in a mixed-mode graph.

Q2A) When the rocket reaches the point A of abscissa Xa, its travel time measured with the rocket time is τ_A. The time of this event A*(point A; time τ_A) seen from the ground frame is denoted by Ta. Give a qualitative answer as to the way of determining Ta. Then compare the time of the event A* seen from the ground and from the Rocket.

Q2B) Find a mathematical expression giving the time Ta. The general mathematical expression suffices, it is not required to make the calculation.

Problem 2.9) The Event Horizon and the Rocket Trajectory – Part 1

Mathematically, there are many possible curves that match Bob's rocket trajectory as shown in Figure 2.6. Relatively simple ones are hyperbolas, knowing that these curves have an asymptote. Let's then consider the case where Bob's trajectory is a hyperbola. We denote by θ the curve parameter, and we use natural units, with distances and time expressed in light-years.

In order to facilitate unit conversions and to have arguments of the functions that are dimensionless, we introduce the constant λ such that when x = 1 light-year, we have: $x/\lambda = 1$ without dimension.

We set the following hyperbolic trajectory: $x/\lambda=\cosh(\theta) - 1$ and: $ct/\lambda = \sinh(\theta)$.

Q1A) Express the value and the dimension of λ using the international system of units.

Q1B) Express the equation of the curve, and then determine its asymptote.

Q1C) Determine the slope of the asymptote of the trajectory.

Q2A) We want to express Bob's trajectory with his proper time as the parameter, in the usual way. To this end, we will express the Lorentzian norm of a small space–time interval dM along Bob's trajectory as a function of θ. Then express τ as a function of θ, setting = 0 when Bob is at the point O.

Q2B) Determine the equation of the asymptote of the trajectory.

Q3) Express the rocket speed as a function of τ. What is the limit of the rocket speed when τ tends to the infinite?

Q4) Express the rocket acceleration $\frac{dV}{dt}$ as a function of τ.

Q5) Calculate the acceleration seen by Bob in his inertial tangent frame, denoted by $A'(\alpha)$. Please comment on the result.

Problem 2.10) The Event Horizon and the Rocket Trajectory – Part 2

Q1) We want Bob's acceleration to be a constant, denoted by a′, as seen from his inertial tangent frame. We choose a value of a′ that is easily bearable by Bob: $a'=9.8\ m/s^2$. Give the parametric equations of Bob's trajectory, the parameter being his proper time.

Q2) Express these equations with natural units, where distances are expressed in light-years and the times in years. Make the calculations of x and t for Bob's proper time duration of 1 year.

Q3) Make the calculations of x and t for Bob's proper time duration of 10 years, and then comment on the difference with Q2).

Q4) What is the year after which Bob cannot receive any information from Alice who stays at point O?

Problem 2.11) The Line of Simultaneity from a Uniformly Accelerated Spaceship

Bob is with his twin Alice at the point O on the ground (assumed inertial) and starts a journey in the cosmos on the day of their 30th birthday. This departure defines the event O*, with Bob's and Alice's clocks being set to zero: Tb=Ta=0. Bob's spaceship has a constant acceleration relative to its inertial tangent frame (ITF), which is: $a' = 9.5\ m/s^2$. Then Bob, after 1 year of his time, wonders: is it also Alice's birthday today? Or is she older or younger than me? Before answering these questions, there are some intermediate questions:

Q1) We denote by A* the event corresponding to Bob, 1 year after his departure as seen with his proper time, denoted by τ.What is Alice's age at this event A*?

For the sake of simplicity, we will make the approximation that $c/a'=31\ 557\ 101$ seconds ≈ 1 year.

Q2A) What is the distance of Bob at the event A*, as seen by Alice?

Q2B) Make a Minkowski diagram with K being Alice's frame, and show the point representing the event A. Use the light-year as a natural unit.

Q3A) What is Alice's age at this event A*, as considered by Bob? Make a qualitative answer using the previous Minkowski diagram.

Q3B) Make an exact calculation of Alice's age at this event A*.

Problem 2.12) The Proper Acceleration

Demonstrate that the proper acceleration is:

$$\frac{\overrightarrow{dS}}{d\tau}=\left(\frac{dc\gamma}{d\tau},\frac{d\vec{v}\gamma}{d\tau}\right)=\left[\gamma^4\left(\vec{a}\cdot\vec{\beta}\right),\gamma^4\left(\vec{a}\cdot\vec{\beta}\right)\vec{\beta}+\gamma^2\vec{a}\right],\ \text{where}\ \vec{\beta}=\frac{\vec{v}}{c}\ \text{and}$$

$\vec{a}=\frac{\overrightarrow{dv}}{dt}$. (Hint: use similar reasoning as in Section 1.5.3.1.)

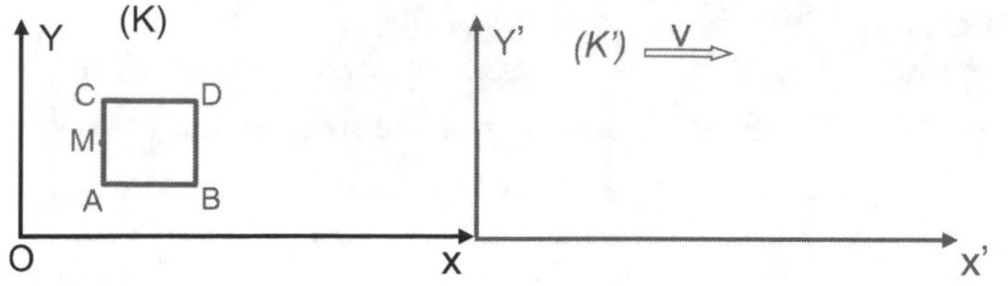

FIGURE 2.14
How is a cube in K seen in K′?

Problem 2.13) Actual Appearance of a Cube Seen from a Fast Spaceship

This problem is a continuation of Problem 2.1) of Volume I, and it is a simplified version of the "Penrose-Terrell effect". Imagine that you are in an extremely fast spaceship, represented by the frame K′. You pass near a space station that has a cubic shape in its rest frame, with each side measuring 10 m. Your velocity, v, relative to this cubic station is 0.9 c (Figure 2.14).

Q1) At the time T_0' of K′, you see in front of you the middle M of the segment AC of the station. What shape has the side AC that you actually see, taking into account the differences in the distances covered by the photons emitted from different points of the station? Indicate the general shape only.

Q2) Still at the time T_0' of K′, your eye is at the point denoted by E, and the distance EM is 20 m as measured in K′. The difference of the abscissas in K′ of the points M and C is denoted by x. Express the relation giving x, and check that the value x=0.561 m satisfies this relation.

Q3) Give the abscissa in K′ of a random point N in the segment AC as seen by your eye at T_0'. This abscissa difference is again denoted by x, and the proper distance MN is denoted by y.

Q4) Still at the time T_0' of K′, are some points of the front side of the station visible to you?

Problem 2.14) Using Hyperbolic Trigonometry, Calculate the Determinant of the Lorentz Matrix

Notes

1 Cf. Section 5.1.1.
2 The corresponding demonstration is in the Mathematical Complement Section 2.8.1.
3 An example is given with Problem 2.8.
4 This point will be further discussed in Section 3.3.
5 The word "light-year" is used to facilitate realizing the magnitude of the distances, but the word "year" is more appropriate to mean that the natural units are used.
6 Cf. the Inertial Tangent Frames Volume I Section 3.5.3.
7 Cf. Volume I Section 3.5.3.
8 Cf. Volume I Section 3.5.2.

3

The Relativistic Laws of Dynamics

Introduction

The relativistic Momentum and energy will be demonstrated with a 3D approach, showing that both are derived from a single 4D conservation law. Then, the important novelties brought by the relativistic energy-momentum will be presented, including nuclear reactions. These will be examined from both a theoretical perspective and the description of important applications, such as thermonuclear fusion in the Sun and Uranium fission. Then, an in-depth presentation of relativistic Forces will be given and their important properties will be demonstrated. An update on the potential energy will follow, including its adaptation to Relativity. Then, the solution brought by Relativity with regard to Maxwell's laws will be explained: the electrical force and the magnetic one actually are one single entity. The Lagrangian variational approach will then be presented, as it is another important method to demonstrate the relativistic Momentum and energy, as well as other laws. Then, other applications and case studies will be presented, such as the Compton effect and Einstein's photon box thought experiment.

3.1 Energy and Momentum from 3D to 4D

We will show that the two fundamental 3D conservation laws for energy and momentum actually stem from a single 4D conservation law. To this end, we will show that each 3D law separately implies the other, and that both merge into a single 4D law.

3.1.1 The Conservation of the 3D Quantity, $\vec{p} = m\vec{v}\gamma_v$, Implies the Conservation of a Scalar Quantity: Energy

Consider an isolated system composed of n objects. A point object M of mass m, is seen in an inertial frame K moving at the velocity $\vec{v} = \left(\frac{dx}{dt}, \frac{dy}{dt}, \frac{dy}{dt}\right)$.

DOI: 10.1201/9781003201359-3

We saw that the 3D expression, $\vec{\mathbf{p}} = \mathbf{m}\vec{\mathbf{v}}\boldsymbol{\gamma}$, satisfies the system Momentum conservation law in any inertial frame (cf. Volume I Section 4.2.2). Thus:

$$\overrightarrow{p_s} = \sum_{i=1}^{n} \overrightarrow{p_i} = \overrightarrow{cst}.$$

We will now demonstrate that this implies the existence of a scalar quantity attached to the object M, denoted by e, which is such that the sum of the n quantities e_i *of the n objects of the isolated system is constant. This will lead to the relativistic definition of energy, and to the fourth dimension of the 4-Vector Momentum.*

Knowing that: $\gamma = \frac{dt}{d\tau}$, the above 3D relativistic impulse $\vec{P}$ becomes:

$$\vec{p} = m\vec{v}\gamma = m\left(\frac{dx}{dt}\frac{dt}{d\tau}, \frac{dy}{dt}\frac{dt}{d\tau}, \frac{dy}{dt}\frac{dy}{d\tau}\right) = m\left(\frac{dx}{d\tau}, \frac{dy}{d\tau}, \frac{dy}{d\tau}\right).$$

This 3D expression invites us to consider 4D quantity $\vec{Q}$ defined as: $\vec{Q} = m\left(\frac{dt}{d\tau}, \frac{dx}{d\tau}, \frac{dy}{d\tau}, \frac{dz}{d\tau}\right) = m\frac{\overrightarrow{dM}}{d\tau}$, because this 4-vector $\vec{Q}$ is contravariant[1], meaning that when changing frame from K to K′, we have: $\overrightarrow{Q'} = (\Lambda)\vec{Q}$ with (Λ) being the Lorentz transformation.

Let's denote the time part of $\overrightarrow{Q'}$ by e, and so we have: $\overrightarrow{Q'} = (e, \vec{p})$, with: $e = m\frac{dt}{d\tau} = m\gamma_v$.

We want to show that Σe_i is constant; to this end, we will show that the quantity: $\overrightarrow{Q_s} = \sum_{i=1}^{n} \overrightarrow{Q_i}$ is constant: We have:

$$\overrightarrow{Q_s} = \Sigma(e_i, \overrightarrow{p_i}) = \left(\Sigma e_i, \sum_{i=1}^{n} \overrightarrow{p_i}\right) = (\Sigma e_i, \overrightarrow{p_s}).$$

Consider two states of this system seen in K, denoted by S_1 and S_2: $\overrightarrow{Q_{s1}} = (\Sigma e_{i1}, \overrightarrow{p_s})$ and $\overrightarrow{Q_{s2}} = (\Sigma e_{i2}, \overrightarrow{p_s})$. The system being isolated, $\overrightarrow{p_s}$ remains constant.

We will show that the fact that $\overrightarrow{p_s}$ is constant whatever the frame where $\overrightarrow{Q_s}$ is considered, implies the constancy of its time part, which is Σe_i. To this end, we will use a Minkowski diagram with two frames K and K′, and take advantage of the contravariance of $\overrightarrow{Q_s}$, meaning that this vector is represented by the same point in both K and K′, like an event.

For the sake of clarity, we will only represent the component of $\overrightarrow{p_s}$ which is along the X-axis, denoted by Xs in K and X′s in K′, knowing that the same reasoning applies to the other components. Likewise, Σe_{i1} will be denoted by T_1 in K, and Σe_{i2} we will denoted by T_2. In K′, these quantities will be denoted by T_1' and T_2'.

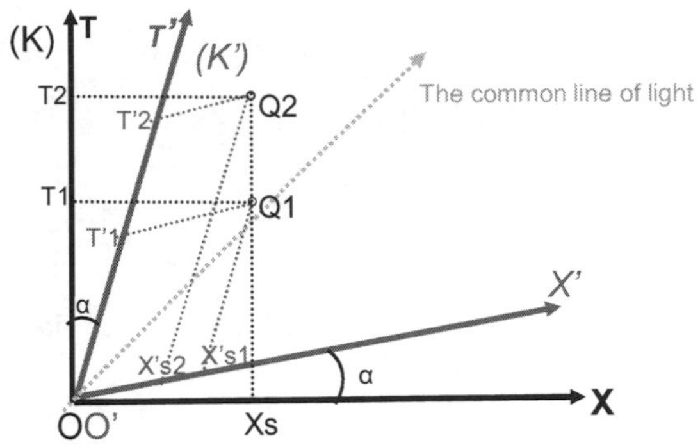

FIGURE 3.1
Minkowski Diagram showing the System Momentum in K and in K′. The equality of the spatial parts of Q1 and Q2 in K implies that their time parts are different in K′.

Thus, the vectors $\overrightarrow{Q_{s1}} = (\Sigma e_{i1}, \overrightarrow{p_s})$ and $\overrightarrow{Q_{s2}} = (\Sigma e_{i2}, \overrightarrow{p_s})$ are, respectively, represented by the points $Q1(T_1, X_{s1})$ and $Q2(T_2, X_{s2})$, as if they were events.

The beauty of the Minkowski diagram is that in the frame K′, the vectors $\overrightarrow{Q'_{s1}}$ and $\overrightarrow{Q'_{s1}}$ are represented by the same points Q1 and Q2; their components in K′ are: $Q1(T'_1, X'_{s1})$ and $Q2(T'_2, X'_{s2})$. Figure 3.1 clearly shows that X'_{s1} is different from X'_{s2}, which contradicts the system Momentum conservation in K′. The only possibility for having the same abscissas in K′ is that points Q1 and Q2 also have the same ordinates, meaning that Σe_i is constant. ■

Consequently, the scalar $e = m\dfrac{dt}{d\tau} = m\gamma_v$ corresponds to the relativistic energy, and the 4D quantity $\vec{Q} = (e, \vec{p})$ to the relativistic Momentum, as shown in Volume 1 Section 5.1.1.

Conversely, we will now see that the same reasoning enables us to show that the system energy conservation law, i.e., the first principle of thermodynamics, implies the system Momentum conservation law, using the 3D momentum expression: $\vec{p} = m\,\vec{v}\,\gamma_v$.

3.1.2 Energy Conservation Implies the Conservation of the 3D Momentum $\vec{p} = m\vec{v}\gamma_v$

We will show that if the quantity, $e = m\,\gamma_v$*, associated to an object of mass m and moving at the speed v in K, satisfies the closed system energy conservation law,*[2] *then the 3D quantity,* $\vec{p} = m\vec{v}\gamma_v$*, satisfies the closed system Momentum conservation.*

Moreover, this will imply the existence of a 4D quantity matching the requirements of the relativistic Momentum: $\vec{P} = m\,\dfrac{\overrightarrow{dM}}{d\tau}$.

We notice that: $e = m\gamma_v = m\,\frac{dt}{d\tau}$. Hence, we are again induced to define the 4D contravariant quantity $\vec{Q}$ such that:

$$\vec{Q} = m\left(\frac{dt}{d\tau}, \frac{dx}{d\tau}, \frac{dy}{d\tau}, \frac{dy}{d\tau}\right) = (e, m\vec{v}\gamma_v) = (e, \vec{p}).$$

Let's then have a closed system in K, seen at two different states: $\overrightarrow{Q_{s1}} = (\Sigma e_i, \overrightarrow{p_{s1}})$ and $\overrightarrow{Q_{s2}} = \left(\Sigma e_i, \overrightarrow{p_{s2}}\right)$ with $\overrightarrow{p_{S1}} = \sum_{i=1}^{n} m_{1i}\overrightarrow{v_{1i}}\ \gamma_{1i}$ and $\overrightarrow{p_{S2}} = \sum_{i=1}^{n} m_{2i}\overrightarrow{v_{2i}}\ \gamma_{2i}$.

This time, it is the sum of the energies that is constant whatever the inertial frame, and we want to show that the sum of the 3D momenta is also constant, meaning that the spatial part of $\vec{Q}$ is constant.

Let's again make a Minkowski diagram, and represent along the OX-axis of the diagram the first component of the 3D momenta, $m\frac{dx}{d\tau}$, knowing that the same reasoning applies to the two other spatial components (Figure 3.2).

We use the same notations as previously, so that the two states of the system are represented by the points Q1 and Q2, with this time the sum of the energies being constant: T1=T2. We can see that T'1 is different from T'2, which contradicts the system energy conservation law. The only possibility to respect this law is that Q1 and Q2 also have the same abscissas, meaning that the component along the X-axis of the system Momentum is constant ∎

These two demonstrations show that neither of the two fundamental 3D conservation laws, respectively, relating to energy and momentum is more primitive than the other. This is consistent with Minkowski's finding that we are in a 4D space–time universe, and consequently, the physical laws are

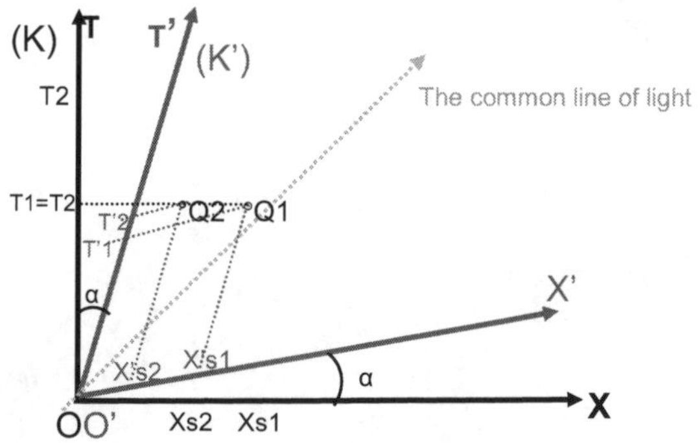

FIGURE 3.2
Minkowski Diagram The equality of the time parts of Q1 and Q2 in K implies that their spatial parts are different in K'.

4D by essence. This further justifies the 4D approach used in the Momentum presentation of Volume I Section 4.2.1.

3.1.3 Geometrical Representation of the System Momentum Conservation and of the CoM Frame

The 4-Vector Momentum can be represented by a space–time separation 4-vector in a Minkowski frame, with its abscissa being its spatial momentum, $mv\gamma$, and its ordinate being its energy. For the sake of simplicity, we will skip the arrows on the 4-vectors as is commonly done (Figure 3.3).

In Figure 3.3, the Momenta in K of two objects are represented by the 4-vectors $P_1(E_1, p_1)$ and $P_2(E_2, p_2)$. The system Momentum is expressed by the vectorial sum of the Momenta, $P_1 + P_2$. This means that after any collision or transformation, the isolated system Momentum conservation implies that the objects after their evolution will always have the sum of their Momenta unchanged: $P_1 + P_2 = \text{cst}$.

The properties of the Minkowski diagram still apply, the Momentum being contravariant. Hence, if we have a second inertial frame K′ whose Energy axis is along the 4-vector $P_1 + P_2$, then in K′ the spatial part of the $P_1' + P_2'$ is null and this means that K′ is the Center of Momentum frame. For instance, if the two objects merge into one object, this object will be fixed in K′. Also, the slope of the Momentum vector is: $\dfrac{m\,c\,\gamma}{m\,v\,\gamma} = \dfrac{1}{\beta}$, like with the Minkowski diagram representing events.

We will now see some important applications of these laws, knowing that most of them were not conceivable at the time they were formulated.

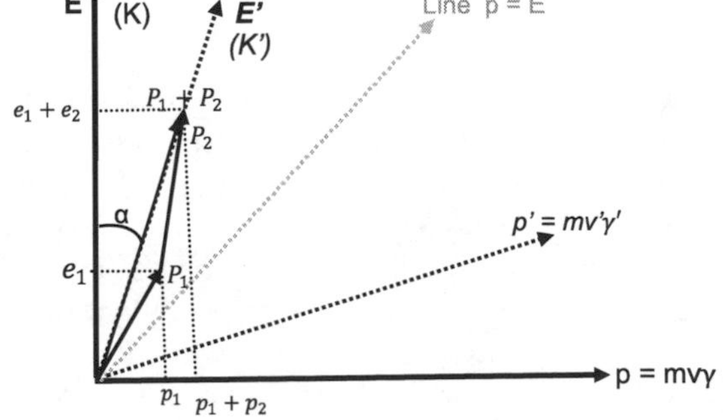

FIGURE 3.3
Minkowski diagram showing the System Momentum and the Center of Momentum frame (K′).

3.2 Nuclear Reactions

Radioactivity means disintegration of the nucleus, and some nucleus indeed spontaneously disintegrate, but not into all its nucleons. We will present three main types of decay that some nuclei undergo. Then, we will address the important nuclear reactions of fission and fusion, first from a theoretical perspective, then with a qualitative description of the proton-proton chain. We will end with additional theoretical considerations.

3.2.1 The Three Main Types of Decay

3.2.1.1 Alpha Decay

The α decay consists of fission with the emission of a nucleus of Helium-4:

$$(A, Z) \rightarrow (A-4, Z-2) + {}^{4}\text{He}.$$

The Helium-4 nucleus has 2 protons and 2 neutrons, and is very stable due to its high binding energy. It is called an "α particle". Helium-4 is nearly as abundant as Hydrogen, and present in the cosmic radiation.

The α decay mainly concerns heavy atoms and involves the strong interaction (strong force), which is responsible for binding the nucleons into atomic nuclei. It is very strong but with an extremely short range: at 10^{-15} meter, it is approximately 137 times stronger than electromagnetism; at longer distances, it fades rapidly.

Some history: in 1896, the French scientist H. Becquerel discovered when he studied the phosphorescence of Uranium salts, that some elements spontaneously emit light: they indeed exposed a photographic plate even when there was no light. A few years later, the French scientists Marie and Pierre Curie worked on such luminescent elements, which they called "radioactive," and they succeeded in identifying the first ones: Polonium and Radium. Radium was transformed into Radon gas.

3.2.1.2 Beta Decay

The beta decay transforms a neutron into a proton with the emission of an electron and an anti-neutrino. Alternatively, a proton is converted into a neutron with the emission of a positron and a neutrino. This only incurs within a nucleus having a proton excess, since the proton itself is stable.

Beta particles are very high-speed electrons or positrons. The beta decay involves the weak interaction, which has an even shorter range than the strong force (about 10^{-17}m), and its strength is about 100 times less.

An isolated neutron has a short lifetime, approximately 15 mn, and transforms itself into one proton, one electron and one anti-neutrino.

Some history: The neutrino existence was first postulated in 1930 by W. Pauli after making a very precise analysis of the energy balance of beta decay. The neutrino is difficult to detect as it interacts only via gravity and the weak interaction. They were first observed in 1956 near a nuclear reactor (where they are produced in great numbers), and due to their interaction with protons.

3.2.1.3 Gamma Decay

The gamma decay consists of the emission by a nucleus of one or several very high energy photon(s): gamma rays, which are more powerful than X-rays. The nucleus simultaneously reduces its energy state. Alternatively, a nucleus can absorb such photon(s) while increasing its energy state. In both cases, the energy states after the decay are well defined for each nucleus.

Some history: Gamma radiation was discovered in 1900 by the French scientist P. Villard while studying the radiation emitted by Radium.

3.2.2 Fusion and Fission: Theoretical Conditions

3.2.2.1 Nuclear Binding Energy

Consider a particle denoted by "a" which disintegrates, giving birth to other particles: $a \rightarrow b+c+d+\ldots$ The system energy conservation in the frame where the particle "a" was at rest yields: $m_a c^2 = m_b c^2 \gamma_b + m_c c^2 \gamma_c + m_d c^2 \gamma_d + \ldots$

Consequently, for a spontaneous disintegration to be possible, we must have: $m_a > m_b + m_c + m_d + \ldots$. We notice that there is a mass defect in the resulting elements.

Conversely, a stable particle satisfies:

$$\mathbf{m_a < m_b + m_c + m_d + \ldots} \tag{3.1}$$

Thus, a stable element has a mass defect compared to the sum of the masses of its components, and it is all the more stable as its mass defect is important.

A nucleus of mass number A is composed of A nucleons, of which Z protons and (A – Z) neutrons. A nucleus does not spontaneously disintegrate into all its nucleons, hence the above condition (3.1) yields:

$$\mathbf{m(A, Z) < Zm_p + (A - Z)m_n}$$

where m_p and m_n are, respectively, the proton and the neutron mass.

Let's multiply by c^2 to represent the energy, and this condition becomes:

$$0 < Zm_p c^2 + (A - Z)\, m_n c^2 - m(A, Z)c^2.$$

Consequently, the nuclear binding energy was defined as:

$$\mathbf{B(A, Z) = Zm_p c^2 + (A - Z)\, m_n c^2 - m(A, Z)c^2.} \tag{3.2}$$

The greater the nuclear binding energy of a nucleus, the more stable it is, meaning the more energy must be provided to perform its disintegration. The term B/c^2 is the nucleus mass defect: it is the mass quantity that the nucleons had to change into binding energy when forming the resulting nucleus, and which was released then.

3.2.2.2 Fusion and Fission Conditions

Consider the general case of a reaction in an isolated system composed of several particles, such that all protons and neutrons of the different nuclei are conserved but arranged differently after the reaction. The initial set of atoms is marked with i and the final one with f.

We thus have: $A_{i1} + A_{i2} + \cdots \rightarrow A_{f1} + A_{f2} + \cdots$

For any element k, its energy E_k can be approximated by the relation: $E_k = m_K c^2 + \frac{1}{2}\, m_K v_k^2$.

Hence, the system energy conservation law yields:

$$\sum (m_i\, c^2 + \tfrac{1}{2}\, m_i\, v_i^2) = \sum \left(m_f\, c^2 + \tfrac{1}{2}\, m_f\, v_f^2\right).$$

We then have: $\sum m_i\, c^2 - \sum m_f\, c^2 = \sum \frac{1}{2}\, m_f\, v_f^2 - \sum \frac{1}{2}\, m_i\, v_i^2$.

Besides, from equation 3.2, we have: $\sum\, m_i\, c^2 - \sum\, m_f\, c^2 = \sum B_f - \sum B_i$ since in this nuclear reaction, all protons and neutrons are conserved but arranged differently. Hence:

$$\sum \tfrac{1}{2}\, \mathbf{m_f}\, \mathbf{v_f^2} - \sum \tfrac{1}{2}\, \mathbf{m_i}\, \mathbf{v_i^2} = \sum \mathbf{B_f} - \sum \mathbf{B_i}\,. \qquad (3.3)$$

Thus, we can see that the reaction is exoenergetic if:

$$\sum \mathbf{B_f} - \sum \mathbf{B_i} > 0 \qquad (3.4)$$

The graph in Figure 3.4 indicates the ratio B/A as a function of the atom mass number A. This ratio B/A gives the binding energy per nucleon since A=N+Z.

A nucleus cannot spontaneously undergo fission toward elements with lower B/A, because the resulting system would have lower binding energy; consequently from equation 3.3 its kinetic energy would be negative, which is absurd.

In Figure 3.4, we notice that Iron (Fe) has the highest B/A ratio. Hence, elements that are lighter than Iron cannot spontaneously undergo fission because the sum of the binding energies of the resulting elements will be lower than the initial one. However, fusions are possible between these light elements, resulting in elements having greater binding energy: a famous example is the fusion of Hydrogen that we will see in Section 3.2.3.

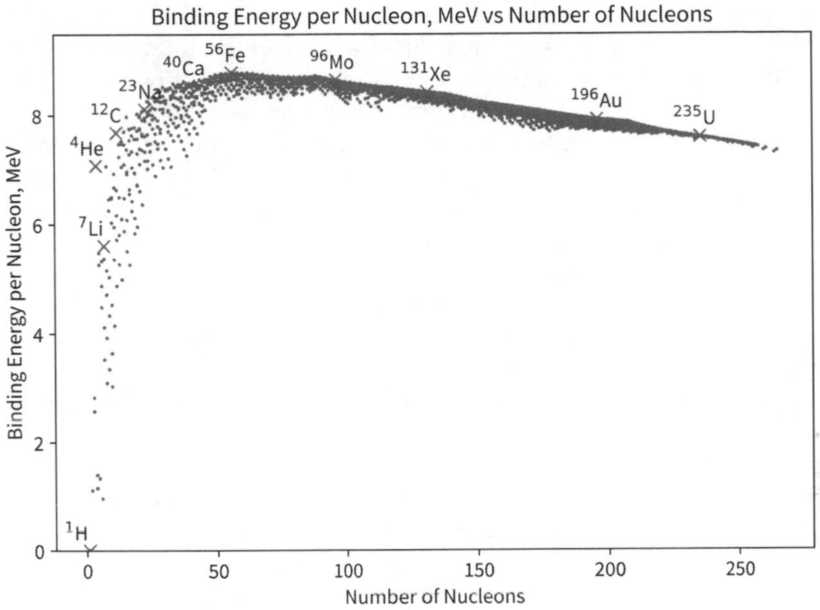

FIGURE 3.4
Variation of binding energy with mass number. (Image from Shutterstock.)

The only elements that may undergo spontaneous fission are then heavier than Iron. The latter is stable, and so are the nuclei having a mass number close to Iron. Fission concerns particularly some very heavy nuclei, such as Uranium 235 or Plutonium 239. Heavy atoms have a high number of protons; these exert a mutually repulsive electrostatic force, whose effect increases with their number. Some very heavy atoms can thus undergo spontaneous fission.

In the case of Uranium 236, which is very unstable, fission can also be triggered by absorption of a neutron. However, the neutron must not be too energetic to be absorbed; then, the Uranium 236 fission liberates neutrons and a chain reaction can follow. When an atom undergoes fission, it produces different sets of atoms while providing them with a significant amount of kinetic energy: for instance, the fission of an atom of Uranium 235 generates energy in the range of 200 MeV, most of which, about 80%, is in the form of kinetic energy endowed to the resulting smaller atoms (Krypton+Baryum+3n) or (Strontium+Xenon 140+2n), making them reach speeds in the range of 10,000 km/s. In the former case, the reaction is:

$$^{235}_{92}\mathrm{U} + 1\mathrm{n} \rightarrow {}^{236}_{92}\mathrm{U} \rightarrow {}^{141}_{56}\mathrm{Ba} + {}^{92}_{36}\mathrm{Kr} + 3\mathrm{n}$$

As these fast moving atoms hit other atoms, their kinetic energy is transformed into thermal energy. However, about 5% of the total energy is given to neutrinos, which pursue their way without any interaction with anything.

The energy produced by the fission of one atom is much greater than the energy involved by chemical reactions, which are in the range of the eV per atom. The reason is that chemical reactions only involve electrons and electromagnetic interaction, which is much weaker than the nuclear interactions involved in fission and fusion.

Fusion: Only elements that are lighter than Iron can incur an exoenergetic reaction by fusion, whereby the resulting binding energy of the system is greater than the initial one. A famous example is the fusion of protons occurring in the core of the Sun, leading to the formation of Helium-4, while generating an enormous amount of energy, mainly through radiation. This phenomenon, called the "Proton–Proton chain (p-p Chain)," is described in the next section.

Remark 3.1: We mentioned that atoms which are heavier the Iron cannot be created by exoenergetic fusion, but they can by endoenergetic reactions. It is assumed to be the case during huge explosions of supernovae, whereby heavy atoms are synthetized and hurled across space.

3.2.3 The Proton–Proton Chain

This part is mainly qualitative and describes the main mechanism by which the Sun transforms part of its mass into radiation.

There are several mechanisms by which the Hydrogen, which is the main component of stars, is transformed into Helium-4 while producing energy mainly through radiation. In stars with masses in the range of the Sun or less, the p-p chain is dominant, whereas in more massive stars, the CNO cycle[3] (Carbon–Nitrogen–Oxygen) is dominant.

There are actually several p-p chains, and the most frequent one in the Sun is presented hereafter: the p-p chain is a series of reactions, the product of one reaction being the starting material of the next one. It starts with the fusion of two protons (i.e., Hydrogen atom), which must break the electrostatic barrier between two positive charges, and this requires the protons to have an extremely high kinetic energy, corresponding to temperatures beyond 10 million Kelvin. Such circumstances occur in the core of the stars, where the pressure due to gravity is extreme, resulting in extremely high temperatures (i.e., kinetic energy). However, the probability of a proton to succeed in fusing with a partner is extremely low, this process being mediated via the weak interaction. Indeed, an average proton waits for 9,000 million years before such an event to occur. This explains why the Sun, fortunately, does not burn all its fuel at once. This p-p fusion forms an atom of Deuterium, while emitting a positron and a neutrino: $p + p \rightarrow {}_1^2D + e^+ + \nu_e$

The positron will rapidly meet with an electron: they will annihilate while emitting 2 gamma rays.

At this step, the reaction has produced 1.442 MeV, a small part of it being in the neutrino energy.

Then, the Deuterium will rapidly fuse with another proton (within 1 second), forming an atom of Helium-3, while emitting a gamma ray with an energy of 5.493 MeV: $p + {}^2_1D \rightarrow {}^3_2H + \gamma$. The rapidity of this reaction is due to the mediation by the strong interaction.

Then, there are several processes by which Helium-3 is transformed into Helium-4, the most frequent in the Sun being the "p-p one" whereby two atoms of Helium-3 merge, forming one atom of Helium-4 and two protons, while releasing 12.859 MeV: ${}^3_2H + {}^3_2H \rightarrow {}^4_2He + 2\,{}^1_1H$.

The total energy produced by this complete chain is 26.732 MeV, with 2% being in the form of neutrinos. About 0.7% of the initial masses is converted into radiation. (Figure 3.5)

Given the fact that the Sun radiates an overall energy estimated to be 3.8×10^{28} W, the number of atoms of Helium-4 produced per second is $\approx 9.1 \times 10^{37}$. In addition, twice this number of neutrinos is emitted every second, of which about 7×10^{14} neutrinos arrive on Earth every second on each square meter.

These fusion reactions induce a mass loss of the Sun of about 4.2×10^9 kg every second, which fortunately is negligible compared to its mass, which is $\approx 2 \times 10^{30}$ kg. The latter is composed of 73%, Hydrogen, 25% Helium and the rest of heavier elements, such as Oxygen, Carbon, Neon and Iron. The part of the Hydrogen in the core that has the temperature required for fusion accounts for approximately 10% of the Sun's mass. Hence, the Sun will

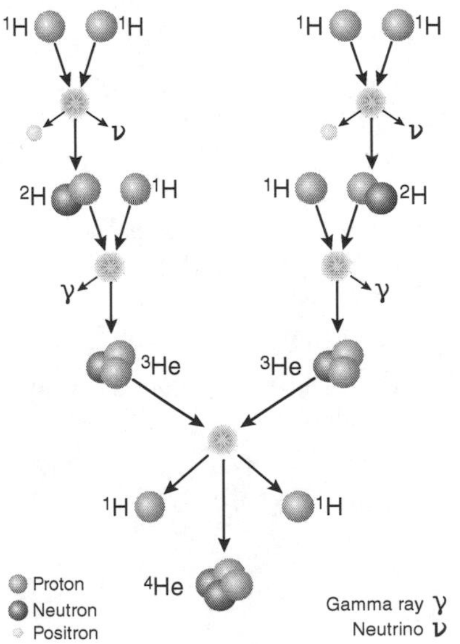

FIGURE 3.5
The Proton–Proton chain. (Sarang, Public Domain, via Wikimedia Commons.)

continue to radiate for about 5 billion years, which means that we are in the middle of its lifetime (it was formed some 4.6 billion years ago). Then, the Sun will start to burn Helium, which will cause the outer layers to expand, reaching even the Earth.

Gamma rays are produced in the core, whose radius is about 20% of the Sun's radius. These rays do not reach us directly (happily), but they radiate in the matter surrounding the core. Their spatial Momentum prevents this matter from falling into the center due to gravity. Then, the heat is transferred by convection to the outer part. When a photon reaches us, it is estimated that the initial proton fusion that led to its creation occurred more than 10,000 years before, due to the number of reactions in the intermediate layers. Note that the rate at which the protons undergo fusion is very temperature sensitive. The reaction becomes sustaining because the energy released is high (mainly the energy released by the follow-on reactions, which are faster and more energetic) and because there are a lot of protons in the Sun. That energy release provides pressure to resist further contraction and heating. The bigger the star, the higher the temperature at which the balance occurs; for very massive stars, the lifetime drops from billions to millions of years. They evolve so fast that their cores become unstable after using only a small fraction of their fuel supply.

3.2.4 Relations Concerning Mass and Energy in Disintegrations

3.2.4.1 Disintegration into Two Bodies

Consider the case of a body that spontaneously disintegrates into two bodies: $A \rightarrow B + C$. We will show the relationship between the masses of these three bodies and their energies as well as their Momenta:

Let's consider the Momenta of A, B and C in the Center of Momentum (CoM) frame, which is the rest frame of the initial body: We have: $(E_B/c, \overrightarrow{p_B})$ and $(E_C/c, \overrightarrow{p_C})$.

The system Momentum conservation gives in the CoM:

$E_A = E_C + E_B = m_A.c^2$ and: $\overrightarrow{p_B} + \overrightarrow{p_C} = 0$. We also have: $E_B^2.p_B^2 = m_B^2.c^4$ and: $E_C^2.p_C^2 = m_C^2.c^4$. Hence: $E_C^2.E_B^2 = (m_C^2.m_B^2)c^4$.

On the other hand, $\left(E_C^2.E_B^2\right) = \left(E_C.E_B\right)\left(E_C + E_B\right) = \left(E_C.E_B\right)E_A$.

We can find E_C and E_B thanks to the two following equations:

$(E_C - E_B).E_A = (m_C{}^2 - m_B{}^2).c^4$ and: $E_A = E_C + E_B$.

Indeed, $E_C = E_A - E_B$, then: $(E_A - 2\,E_B).\,E_A = (\,m_C{}^2 - m_B{}^2).c^4$, and:

$2\,E_B.E_A = E_A{}^2 - (m_C{}^2 - m_B{}^2).c^4$. Finally, from $E_A = m_A.c^2$, we obtain:

$$E_B = \frac{m_A^2 + m_B^2 - m_C^2}{2m_A} \cdot c^2 \text{ and } E_C = \frac{m_A^2 + m_C^2 - m_B^2}{2m_A} \cdot c^2. \qquad (3.5a) \blacksquare$$

Let's calculate the common value of p_B and p_C: $c^2p_B{}^2 = E_B{}^2 - m_B{}^2.c^4$. Then:

$$p_B^2 = p_C^2 = \frac{\left(m_A^2 + m_B^2 - m_C^2\right)^2}{4m_A^2} \cdot c^2 - m_B^2 c^2$$

$$= \frac{m_A^4 + m_B^4 - m_C^4 - 2m_A^2 m_B^2 - 2m_A^2 m_C^2 - 2m_B^2 m_C^2}{4m_A^2} \cdot c^2. \qquad (3.5b) \blacksquare$$

3.2.4.2 *The Sum of the Resulting Masses Is Less Than E/c² in the CoM*

Consider a reaction in an isolated system composed of several particles; the initial set of atoms are marked with i and the final one with f. We thus have: $A_{i1} + A_{i2} + \cdots \rightarrow A_{f1} + A_{f2} + \cdots$

The system Momentum constancy reads:

$$\Sigma P_i = \Sigma P_f. \qquad (3.6)$$

Let's apply the Lorentzian norm triangle inequality (Section 2.6.1) to $\Sigma P_f : \sqrt{(\Sigma P_f)^2} > \Sigma \sqrt{P_f^2}$. *We used the common convention to denote the Lorentzian norm², ds²(P), by P².*

Besides, for any element j of the final state, we have: $P_{fj}^2 = m_{fj}^2 \cdot c^2$.

Then, applying equation 3.6 we obtain: $\sqrt{(\Sigma P_i)^2} > c\Sigma m_f$. In the Center of Momentum frame, we have: $\sqrt{(\Sigma P_i)^2} = E/c$, with E being the system Energy in the CoM. So finally:

$$\Sigma m_f < E/c^2 \qquad (3.6a) \blacksquare$$

3.2.5 Ultra-Relativistic Particles

At the CERN particle accelerator, electrons could reach a speed of 0.999 999 999 983 c. More generally, when a particle goes at a speed close to c, its gamma factor is very important. Its ratio $\beta = v/c$ is very close to 1; hence, we introduce: $\boldsymbol{\varepsilon = 1 - \beta}$. We can then approximate the gamma factor and subsequently the energy of such particles as follows:

$$\gamma = \sqrt{\frac{1}{1-\beta^2}} = \sqrt{\frac{1}{1-(1-\varepsilon)^2}} = \sqrt{\frac{1}{2\varepsilon - \varepsilon^2}} = \sqrt{\frac{1}{2\varepsilon(1-\varepsilon/2)}} \approx \sqrt{\frac{1}{2\varepsilon}} \qquad (3.7)$$

Consequently,

$$E = mc^2\gamma \approx \frac{mc^2}{\sqrt{2\varepsilon}} \text{ and: } \varepsilon \approx \frac{m^2c^4}{2E^2}. \qquad (3.8)$$

For the electron moving at 0.999 999 999 983 c, the value of ε is: 17×10^{-12}. Its rest mass being 0.51 MeV/c², its energy is: $E \approx \frac{0.51 \times 10^5 \text{MeV}}{\sqrt{2 \times 17 \times 10^{-12}}} = 87\,\text{GeV}$.

We can see that the rest mass accounts for a negligible part of the energy of such particles (in the range of 10^{-6}), hence the name ultra-relativistic for such particles.

In nature, cosmic rays comprise photons, some being extremely energetic, and also various ultra-relativistic particles, such as protons, neutrinos and even alpha particles (Helium-4 nuclei). These were produced by nuclear reactions in the stars. Fortunately, most of these very energetic particles do not reach the Earth's surface because of reactions occurring in the outer layers of the atmosphere. As a result of these reactions, new particles that are less energetic are created and reach us, such as muons. Neutrinos also reach us since they interact with almost nothing, but their energy is extremely low.

3.3 The Force in Relativity

There are two Forces in Relativity, as seen in Volume I Section 5.3. We will see their properties and the laws involving them.

3.3.1 The Newtonian-Like Force and the Force of Minkowski

3.3.1.1 The Newtonian-Like Force

The Newtonian-like Force is the time rate of change of the relativistic Momentum incurred by an object submitted to a force, the time considered being the time t of the inertial frame K of the observer who considers this Force. Hence, it is: $\vec{F} = \frac{d\vec{P}}{dt} = \left(\frac{d(mc\gamma)}{dt}, \frac{d(m\vec{v}\gamma)}{dt}\right)$.

- The time part is: $F_t = \frac{d(mc\gamma)}{dt} = \left(\frac{dE}{cdt}\right) = \frac{\Pi}{c}$ with Π being the power of the Force.

  From equation 1.19, we have: $F_t = \frac{d(mc\gamma)}{dt} = mc\gamma^3\left(\vec{\beta}\cdot\vec{\beta}'\right)$

- The spatial part is in the context of constant mass:

$$\vec{F_S} = \frac{d(m\gamma\vec{v})}{dt} = m\gamma\vec{a} + \frac{d(m\gamma)}{dt}\vec{v} = m\gamma[\vec{a} + \vec{v}\gamma^2\left(\vec{\beta}\cdot\vec{\beta}'\right)] \tag{3.9}$$

We can see that the Newtonian-like Force is not the same in all inertial frames, unlike the classical one. Note that the Newtonian-like Force still respects the inertial frame equivalence postulate, its expression in K′ being $\vec{F}' = \frac{d\vec{P}'}{dt'}$.

3.3.1.2 The Force of Minkowski

The Force of Minkowski, also called the Four-Force, is the particular case of the Newtonian-like Force where the observer is in the Inertial Tangent Frame (ITF) of the object. During a small time lapse, the time of the ITF is the same as the object proper time. Hence, the Four-Force is:

$$\vec{\Phi} = \frac{d\vec{P}}{d\tau} = \left(\frac{dE}{cd\tau}, \frac{d(m\vec{v}\gamma)}{d\tau}\right).$$

- The Force of Minkowski is contravariant (since P is contravariant and τ is intrinsic).
- There is a relation between the two Forces:

$$\vec{\Phi} = \frac{d\vec{P}}{d\tau} = \frac{d\vec{P}}{dt}\frac{dt}{d\tau} = \vec{F}.\gamma. \tag{3.10}$$

- In case of constant mass, equation 2.15 yields:

$$\vec{\Phi} = \left[\frac{d(mc\gamma)}{d\tau}, \frac{d(m\gamma\vec{v})}{d\tau}\right] = m\left[\gamma^4\left(\vec{a}.\beta\right), \gamma^4\left(\vec{a}.\vec{\beta}\right)\vec{\beta} + \gamma^2\,\vec{a}\right]. \tag{3.10a}$$

Notably, in the ITF, the spatial parts of the two relativistic Forces are also equal to the classical Newtonian force: indeed in the ITF, $v=0$ then $\gamma = 1$ and $\beta=0$. Hence, relation (3.9) yields: $\vec{F_S} = m\vec{a}$.

Physically, the Momentum actually represents the global energy of an object. Hence, the relativistic Forces, being the time rate of change of the Momentum, represents the global power.

In classical physics, we had a very important relation: the work of the force equals the kinetic energy variation. However, Relativity showed that the Newtonian-like Force is not identical to the classical Newtonian one, and that the kinetic energy is no longer equal to ½ mv². Hence, it is important to know if the work of the relativistic Force still gives the relativistic energy variation, all the more as the work of the force is the basis on which the potential energy is defined.

3.3.2 The Work of the Force in the Context of Constant Mass

We will skip the arrows on the 4D vectors when there is no risk of ambiguity, and the vectors will be marked with bold characters.

The Force is the derivative of the Momentum **P**, and we saw an important relation that impacts the derivative of **P**, indeed: $ds^2(\mathbf{P}) = m^2c^2$. As previously mentioned, $ds^2(\mathbf{P})$ is also written $\mathbf{P}^2$. Hence, in the context where the mass is constant, the derivative of $\mathbf{P}^2$ is null.

Besides, the Lorentzian norm, like any norm, is a scalar product of a vector with itself.[4] In the case of the Lorentzian norm, the corresponding scalar product is the Lorentzian scalar product, denoted by • in order to avoid confusion[5] with the classical Euclidean scalar product, and defined as follows:

Consider two 4D vectors expressed with homogeneous units: $\mathbf{V}(ct, x, y, z)$ and $\mathbf{V}'(ct', x', y', z')$. The Lorentzian scalar product between these vectors is defined as: $\mathbf{V} \bullet \mathbf{V}' = c^2 t\, t' - (x\, x' + y\, y' + z\, z')$.

We can see that the Lorentzian norm² is the Lorentzian scalar product of **V** with itself: $ds^2(\mathbf{V}) = \mathbf{V}^2 = \mathbf{V} \bullet \mathbf{V}$. Like the Lorentzian norm, the Lorentzian scalar product has the important property of being invariant (cf. Volume II Section 2.8.2), meaning that it gives the same result in all inertial frames.

We thus have: $\mathbf{P}^2 = \mathbf{P} \bullet \mathbf{P} = m^2c^2$. Consequently, the mass constancy implies that the derivative of $\mathbf{P} \bullet \mathbf{P}$ is null. This derivative can be made relative to the time t of the frame K where the Newtonian-like Force is considered, or to the object proper time τ. We thus have: $0 = \frac{d(\mathbf{P} \bullet \mathbf{P})}{dt} = 2\mathbf{P} \bullet \frac{d\mathbf{P}}{dt}$ (cf. more in Section 3.6.5.1), meaning: $\mathbf{P} \bullet \mathbf{F} = 0$.

Similarly, the derivative relative to the object proper time τ leads to: $\mathbf{P} \bullet \mathbf{\Phi} = 0$.

We recall that: $\mathbf{P} = m\frac{\mathbf{dM}}{d\tau}$, where **dM** is the small space–time separation vector of the object in its trajectory during dτ. We then have:

$$\mathbf{dM} \bullet \mathbf{\Phi} = 0 \text{ and } \mathbf{dM} \bullet \mathbf{F} = 0 \tag{3.11}$$

which means that the relativistic Forces are perpendicular to the object trajectory, the perpendicularity being relative to the Lorentzian scalar product (and not the classical Euclidean one).

Let's expand: $\mathbf{dM} \bullet \mathbf{F} = 0$. It is: $(cdt, dx, dy, dz) \bullet (F_t, F_x, F_y, F_z) = 0$.

Besides, we have: $F_t = dE/cdt$. Hence: $0 = dE - (F_x dx, + F_y dy + F_z dz)$ or: $dE = F_x.dx + F_y.dy + F_z.dz$. ■

This relation can be written:

$$dE = \overrightarrow{F_s} \cdot \overrightarrow{dM} \;\blacksquare \tag{3.12}$$

In the context of mass constancy, the energy variation is equal to the kinetic energy variation. Hence, relation (3.12) shows that the kinetic energy variation is the classical Euclidean scalar product between the spatial part of the Newtonian-like Force and the object spatial displacement.

3.3.3 Other Properties of the Newtonian-Like Force

3.3.3.1 The Power of the Newtonian-Like Force

If we divide both terms of relation (3.12) by dt, we obtain: $\Pi=\vec{F}_s\cdot\vec{v}$, meaning that the power of the Newtonian-like Force is, like in classical physics, the classical Euclidean scalar product between the object velocity and the spatial part of the Newtonian-like Force.

3.3.3.2 The Acceleration Due to the Newtonian-Like Force

We saw that: $\vec{F_S}=\dfrac{d(m\gamma\vec{v})}{dt}=m\gamma\vec{a}+\dfrac{d(m\gamma)}{dt}\vec{v}$. We have from equation 3.12: $\dfrac{d(m\gamma)}{dt}=\dfrac{d(E)}{c^2dt}=\dfrac{\vec{F}_s\cdot\vec{dM}}{c^2dt}=\dfrac{\vec{F}_s\cdot\vec{v}}{c^2}$. Then:

$$\vec{F}_s=m\gamma\vec{a}+\frac{\vec{F}_s\cdot\vec{v}}{c^2}\vec{v} \tag{3.13}$$

and

$$\vec{a}=\frac{1}{m\gamma}\vec{F}_s-\frac{\vec{F}_s\cdot\vec{v}}{m\gamma c^2}\vec{v}. \tag{3.14}$$

We can see that if the Force is perpendicular to the object velocity, so is its acceleration: the object has a circular trajectory. In three spatial dimensions, it is generally a helix.

In the case where the Force is parallel to the velocity, equation 3.13 becomes:

$$F_s=m\gamma a\left(1+\frac{\gamma^2v^2}{c^2}\right). \tag{3.15}$$

We can clearly see that for the same acceleration, the Force required increases with the speed. Note that the last term, $\dfrac{\gamma^2v^2}{c^2}$, is important for ultra-relativistic particles only.

3.3.3.3 The Newtonian-Like Force Complies with Force Reciprocity

The demonstration of Volume I Section 4.1.1.4.1 indeed applies for the Newtonian-like Force, but not for the Four-Force as we will see.

From the constancy of $\mathbf{P_A}+\mathbf{P_B}$ in an isolated system, if we differentiate $(\mathbf{P_A}+\mathbf{P_B})$ relative to the time of an inertial frame K, we obtain: $\dfrac{d\mathbf{P_A}}{dt}=-\dfrac{d\mathbf{P_B}}{dt}$, which expresses the Newtonian-like Force reciprocity (also called the 3rd Newtonian law).

Concerning the Four-Force reciprocity: The Four-Force incurred by the objects A and B, respectively, are $\frac{\mathbf{dP_A}}{d\tau_A}$ and $\frac{\mathbf{dP_B}}{d\tau_B}$. From the previous relation: $\frac{\mathbf{dP_A}}{dt} = -\frac{\mathbf{dP_B}}{dt}$, we have: $\frac{\mathbf{dP_A}}{d\tau_A}\frac{d\tau_A}{dt} = -\frac{\mathbf{dP_B}}{d\tau_B}.\frac{d\tau_B}{dt}$, which shows that we generally don't have: $\frac{\mathbf{dP_A}}{d\tau_A} = -\frac{\mathbf{dP_B}}{d\tau_B}$ since $\frac{d\tau_A}{dt}$ is generally different from $\frac{d\tau_B}{dt}$ (except if $v_A = v_B$). ■

3.3.4 The Four-Force in Case of Mass Variation

We saw that $ds^2(\mathbf{P})=m^2c^2$. Let's then differentiate this relation:

$$2\mathbf{P}\bullet\frac{d\mathbf{P}}{d\tau}=2\,\mathbf{P}\bullet\mathbf{\Phi}=2mc^2\frac{dm}{d\tau}.$$

Using $\mathbf{P} = m\frac{\mathbf{dM}}{d\tau}$, we have: $\frac{\mathbf{dM}}{d\tau}\bullet\mathbf{\Phi}=c^2$, so:

$$\mathbf{dM}\bullet\mathbf{\Phi}=c^2\,dm. \tag{3.16}$$

This means that the "4D Lorentzian Work" of the Four-Force represents the mass variation.

More generally, let's show that: $\mathbf{dM}\bullet\mathbf{\Phi} = dE$. The Lorentzian scalar product between two intrinsic 4-vectors, such as $\mathbf{dM}$ and $\mathbf{\Phi}$, is invariant: it gives the same result in all inertial frames. Hence, let's calculate $\mathbf{dM}\bullet\mathbf{\Phi}$ in the object ITF, where it is relatively simple, and we obtain:

$$\mathbf{dM}\bullet\mathbf{\Phi}=(cd\tau,0,0,0)\bullet\left(\frac{dE}{cd\tau},\Phi_x,\Phi_y,\Phi_z\right)=dE$$ ■

3.3.5 Potential Energy Update and Applications

The concept of potential energy is even more important in Relativity and modern physics than it was in classical physics. We will first present the potential energy in the context of a gravitational field using the classical Newtonian gravitation law. We will then generalize this concept.

Consider an object of mass m at a point A of altitude Z_a relative to the center of a great mass M. Intuitively, this object has a potential energy because if it falls to a lower point B of altitude Z_B, it will acquire a velocity and thus a kinetic energy, which is: $K=\frac{1}{2}mv^2$ in classical physics, but in relativity: $K = mc^2(\gamma - 1)$.

The potential energy was thus defined so that the sum of the kinetic energy and the potential energy remains constant for a free-falling object, in accordance with the energy conservation law of isolated systems.

For the sake of simplicity, we first assume that the gravity field is constant between Z_A and Z_B and is equal to g. This provides a good approximation for the usual cases of a few tens of meters, and we will further see the case of long vertical distances. We saw that the Coriolis law stating the equality between the relativistic kinetic energy variation, ΔK, and the Work done by the gravitational force, is still valid in Relativity:

Thus, $\Delta K = K_b - K_a = mg(Z_a - Z_B)$.

This leads to the potential energy definition: E_p=mgZ, with Z being the object altitude, and assuming g is constant. We thus have:

$$K_B + E_{pB} = K_A + E_{pA} = cst \tag{3.17}$$

with $E_{pA} - E_{pB} = mg(AB)$.

We can indeed see that the increase of kinetic energy acquired by the falling object when reaching B is equal to the decrease of its potential energy. Thus, the higher the object, the greater its potential energy. We can then model a gravitational field by a field of scalar numbers Φ, such that:

$$\Phi_A - \Phi_B = \vec{g}.\overrightarrow{AB}\,. \tag{3.18}$$

hence: $E_{pA} - E_{pB} = m.(\Phi_A - \Phi_B)$.

If g is not constant, we will consider the distance AB as a succession of small segments dz where g is constant:

Mathematically, relation (3.18) can be written in a more general form:

$$d\Phi = -\vec{g}\cdot\overrightarrow{dM}\,. \tag{3.19}$$

where $\vec{g}$ is the 3D gravitational acceleration vector at the object location, and $\overrightarrow{dM}$ the small displacement of this object. In turn, the term dΦ can be expressed using the partial derivatives of Φ relative to the directions dx, dy and dz: this is done via the gradient[6] operator:

$$d\Phi = \overrightarrow{grad}(\Phi)\cdot\overrightarrow{dM}, \text{ with: } \overrightarrow{grad}(\Phi) = \left(\frac{\partial\Phi}{\partial x}, \frac{\partial\Phi}{\partial y}, \frac{\partial\Phi}{\partial z}\right).$$

Together with equation 3.19, we deduce: $\overrightarrow{grad}(\Phi) = -\vec{g}$ and:

$$\vec{F} = -\, m.\overrightarrow{grad}(\Phi). \tag{3.20}$$

These relations are valid with both Newtonian gravity and that of General Relativity, but the values of Φ are different (they are very close in the case of the Earth).

3.3.5.1 The Potential Energy of the Gravity Field in the General Case

At the altitude Z, the classical Newtonian gravitational force is: $\frac{mMG}{Z^2}$, with M being the Earth's mass and Z being relative to the Earth Center. When falling from A to B the work of the gravitational force is:

$$\int_A^B \frac{mMG}{Z^2} dz = \frac{mMG}{Z_B} - \frac{mMG}{Z_A}.$$

The equality between the kinetic energy variation and the work of the Force then gives: $K_B - K_A = \frac{mMG}{Z_B} - \frac{mMG}{Z_A}$. We still have:

$$K_B - K_A = E_{pA} - E_{pB} = m.(\Phi_A - \Phi_B).$$

Hence:

$$\mathbf{\Phi_A = -\frac{MG}{Z_A}} \tag{3.21}$$

We can see that the gravitational potential is negative, but still, the higher the point, the greater its potential energy. Note that the consistency with equation 3.20 is ensured: $\vec{F} = -m\,\overrightarrow{grad}(\Phi)$ so: $\overrightarrow{grad}(\Phi) = \left(0, 0, \frac{d\left(-\frac{MG}{Z}\right)}{dz}\right) = \left(0, 0, \frac{MG}{Z^2}\right)$.

For an object moving along the radial axis, relation (3.17) becomes:

$$\mathbf{\tfrac{1}{2}\dot{Z}^2 - \frac{MG}{Z} = cst} \blacksquare \tag{3.21a}$$

with the notation $\dot{Z} = \frac{dZ}{dt}$, and considering the classical kinetic energy.

In the case of the Earth, we have international units: $G = 6.67408 \times 10^{-11}\ m^3 kg^{-1} s^{-2}$ and $M = 5.972 \times 10^{24}$ kg, which yields on the Earth's surface: $g = 9.806\ m/s^2$.

The gravitational potential on the Earth's surface is: -6.2×10^7 J/kg ($=m^2/s^2$).

The concept of potential energy also applies to various domains, such as electromagnetism with Maxwell's laws (classical and relativistic), or the nutritional domain: a nutrient indeed has a potential energy, which is acquired by our body after digestion. Moreover, the concept of potential energy is fundamental to the Lagrangian approach.

Additional considerations on the potential energy are given in Section 3.6.6, including the orbital potential energy.

3.4 The Unity between Electrical and the Magnetic Forces

Einstein: "What led me more or less directly to the special theory of relativity was the conviction that the electromotive force acting on a body in motion in a magnetic field was nothing else but an electric field."

We will see how Relativity solved the issue and achieved the unity of the electrical and the magnetic forces.

When an electrically charged particle is in an electromagnetic field, it incurs a force that was in classical physics the Lorentz force:

$$\vec{f} = q\vec{E} + \vec{B} \times \vec{v} \tag{3.22}$$

where

- q is the electrical charge of the particle incurring the considered force.
- $\vec{v}$ is the 3D speed vector of this particle in the inertial frame K where the force is considered.
- $\vec{E}$ and $\vec{B}$, respectively, are the 3D electrical and magnetic field vectors at the point of this particle.
- $\vec{B} \times \vec{v}$ means the vector product: $\vec{B} \times \vec{v}$ is a new vector that is perpendicular to both $\vec{B}$ and $\vec{v}$ (cf. more in Section 5.1.6.1).

Relativity showed that the 3D classical laws actually are derived from 4D laws. We are then induced to find the 4D law which relates the relativistic Force experienced by a particle under an electrical and a magnetic field.

We will use capital letters for 4D quantities, without an arrow on the top. Conversely, we will still place arrows on the 3D vectors representing the classical electrical and magnetic field vectors.

We will assume the particle mass is constant. We saw with relation (3.11) that this implies that both relativistic Forces are always perpendicular to the object trajectory, this perpendicularity being considered according to the Lorentzian scalar product.

To avoid confusion between the Lorentzian scalar product and the classical one, the latter will be denoted by A·B (as in classical physics) and the former A•B. Thus, this perpendicularity means:

$$\mathbf{V} \bullet \mathbf{F} = 0 \tag{3.23}$$

with **V** being the 4D speed of the particle, and **F** the 4D Newtonian-like Force. This 4D speed, expressed with homogeneous units, is:

$$\mathbf{V} = \mathbf{dM}/dt = (cdt/dt,\ dx/dt,\ dy/dt,\ dz/dt) = (c,\ v_x,\ v_y,\ v_z).$$

The Lorentzian scalar product between two 4D vectors expressed with homogeneous units: $V(c, v_x, v_y, v_z)$ and $F\,(F_t, F_x, F_y, F_z)$ is:

$$V \bullet F = c.F_t - v_x F_x - v_y F_y - v_z F_z. \tag{3.24}$$

We can obtain the same result by using the classical scalar product between V* and F, with $V^* = (c, -v_x, -v_y, -v_z)$. We indeed have: $V \bullet F = V^* \cdot F$. Hence, relation (3.24) becomes:

$$V^* \cdot F = 0. \tag{3.25}$$

The fact that this scalar product is null whatever V* implies that the Force F is a function of V*. Let's further characterize F: relation (3.22) shows that the classical force is linear in the electrical charge q, and that there is another element of linearity that is the particle speed (since $\vec{v} \times \vec{B}$ is linear relative to $\vec{v}$). We can then deduce that the relativistic Force F is a linear function of both V* and q, which will be further confirmed.

With this assumption, the 4D vector F is the result of a linear function applied to a 4D vector having $\vec{v}$ in its spatial part. This vector can be V or V*: we are induced to choose the latter since it is involved in relation (3.25). This choice is not compulsory, but it simplifies the calculations. Any function giving the 4D vector F from the 4D vector V* can be expressed by a 4×4 matrix; hence, we will call this matrix the electromagnetic matrix (EM).

Thus, the relativistic Electromagnetic Force is given by:

$$F = q\,(EM)(V^*). \tag{3.26}$$

With the matrix format, relation (3.25) reads[7]:

$$\left(V^{*T}\right)(F) = 0 = q\left(V^{*T}\right)(EM)\left(V^*\right).$$

This relation implies that the (EM) matrix is antisymmetric, as shown in Section 5.1.6.6. Consequently, the (EM) matrix has only 6 degrees of freedom instead of 16, and this simplification is due to the choice of V* in relation (3.26), which reads with the matrix format:

$$\begin{pmatrix} F_t \\ F_x \\ F_y \\ F_z \end{pmatrix} = q \cdot \begin{pmatrix} 0 & -E_x/c & -E_y/c & -E_z/c \\ E_x/c & 0 & -B_z & B_y \\ E_y/c & B_z & 0 & -B_x \\ E_z/c & -B_y & B_x & 0 \end{pmatrix} \begin{pmatrix} c \\ -v_x \\ -v_y \\ -v_z \end{pmatrix} \tag{3.26a}$$

The names of the components of the (EM) matrix have been chosen by anticipation and we will now see why. Let's first expand the spatial parts of relation (3.26a):

The first spatial component gives: $F_x = dP_x/dt = q(E_x + B_z v_y - B_y v_z)$. We recognize that the last two terms correspond to the 3D vector product $\vec{B} \times \vec{v}$, so that the 3D spatial part of the Force is: $\vec{F}_S = d\vec{P}_S/dt = q\vec{E} + \vec{B} \times \vec{v}$. This relation matches with the classical Lorentz force 3.22.

Let's now expand the time part of relation (3.26): we first recall that the time part of the Newtonian-like Force, dP/dt, is equal to the power, $d\varepsilon/dt$, with ε being the energy of the particle (the letter ε was chosen to avoid confusion with the electrical field E). We obtain: $F_t = d\varepsilon/dt = q\,\vec{E} \cdot \vec{v}$, where the 3D classical scalar product is used. We notice that this relation is the same as in classical physics where the energy variation is equal to the work of the Lorentz force: $d\varepsilon = f_x dx + f_y dy + f_y dy$. Let's then divide by dt and we have: $d\varepsilon/dt = q\vec{F}_s \cdot \vec{v}$; then: $d\varepsilon/dt = (q\vec{E} + \vec{B} \times \vec{v}) \cdot \vec{v} = q\vec{E} \cdot \vec{v}$ since $(\vec{v} \times \vec{B})$ is perpendicular to $\vec{v}$. Thus, the classical energy acquired by the particle is equal to the relativistic one. Similarly, the power of the relativistic Force F is identical to that of the Lorentz force.

Conclusion: Relativity showed that the Electric and Magnetic fields are one single entity characterized by the EM matrix. This unity is logical since we knew that a fixed electrical charge creates an electric field, whereas a mobile charge creates a magnetic field, but Relativity stresses that (i) there is no fixed or mobile charge in the absolute sense (it all depends on the frame where it is considered), and (ii) all physical laws are the same in all inertial frames.

The Relativistic Maxwell's laws are presented in Section 5.1. We will see in particular that when changing the frame, the Electric and Magnetic fields change differently than in classical physics.

3.5 The Lagrangian Approach for Momentum and Energy

3.5.1 Lagrangian History and Basic Considerations

The general idea behind the Lagrangian approach (also called variational) is that nature follows the least action: In the 1600s, the French scientist P. Fermat postulated that "*light travels between two given points along the path of shortest time.*" Then, in the early 1700s, P. L. Maupertuis added: "Nature is thrifty in all its actions," and gave a first physical definition of the action that must be minimized between two states of a system: it was the integral of the "living force," which was the ancestor of kinetic energy.

At the same time, L. Euler stated that the action to be minimized between the times t1 and t2 was the integral of the momentum: $S = \int_{t1}^{t2} p\,dt$, with p(t)=mv(t). Minimum action means dS=0. Any other trajectory having the same states at t1 and t2, but differing in between by very small differences regarding the position and the speed of the object, would have an action S′, which is equal to S in the first order (dS = 0), and greater due to higher orders. This approach has also the interest of being applicable to systems composed of several objects.

3.5.1.1 The Lagrangian Theory

Later in 1760, J-L. Lagrange and W. Hamilton pursued this approach by taking into account forces that are derived from a potential V, such as gravity. They found that the action to be minimized was the integral of the function L=T−V, with T being the kinetic energy (1/2 mv^2) and V the potential energy. The function L was called the Lagrangian, and we can see that it has the dimensions of energy. The trajectories chosen by objects submitted only to gravity are such that the integral of this Lagrangian is minimized. Note the consistency with the Coriolis law stating that the work of the force of gravity (mgh) equals the kinetic energy variation.

In a more general perspective, if a system has n degrees of freedom with the positions of its elements being defined by n coordinates $q_i(t)$, then the existence of a Lagrangian function $L[q_i(t), \dot{q}_i(t), t]$ is postulated such that its integral $S = \int_{t1}^{t2} L\,dt$ is extremized: dS=0. $\left(\text{We use the usual notation: } \dot{q}_i(t) = \frac{dq_i}{dt}\right)$. Euler and Lagrange showed that this implies that:

$$\frac{\partial L}{\partial q_i} - \frac{d}{dt}\left(\frac{\partial L}{\partial \dot{q}_i}\right) = 0. \tag{3.27}$$

This relation is useful; its mathematical demonstration is presented in Section 3.5.2.

Let's apply this relation 3.27 in the case of an object of mass m in a gravity field: the above Lagrangian, L=T−V, is in the general case: $L = \frac{1}{2}\, m \sum_{i=1}^{3} \dot{x}_i^2(t) - V(x_i, t)$.

We then have: $\frac{\partial L}{\partial x_i} = -\frac{\partial V(x_i, t)}{\partial x_i}$, and

$$\frac{d}{dt}\left(\frac{\partial L}{\partial \dot{x}_i}\right) = \frac{d}{dt}\left(\frac{\partial\left(\frac{1}{2} m \sum_{i=1}^{3} \dot{x}_i^2(t)\right)}{\partial x_i}\right) = \frac{d\left(m \sum_{i=1}^{3} \dot{x}_i(t)\right)}{dt}.$$

Hence, $m\frac{d(\vec{v})}{dt} = -\overrightarrow{grad}(V)$, which matches with the Newtonian laws.

Remark 3.2: Different Lagrangian functions can lead to the same objects trajectories.

3.5.1.2 The Famous Noether Theorem

Much later in 1915, the mathematician Emmy Noether showed that the universal homogeneity postulate has important consequences on the Lagrangian: it implies the existence of conserved quantities, indeed:

- The invariance of the Lagrangian under translation in space implies the invariance of the quantity: $\frac{dL}{dv}$, which matches with the spatial part of the Momentum.
- The invariance of the Lagrangian under direction (isotropy) implies the invariance of the angular momentum.
- The invariance of the Lagrangian under translation in time implies the invariance of the quantity: $\frac{dL}{dv} - L$, which matches with the energy (Time part of the Momentum).

This theorem was ascribed by Einstein: "A monument of mathematical thinking." The mathematics involved is out of the scope of this document, but its application to Relativity shows that we can rapidly derive the relativistic expression of momentum and energy, as we will see.

3.5.2 Application to Relativity

The basic idea of the Lagrangian still applies, but the above classical Lagrangian, L=T−V, cannot fit because T does not match with the relativistic kinetic energy, and moreover V refers to gravitation, but the principle of equivalence states that gravitation does not exist (but is due to deformations of our space–time universe).

Hence, if we want to find the Lagrangian of a free-falling particle, we should return to the fundamentals and recall that this particle follows the path that maximizes the (Lorentzian) distance covered (cf. Volume I Section 3.5.4). We also recall that the Lorentzian distance is the sole distance in our space–time universe, and that for time-like intervals, this distance, ds, is equal to c.dτ, with τ being the proper time of the particle. In other words, a free-falling particle chooses the path which maximizes its proper time, and this property induced us to define the relativistic action S between two instants t1 and t2 as:

$$S = -K \int_{t1}^{t2} ds = -K\,c \int_{t1}^{t2} d\tau.$$

with K being a positive constant, and Kc having the dimension of energy.

The trajectory that extremizes the (Lorentzian) distance is a geodesic (a straight line if there are no space–time deformations), as we will see in section 4.2.6.

Regarding the relativistic Lagrangian L: L is related to the action by: $S = -\int_{t1}^{t2} Ldt.$

Knowing that $dt = \gamma_v d\tau$, we have:

$$S = -\int_{t1}^{t2} L\gamma_v d\tau.$$

Hence: $-Kc = L\gamma_v$, so finally:

$$L = -Kc/\gamma_v \blacksquare$$

Regarding K: L is specific to the object and represents an energy; hence, we deduce that K is proportional to the object mass; we saw that in Relativity the natural dimension of the energy is the mass, but with our usual international system of units, the energy is a mass multiplied by c^2.

Hence, K=mc, meaning that $L = -Mc^2/\gamma_v = -mc^2\sqrt{1-\frac{v^2}{c^2}}.$

Subsequently, we can apply the Noether theorem, and obtain:

- the momentum: $p = \frac{dL}{dv} = mv\left(\sqrt{1-\frac{v^2}{c^?}}\right)^{-1} = mv\gamma_v$

- the energy:
  $E = v\frac{dL}{dv} - L = mv^2\gamma_v + mc^2\gamma_v^{-1} = m\gamma_v[v^2 + c^2(1 - v^2/c^2)] = m\gamma_v c^2.$
  which, respectively, match with the spatial part and the time part the relativistic Momentum. ■

3.5.3 The Euler-Lagrange Equations

We will demonstrate the Euler-Lagrange equations that play a key role in the variational Lagrangian approach.

Any system can be characterized at a given time t by n independent variables which evolve with the time: $q_i(t)$. This system thus has n degrees of freedom. Its evolution is assumed to be characterized by a function of these variables and their derivatives, $\frac{dq_i(t)}{dt}$, also denoted by $\dot{q}_i(t)$. Such a

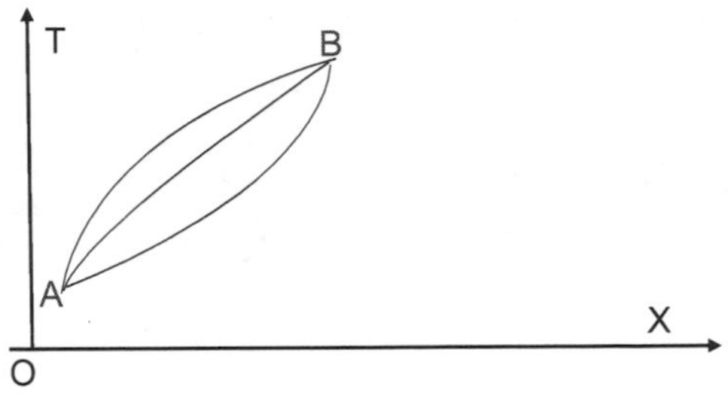

FIGURE 3.6
Different paths between A and B.

function, called Lagrangian, has the form: $L\left[q_1(t),\ldots,q_n(t),\dot{q}_1(t),\ldots,\dot{q}_n(t),t\right]$,

also written: $L\left[q_i(t),\dot{q}_i(t),t\right]$. This function thus has 2n+1 variables.

Then, the "action" was defined as the integral of the Lagrangian between two times, meaning two states of the system: $S\left[q_i\right]=\int_{t1}^{t2} L\left[q_i(t),\dot{q}_i(t),t\right]dt$.

The least action principle states that the natural evolution of a system is such that this action is extremized. This means that for any small variations of the $q_i(t)$ during the system evolution from t_1 to t_2, the action variation dS is null at the first order (Figure 3.6).

Consider a small variation of the action resulting from small variations of the $q_i(t)$ during the system evolution, but not at the extremities in t_1 and t_2, the system being precisely defined and fixed at its extremities.

We then have: $S\left[q_i+\delta q_i\right]=\int_{t1}^{t2}\mathcal{L}\left[q_i(t)+\delta q_i,\frac{dq_i(t)}{dt}+\frac{d}{dt}\delta q_i(t),t\right]dt$.

Let's make a Taylor development to the first order of this Lagrangian:

$$L\left[q_i(t)+\frac{dq_i(t)}{dt},t\right]+\sum_i\frac{\partial L}{\partial q_i}\delta q_i(t)+\sum_i\frac{\partial L}{\partial\left(\frac{dq_i(t)}{dt}\right)}\frac{d\delta q_i(t)}{dt}.$$

The small variation of the action is then:

$$\delta S=0=S\left[q_i+\delta q_i\right]-S\left[q_i\right]=\int_{t_1}^{t_2}\left[\sum_i\frac{\partial\mathcal{L}}{\partial q_i}\delta q_i(t)+\sum_i\frac{\partial\mathcal{L}}{\partial\left(\frac{dq_i(t)}{dt}\right)}\frac{d\delta q_i(t)}{dt}\right]dt$$

We can permute the summation and the integration:

$$\delta S = \sum_i \int_{t_1}^{t_2} \left[\frac{\partial \mathcal{L}}{\partial q_i} \delta q_i(t) + \frac{\partial \mathcal{L}}{\partial (\frac{dq_i(t)}{dt})} \frac{d\delta q_i(t)}{dt} \right].$$

Let's integrate by parts the second term:

$$\int_{t_1}^{t_2} \frac{\partial \mathcal{L}}{\partial \left(\frac{dq_i(t)}{dt} \right)} \frac{d\delta q_i(t)}{dt} dt = \frac{\partial \mathcal{L}}{\partial \left(\frac{dq_i(t_2)}{dt} \right)} \delta q_i(t_2) - \frac{\partial \mathcal{L}}{\partial \left(\frac{dq_i(t_1)}{dt} \right)} \delta q_i(t_1) - \int \frac{d}{dt} \frac{\partial \mathcal{L}}{\partial \left(\frac{dq_i(t)}{dt} \right)} \delta q_i(t) dt$$

We have $\delta q_i(t_2) = \delta q_i(t_1) = 0$ because the system is precisely defined and fixed at the times t_1 and t_2.

We then have: $\delta S = 0 = \sum_i \int_{t_1}^{t_2} \left[\frac{\partial \mathcal{L}}{\partial q_i} - \frac{d}{dt} \left(\frac{\partial \mathcal{L}}{\partial \left(\frac{dq_i(t)}{dt} \right)} \right) \right] \delta q_i(t) dt$. For sum of the integrals to be null whatever the $dq_i(t)$ and their evolutions between t_1 and t_2, we must have: $\forall i \in [1,n]: \frac{\partial \mathcal{L}}{\partial q_i} - \frac{d}{dt} \left(\frac{\partial \mathcal{L}}{\partial \left(\frac{dq_i(t)}{dt} \right)} \right) = 0.$

Indeed, if a part is not null, we can set all the other dq_i to zero, and the whole sum won't be null.

These are the Euler-Lagrange equations, also written: $\frac{\partial L}{\partial q_i} - \frac{d}{dt} \left(\frac{\partial L}{\partial \dot{q}_i} \right) = 0.$

3.6 Complements and Case Studies

3.6.1 The Compton Effect

The American scientist Arthur Compton (Nobel Prize in 1927) explained the collision between a high-energy photon (X-ray) and an electron. This collision

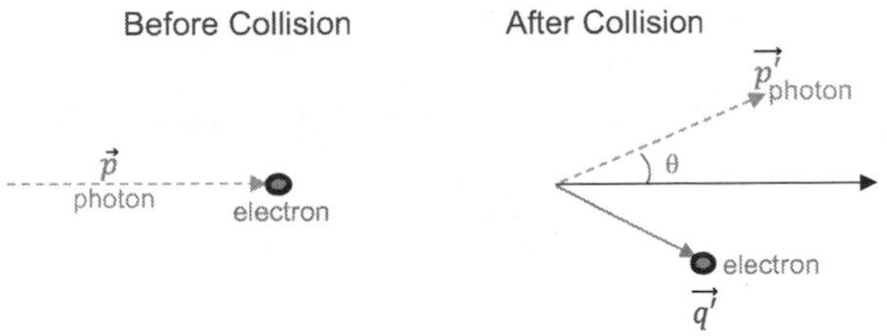

FIGURE 3.7
The Compton effect.

deflects the electron, and simultaneously the photon is reflected with a lower energy (lower frequency): part of the photon energy has been converted into kinetic energy for the electron. This collision is thus inelastic.

We assume the electron is at rest initially, and then it is hit by a photon coming along the X-axis (Figure 3.7).

Let's express the system Momentum conservation law:

Situation before the collision:

The Momentum of the photon is: $\vec{P}_{photon} = (p, \vec{p})$ and we have: $p = h\nu/c$.

The Momentum of the electron is: $\vec{P}_{electron} = (q, \vec{0})$ and we have: $q = mc$ with m being the mass of the electron. The system Momentum before the collision then is: $\vec{P}_{System} = (p + q, \vec{p})$.

Situation after the collision:

The Momentum of the photon is: $\vec{P}'_{photon} = (p', \vec{p}')$ and we have: $p' = h\nu'/c$.

The Momentum of the electron is: $\vec{P}'_{electron} = (E'/c, \vec{q}')$ with $E'=mc^2\gamma$, and $\vec{q}' = m\vec{v}'\gamma$.

The system Momentum after the collision is then: $\vec{P}'_{System} = (p' + E'/c, \vec{p}' + \vec{q}')$.

The system Momentum conservation law yields:

$$p + q = p' + E'/c \quad \text{or} \quad p + mc = p' + E'/c \tag{3.28}$$

$$\vec{p} = \vec{p}' + \vec{q}' \quad \text{or} \quad \vec{q}' = \vec{p} - \vec{p}'. \tag{3.29}$$

In equation 3.28, let's use the relation: $E'^2 = m^2c^4 + q'^2c^2$.

We then have: $\frac{E'^2}{c^2} = \left[(p - p') + mc\right]^2 = m^2c^2 + q'^2$ then:

$$(p - p')^2 + 2mc(p - p') = q'^2. \tag{3.30}$$

With equation 3.29, let's calculate the norm:

$$(\vec{q}')^2 = q'^2 = (\vec{p} - \vec{p}')^2 = p^2 + p'^2 - 2\vec{p}.\vec{p}' = p^2 + p'^2 - 2pp'\cos\theta. \tag{3.31}$$

From equations 3.30 and 3.31, we have:

$$(p-p')^2 + 2mc(p-p') = p^2 + p'^2 - 2pp'\cos\theta.$$

Then: $2mc(p-p') = 2pp'(1-\cos\theta)$ or:

$$p - p' = \frac{pp'}{mc}(1-\cos\theta). \tag{3.32}$$

Let's use the relations concerning the photon: $p = h\nu/c$ and $\lambda = c/\nu$, then $p = h/\lambda$. Relation (3.32) becomes:

$$h\left(\frac{1}{\lambda} - \frac{1}{\lambda'}\right) = \frac{h^2}{\lambda\lambda' mc}(1-\cos\theta), \text{ so finally: } \lambda' - \lambda = \frac{h}{cm}(1-\cos\theta) \ \blacksquare$$

The term $h/(cm)$ is called the Compton wavelength of the electron, and it is close to: 2.43×10^{-12} mm, which is in the range of the X-rays, and explains why the Compton effect mostly concerns X-ray photons.

**

Historically, the Compton effect was an important confirmation of Einstein's foreshadowing assertion that the photon is the quantum of electromagnetic energy, and that it has both particle and wave aspects.

The inverse Compton effect also exists in the cosmos, whereby high-energy electrons give part of their kinetic energies to photons, which then acquire higher frequencies. This is significant because the photons have a much longer range, since the electrons are deflected by magnetic fields.

3.6.2 A Thought Experiment by Einstein Revisited: The Photon Box

Consider an empty box that is not submitted to any force. In one of its interior faces, an atom emits a photon, which goes toward the opposite interior face where it is absorbed. We will first apply classical physics principles and show that the photon must have transferred some mass from its emission location to its absorption location, and that this mass corresponds to the famous relation $E = mc^2$.

We will then review this scenario in the light of Relativity and we will be somewhat disconcerted by the conclusions: the exchange photon-mass still occurs, but the box mass loss doesn't correspond to $E = mc^2$, and the center of gravity is not immobile, nor the center of inertia (Figure 3.8).

Step 0: The box is at rest in its rest frame K. Its length is L and its mass M.

Step 1: A photon is emitted from the left interior side, and moves toward the right interior face.

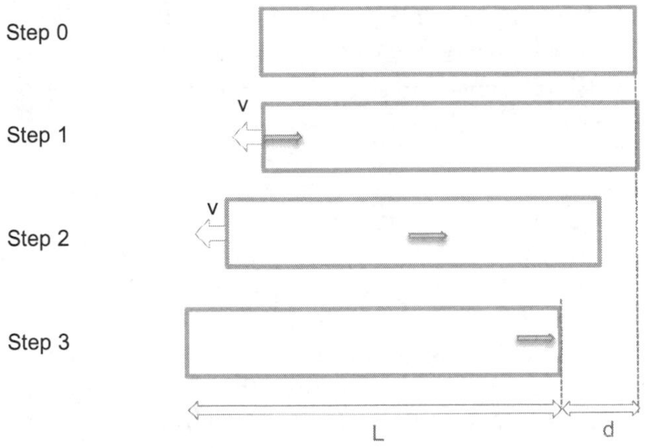

FIGURE 3.8
Einstein's photon box.

Step 2: The photon continues its motion toward the right interior side.

Step 3: The photon has just arrived against the right interior side, and is instantaneously absorbed.

Step 4: No event occurs inside nor outside the box.

3.6.2.1 First Analysis Based on Classical Physics Principles

The classical 3D momentum and the Relativistic one will be, respectively, denoted by $\vec{p}$ and $\vec{P}$.

Step 1: Consider the system made of the photon and the box without the photon: this system being submitted to no external force, its momentum remains constant, meaning null since it was null in step 0. Hence, the momentum of the box without the photon equals: $-\vec{p}$, with $\vec{p}$ being the photon momentum. Consequently, the box moves in the opposite direction from the photon: it thus incurs a recoil with the speed v such that: Mv = p. Hence:

$$v = E/(Mc). \quad (3.33)$$

Step 2: No change occurs regarding the momenta of the photon and the box without the photon. Hence, the box continues moving in the opposite direction of the photon, and with the same speed v.

Step 3: After the photon absorption, the system only comprises the box; as no external force was applied to the system since step 0, its momentum remained unchanged, meaning that the box speed in K is the same as in step 0, meaning null.

Step 4: No force is applied, no particle is emitted: the box speed remains null.

One thing is disconcerting: from step 0 to step 4 the center of gravity of the box has moved to the left, whereas no external force was applied to the box. How can that be?

To comply with the classical physics principle stating that the center of mass (also called the center of gravity) remains fixed in K, we must consider that when the photon was emitted, it resulted in some mass loss, denoted by m, for the atom, which has emitted it. Then, an equivalent mass addition occurred for the atom that finally absorbs the photon. This is the condition for the center of gravity to remain fixed although the box has moved to the left.

Let's calculate the value of this mass loss/addition, using the condition that the center of mass remains constant.

Let d be the distance covered by the box. To calculate the displacement of its center of gravity, we can consider that one small part m of the box mass has moved to the right, covering the distance L, while the rest of the Box (M − m) has moved to the left covering the distance d. The conservation of the center of mass then implies:

$$mL = (M - m)d. \quad (3.34)$$

This relation will enable us to calculate m: Let's call t the duration of step 2. The box moved at the speed v, so that: t=d/v. During the same time, the photon has covered the distance L, so that: t=L/c. We then have:

$$d/v = L/c. \quad (3.35)$$

From equations 3.32 and 3.33, we have:

$$m = (M - m)d/L = (M - m)Lv/Lc = (M - m)v/c. \quad (3.36)$$

The speed v can be obtained from equation 3.33, however, this relation needs to be amended since the mass of the box has slightly decreased: we thus have: (M − m)v = E/c, so that:

$$v = E/\left(c(M - m)\right). \quad (3.37)$$

Coming back to equation 3.36, we have:

$$m = (M - m)E/[c(M - m)c] = E/c^2. \quad (3.38)$$

which gives the famous energy relation: $E = mc^2$.

Comments: It may sound ironic that the photon, which has no mass, actually transferred some mass from the left side of the box to its right side. Moreover, we witness the disconcerting phenomenon that the total mass of the system does not remain constant: in step 2 the total mass indeed is M-m, whereas it was M in step 0. In Relativity, however, this paradox disappears due to the equivalence energy-inertia and to the new system-mass definition (norm of the system Momentum).

3.6.2.2 This Scenario Revisited in the Light of Relativity

We actually made some approximations, both in the calculations and in the fact that we used classical physics laws. In this section, we will assume valid the center of gravity immobility law, and we will discuss this law in Section 3.6.2.3.

Relation (3.33) expresses the classical momentum conservation law, but the correct one is the relativistic one, which we will now apply between steps 0 and 1: the conservation of the spatial part of $\vec{P}$ yields: $0 = M'vg - p$ with M′ being the mass after the photon emission ($M' = M - m$), and γ refers to the box speed v; we thus obtain:

$$M'v\gamma = p = E/c. \tag{3.39}$$

The box speed being very low, the value of γ is very close to 1:

$$v = E/(M' - \gamma c). \tag{3.40}$$

In relation (3.34) expressing the center of gravity conservation, the precise distance covered by the photon was not L, but (L−d). Hence:

$$M'd = m(L - d). \tag{3.41}$$

Similarly, relation (3.35) becomes:

$$d/v = (L - d)/c \tag{3.42}$$

From equations 3.41 and 3.42, we have: $m = M'd/(L - d) = M'v/c$; then with equation 3.40:

$$m = M'E/cM'\gamma c = E/c^2\gamma\,. \tag{3.43}$$

Which gives the correct relativistic Energy expression: $E = mc^2\gamma_v$ ■

The above reasoning is still based on the statement that the center of mass of an isolated system remains fixed (in K), but let's challenge this statement in Relativity.

3.6.2.3 The Center of Gravity in Classical Physics and the Center of Inertia in Relativity

3.6.2.3.1 *The Immobility of the Center of Gravity in Classical Physics*

In classical physics, the immobility of the center of mass (or of gravity) is a direct consequence of the conservation of the system Momentum of isolated systems as we will see. For the sake of simplicity, we assume the objects of a closed system are moving along the X-axis of the frame K; the center of mass is defined as the point having for abscissa:

$$\frac{\sum_{i=1}^{N} m_i x_i}{\sum_{i=1}^{N} m_i}. \tag{3.44}$$

Let's set that in K at t = 0, the center of mass coincides with the origin of frame K. We thus have at the time t = 0: $\sum_{i=1}^{N} m_i x_i = 0$. For this expression to remain constant, let's check its derivative:

$$\sum_{i=1}^{N} m_i \vec{v_i} \tag{3.45}$$

We recognize that this expression is the classical system Momentum, which is constant since the system is isolated. Let's call k this constant, we then have by integrating equation 3.45: $\sum_{i=1}^{N} m_i x_i = kt$. Consequently, the center of mass evolves at a constant velocity $k' = \frac{k}{\sum_{i=1}^{N} m_i}$. Then, if we choose the frame K′ that moves at the speed k′ relative to K, the center of mass (gravity) will be fixed in this frame K′.

Also, the constancy of the sum of the masses is a consequence of the system Momentum conservation in classical physics (cf. Volume I Section 4.4.1).

Thus in classical physics, the immobility of the center of gravity was a consequence of the system Momentum conservation. In Relativity, however, this demonstration does not work as we will see.

3.6.2.3.2 *The Center of Gravity Does Not Fit in Relativity*

Relation (3.45) is not valid in Relativity, but might be replaced by the spatial part of relativistic system Momentum, which is: $\sum_{i=1}^{N} m_i \gamma_i \vec{v_i} = \text{cst}$. However, equation 3.44 then has problems with both the numerator and the denominator, which are not constant in Relativity. Hence, equation 3.44 cannot define a center of mass in Relativity.

3.6.2.3.3 The Center of Inertia in Relativity

This issue invites us to define the center of inertia as:

$$\frac{\sum_{i=1}^{N} m_i \gamma_i x_i}{\sum_{i=1}^{N} m_i \gamma_i} \tag{3.46}$$

We notice that the denominator represents the system energy, and so is constant. If we choose that the origin of K matches with the center of inertia at $t = 0$, we have at $t = 0$:

$$\sum_{i=1}^{N} m_i \gamma_i x_i = 0. \tag{3.47}$$

To check if it remains fixed, let's check its derivative:

$$\sum_{i=1}^{N} m_i \gamma_i \overrightarrow{v_i} + \sum_{i=1}^{N} m_i' \gamma_i \overrightarrow{x_i} + \sum_{i=1}^{N} m_i \gamma_i' \overrightarrow{x_i}. \tag{3.48}$$

We can see that the first term is the system Momentum (spatial part), which indeed is constant. The second term is null if the masses are constant; the third term is null except if some parts have an acceleration. Consequently, the center of inertia has an inertial trajectory in the case where all the parts have a constant mass and a constant speed.

3.6.2.4 Consequence Regarding Our Photon Box Scenario and Alternative Method

These conditions are met during step 2. In steps 1 and 3, which are instantaneous, there are sharp accelerations and rapid mass changes. Thus, there is no guarantee that the center of inertia remains fixed between steps 0 and 4, and the same for the center of mass since it is the same as the center of inertia in the frame K in steps 0 and 4. Consequently, we were not legitimate in our demonstration of the relation $E = mc^2\gamma$ to assume the center of mass immobility between steps 0 and 4. Hence, we are invited to use another method to find the value of the mass loss m:

3.6.2.4.1 Correct Method for Calculating the Box Mass Loss

We will apply the system Momentum conservation law for isolated systems. The time part of the system Momentum is the sum of the inertias. To avoid confusion, we will denote by m′ the mass loss of the box calculated with this method:

Between steps 0 and 1, the constancy of the sum of the inertias of the different parts yields:

- In step 0, the system only comprises the box of mass M; being fixed in K, its inertia is M.
- In step 1, the sum of the inertias is: $(M - m')\gamma E/c^2$ (*E/c^2 being the inertia of the photon).* We thus have:

$$M=(M - m')\gamma + E/c^2. \tag{3.49}$$

This equation shows that m′ is slightly different from the previous mass loss m of equation (3.43) that corresponds to $E = mc^2\gamma$, and we can show that m′ is greater than m: equation (3.49) gives: $m'\gamma = M(\gamma - 1) + E/c^2$, then:

$$m' = \frac{E}{c^2\gamma} + M\frac{(\gamma - 1)}{\gamma} = m + M\frac{(\gamma - 1)}{\gamma}.$$

Since m is the mass that corresponds to the immobility of the center of gravity, we can deduce that the center of gravity has moved to the right, and the same for the center of inertia since both are identical in steps 0 and 4.

3.6.2.5 Conclusive Comments

It was by an error that the mass loss corresponded to the famous relation $E=mc^2$, and even then to $E = mc^2\gamma$. Moreover, this scenario shows that at the microscopic level, the immobility of the center of mass is not respected in Relativity, and the same for the center of inertia.

Besides, the emission of a photon creates a recoil for the atom which has emitted this photon, but this recoil cannot instantaneously set the whole box moving (otherwise there would exist instantaneous distant interactions). Actually, this recoil generates a vibrational wave which propagates along the box from its left side to the right side at the sound velocity. Hence, the photon arrives at the right side far before the vibrational wave: the right side has not moved when the photon reaches it. An illustration that rigid bodies aren't rigid in Relativity.

3.6.3 The Mossbauer Effect

We saw that when an atom emits a photon, it incurs a recoil according to the Momentum conservation. Relation (3.39) indeed gives:

$$M' \, v \, \gamma = p = E/c. \tag{3.50}$$

with M′ being the atom mass after the photon emission: $M' = M - m$, and γ refers to the atom's speed v. High-energy photons are considered, such as gamma rays.

This effect makes it impossible in the general case to have a resonance effect whereby an atom emits a photon, which is then absorbed by a neighboring atom having the same transition energy. Indeed, part of the transition energy given to the photon is lost by the recoil, and another part by the push incurred by the neighboring atom. The resulting energy available to this neighboring atom is thus lower than its transition energy, so that the photon is not absorbed.

However, in some crystals the atoms are bound together so that they don't incur a recoil, it is as if the mass M′ to be considered is the mass of the crystal, or part of it, but in any case, it is much greater than the mass of one single atom. Then, if the crystal is very homogeneous, the transition energies of neighboring atoms are identical, so that a photon emitted by one atom is absorbed by a neighboring one since their transition energies are the same, and there is no energy loss by the recoil and push effects.

When the crystal is not absolutely pure, the transition energies between neighboring atoms differ, and when modulating the energy of the gamma photons, absorption peaks are observed, which are characteristic to the chemical environment of this atom.

3.6.4 Mathematical Demonstration of the 4D-Relativistic Momentum

We saw in Volume I Section 4.2.1 that the relativistic Momentum of an object must be the product of its mass with its velocity using the general relativistic velocity definition, which is the time rate of change of the events characterizing the object in its trajectory. However, we need to find the appropriate time source, and we will do so by placing the condition that the system Momentum is conserved in a collision whatever the inertial frame where it is considered.

We denote by 0 the time of the desired time source. An object trajectory can then be characterized by the succession of events M(θ), seen in the frame K with the coordinates: $M^K(\theta)=[T(\theta),X(\theta),Y(\theta),Z(\theta)]$. The relativistic Momentum is then in K:

$$\overrightarrow{P^K(\theta)}=m\overrightarrow{S^K(\theta)}=m\frac{\overrightarrow{dM^K(\theta)}}{d\theta}=m\left[\frac{dT(\theta)}{d\theta},\frac{dX(\theta)}{d\theta},\frac{dY(\theta)}{d\theta},\frac{dZ(\theta)}{d\theta}\right]. \quad (3.51)$$

The universal homogeneity–isotropy implies that the time source associated to the Momentum of an object cannot be a function of the location of this object nor the direction of its motion. Hence, the time source of the Momentum in K can only be function of the object speed in this frame. We thus have: $d\theta=f(v)dt$. We indeed address the case of objects having a constant velocity. The case of accelerated objects will be addressed in the end.

Let's introduce the object proper time, which will help us when considering the Momentum in different inertial frames: In K, we have: $dt=\gamma_v d\tau$. Consequently, $d\theta=f(v)\gamma_v d\tau$. Then, for the sake of simplicity, we will set: $\varphi(v)=1/f(v)\gamma_v$ so that we have in K:

$$\mathbf{d\theta=d\tau/\varphi(v)}. \quad (3.52)$$

When $v \ll c$, the time source θ must be very close to the time t of the inertial frame K since the classical momentum proved its validity in context. When the object speed is null, the Momentum continuity implies: $d\theta = dt$, and so: $\varphi(0) = 1$.

In order to find the function φ, we will express the system Momentum conservation in K and K′:

3.6.4.1 Application to a Collision Seen from Two Inertial Frames

Let's have a billiard table with two balls A and B incurring an elastic collision as shown in Figure 3.9.

In the inertial frame K where the billiard is fixed, the classical speeds of A and B are, respectively, denoted by Va and Vb before the collision, and then by V2a and V2b after the collision.

Let's express the system Momentum conservation law in K: Before the collision, the appropriate time sources of A and B are denoted by θ_a^K and θ_b^K. Similarly, after the collision, they are denoted by θ_{2a}^K and θ_{2b}^K. The system Momentum conservation law in K then gives:

$$\overrightarrow{P_a^K} + \overrightarrow{P_b^K} = \overrightarrow{P_{2a}^K} + \overrightarrow{P_{2b}^K} \tag{3.53}$$

Let's express the first term: $\overrightarrow{P_a^K} = m_a \dfrac{\overrightarrow{dM_a^K}}{d\theta_a^K} = m_a \left[\dfrac{(dT)}{d\theta_a^K}, \dfrac{dX}{d\theta_a^K}, \dfrac{dY}{d\theta_a^K}, \dfrac{dZ}{d\theta_a^K} \right]$.

We saw that $d\theta_a^K = d\tau_a / \varphi(Va)$; we then have: $\overrightarrow{P_a^K} = m_a \varphi(Va) \dfrac{\overrightarrow{dM_a^K}}{d\tau_a}$.

Let's introduce the contravariant 4-vector: $\overrightarrow{Q_a^K} = m_a \dfrac{\overrightarrow{dM_a^K}}{d\tau_a}$; we thus have:

$$\overrightarrow{P_a^K} = \varphi(Va) \overrightarrow{Q_a^K}. \tag{3.54}$$

We will show that $\overrightarrow{Q_a^K}$ is the correct relativistic Momentum.

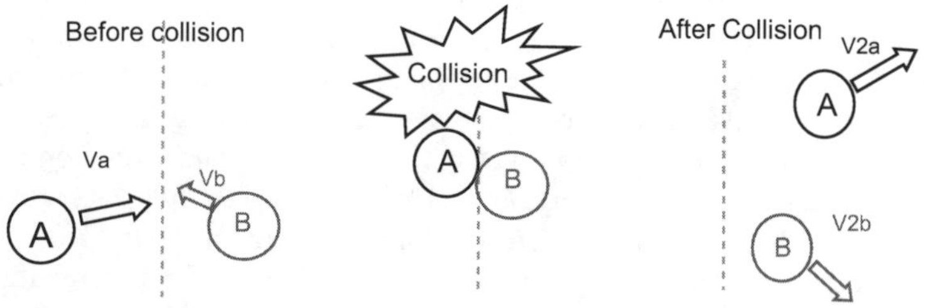

FIGURE 3.9
Collision scenario.

Relation (3.53) becomes using equation 3.54:

$$\varphi(Va)\overrightarrow{Q_a^K} + \varphi(Va)\overrightarrow{Q_b^K} = \varphi(Va)\overrightarrow{Q_{2a}^K} + \varphi(Va)\overrightarrow{Q_{2b}^K} \tag{3.55}$$

Let's now express the system Momentum conservation in another inertial frame K'.
Relation (3.55) is in K':

$$\varphi\left(V_a'\right)\overrightarrow{Q_a^{K'}} + \varphi\left(V_b'\right)\overrightarrow{Q_b^{K'}} = \varphi\left(V_a'\right)\overrightarrow{Q_{2a}^{K'}} + \varphi\left(V_{2b}'\right)\overrightarrow{Q_{2b}^{K'}} \tag{3.56}$$

In order to compare equations 3.55 and 3.56: we know that V'a is the composition of the speed Va with the speed W of K relative to K', so: $V_a' = V_a \oplus W$. Then using the contravariance of the 4-vector $\overrightarrow{Q_a^K}$, relation (3.56) becomes:

$$\varphi(V_a \oplus W)(\Lambda_w)\overrightarrow{Q_a^K} + \varphi(V_b \oplus W)(\Lambda_w)\overrightarrow{Q_b^K} = \varphi(V_{2a} \oplus W)(\Lambda_w)\overrightarrow{Q_{2a}^K} + \varphi(V_{2b} \oplus W)(\Lambda_w)\overrightarrow{Q_{2b}^K}.$$

Let's apply the reverse Lorentz transformation, (Λ_{-w}), to both sides:

$$\varphi(V_a \oplus W)\overrightarrow{Q_a^K} + \varphi(V_b \oplus W)\overrightarrow{Q_b^K} = \varphi(V_{2a} \oplus W)\overrightarrow{Q_{2a}^{K'}} + \varphi(V_{2b} \oplus W)\overrightarrow{Q_{2b}^K} \tag{3.57}$$

The conjunction of equations 3.55 and 3.57, whatever Va, Vb and W, implies that whatever v:

$$\frac{\varphi(v)}{\varphi(v \oplus w)} = k(w). \tag{3.58}$$

We will show that $k(w) = 1$: If we choose $w=-v$, equation 3.58 gives: $k(-v)=\varphi(v)$ since we saw that $\varphi(0)=1$. Besides, $\varphi(v)=\varphi(-v)$ due to the universal isotropy, so: $k(w) = \varphi(-v) = \dfrac{\varphi(v)}{\varphi(v \oplus w)} = \dfrac{\varphi(-v)}{\varphi(v \oplus w)}$. Hence: $\varphi(v \oplus w) = 1$ whatever v and w, which implies that the function φ is always equal to 1. Consequently equation 3.54 yields: $\overrightarrow{P_a^K} = \overrightarrow{Q_a^K} = m_a \dfrac{\overrightarrow{dM_a^K}}{d\tau_a}$ ■

Thus, the appropriate time source for the relativistic Momentum of an object is its proper time.

Note that we took an elastic scenario as an illustration, but this demonstration did not use the conservation of the kinetic energies; hence, it is valid for all types of collisions.

3.6.5 Demonstration of the Famous Energy Relation $E = mc^2\gamma$ from the Work of the Force

If we did not know that the time part of the Momentum, $mc\gamma$, represents energy, but if we assume that the 3D work of the Newtonian-like Force is equal to the kinetic energy variation (like in classical physics), then can demonstrate the famous energy relation $E=mc^2\gamma$. Thus, we now assume that:

$$F_x dx + F_y dy + F_z dz = dE . \quad (3.59)$$

We will use the notation • which means the Lorentzian scalar product (cf. §3.3.2), and we recall that the Lorentzian norm² of **P** is equal to m^2c^2 (cf. Volume I Section 4.3.1): $\mathbf{P}^2=\mathbf{P}\bullet\mathbf{P}=m^2c^2$.

Hence, in the context of constant mass, we have:

$0=\frac{d\mathbf{P}^2}{dt}=\frac{d(\mathbf{P}\bullet\mathbf{P})}{dt}=2\mathbf{P}\cdot\frac{d\mathbf{P}}{dt}$, meaning: $\mathbf{P}\bullet\mathbf{F}=0$.

Then, as: $\mathbf{P}=m\frac{d\mathbf{M}}{d\tau}$, we have: $d\mathbf{M}\bullet\mathbf{F}=0$, which means:

$F_t cdt-\left(F_x dx+F_y dy+F_z dz\right)=0.$

With the above assumption (equation 3.59), we have: $F_t cdt - dE = 0$.

Besides: $F_t=\frac{d(mc\gamma)}{dt}$; hence, $dE=F_t cdt=\frac{d(mc\gamma)}{dt}cdt=d(mc^2\gamma)$, which gives after integration: $\mathbf{E}=\mathbf{mc^2\gamma}$ ■

3.6.5.1 The Derivative of the Lorentzian Scalar Product

We will demonstrate that: $\frac{d\left(\mathbf{P}^2\right)}{dt}=\frac{d(\mathbf{P}\bullet\mathbf{P})}{dt}=\mathbf{2P}\bullet\frac{d\mathbf{P}}{dt}$, which can be written more simply: $(\mathbf{P}\bullet\mathbf{P})'=2\mathbf{P}\bullet\mathbf{P}'$.

The frame is assumed to be orthonormal, i.e., a frame of Minkowski. This result is a consequence of a more general law that we will demonstrate. Let's have any couple of 4D vectors: $V(t)=\left(V_t(t),V_x(t),V_y(t),V_z(t)\right)$ and $W(t)=\left(W_t(t),W_x(t),W_y(t),W_z(t)\right)$. The Lorentzian scalar product between V(t) and W(t) is:

$$V(t)\bullet W(t)=V_t(t)W_t(t)-V_x(t)W_x(t)-V_y(t)W_y(t)-V_z(t)W_z(t).$$

We will show that: $(\mathbf{V}\bullet\mathbf{W})'=\mathbf{V}'\bullet\mathbf{W}+\mathbf{V}\bullet\mathbf{W}'$, which will imply: $(\mathbf{P}\bullet\mathbf{P})'=2\mathbf{P}\bullet\mathbf{P}'$. To this end, let's differentiate:

$\mathbf{V}\bullet\mathbf{W}$: $[V(t)\bullet W(t)]'=\left[V_t(t)W_t(t)-V_x(t)W_x(t)-V_y(t)W_y(t)-V_Z(t)W_Z(t)\right]'$,

and so: $[V(t)\bullet W(t)]'=[V_t(t)W_t(t)]'-[V_x(t)W_x(t)]'-[V_y(t)W_y(t)]'-[V_Z(t)W_z(t)]'$.

Let's expand, knowing that for each term in bracket we have:

$$[V_i(t)W_i(t)]' = V_i'(t)W_i(t) + V_i(t)W_i'(t).$$

Then by regrouping the terms, we obtain:

$$[\mathbf{V(t)\ W(t)}]' = \mathbf{V'(t)\ W(t) + V(t)\ W'(t)} \ \blacksquare \tag{3.60}$$

3.6.6 Supplements to the Potential Energy: The Poisson's Equation and the Orbital Potential Energy

3.6.6.1 The Poisson's Equation

We will demonstrate the Poissons equation, which is: $\nabla^2\Phi = 4\pi G\rho$, where ∇^2 is the Laplacian operator: $\nabla^2\Phi = \frac{d^2\Phi}{dx^2} + \frac{d^2\Phi}{dy^2} + \frac{d^2\Phi}{dz^2}$.

The different operators used are further explained in Section 5.1.6.

The Newtonian law tells us that a small sphere of volume dv, centered at a point X, and with the mass density ρ generates a gravitational field at the distance r whose intensity is: $g = \frac{Gdm}{r^2} = \frac{G\rho dv}{r^2}$.

Let's integrate the outward component of the gravitational field on the surface centered at X and of radius r, knowing that g is the same on this sphere of surface area S: $\oiint \vec{g} \cdot \vec{ds} = -gS = -\frac{4\pi r^2 G\rho dv}{r^2} = -4\pi G\rho dv$.

Remark: this value is independent of the sphere size provided that it is greater than dv. Besides, the negative sign is because we want the *outward* component in order to use the Ostrogradski theorem, which tells us:

$$\oiint \vec{g} \cdot \vec{ds} = \iiint \mathrm{div}(\vec{g})dv'.$$

This integral concerns the volume inside the surface (the sphere in the present case).

Thus: $\iiint \mathrm{div}(\vec{g})dv' = 4\pi G\rho dv$. Besides, we saw that: $\vec{g} = -\overrightarrow{\mathrm{grad}}(\Phi)$, and so:

$$\mathrm{div}(\vec{g}) = -\mathrm{div}(\overrightarrow{\mathrm{grad}}(\Phi)) = \frac{d^2\Phi}{dx^2} + \frac{d^2\Phi}{dy^2} + \frac{d^2\Phi}{dz^2} = \nabla^2\Phi.$$

We thus have: $4\pi G\rho dv = \iiint \mathrm{div}(\vec{g})\,dv' = \iiint \nabla^2\Phi dv'$.

So finally: $\nabla^2\Phi = 4\pi G\rho$ ■

More generally, the flow of $\vec{g}$ through a closed surface is related to the total mass inside this surface, denoted by M_{int}:

$$\oiint \vec{g} \cdot \overrightarrow{ds} = 4\pi G M_{int} . \tag{3.61}$$

3.6.6.2 The Orbital Potential

Similarly as the momentum of an inertial object is constant, the angular momentum of an orbiting object is also constant. This was a principle of classical physics, which could then be demonstrated by the Noether theorem (cf. Section 3.5.1.2). Its applications are diverse and important, ranging from spinning tops, the gyroscope, to orbiting planets, the spin of particles…

In an inertial frame, an object is orbiting with the angular velocity ω. It is located at the distance r from the force center O (e.g., the Sun in the case of a planet). Consequently, it has an angular speed $V_\omega = \omega r$, and an angular momentum, denoted by L, with $L = r\,V_\omega = r^2\omega$ (Figure 3.10).

This object has a centripetal acceleration, $\omega^2 r$, which generates an inertial force in the opposite direction: the centrifugal force, $F = m\,\omega^2 r = m\frac{L^2}{r}$. This force tends to push the object further away from the center O, and as this force is a function of r, we are in a similar situation as for gravity where the altitude determines the force. Hence, the reasoning leading to the gravity potential also applies to the orbital potential:

The variation of the orbital potential is equal to the work of the centrifugal force: when the object moves from r to r+dr, this work is:

$$dW = F \cdot dr = m\,\omega^2 r dr.$$

When going from a point A at the distance r_A from O, to a point B at a greater distance r_B from O, the object has lost a potential energy of:

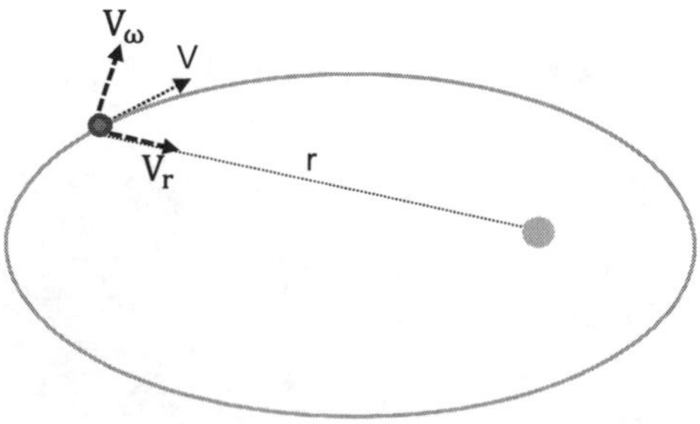

FIGURE 3.10
The angular and the radial components of the velocity of an orbiting planet.

$\int_A^B m\omega^2 r\,dr = 1/2m\omega^2 r_B - 1/2m\omega^2 r_A$. Hence, the orbital potential of an orbiting object located at the distance r from the center is:

$$U_{orb} = 1/2\ \omega^2 r^2 = \frac{L^2}{2r^2}.$$

L being constant, if r is infinite, this potential is null; if r is very small, it tends to the infinite.

Then, when adding the gravitational potential energy, we obtain the total potential, also called "effective potential" of an orbiting body in a gravitational field:

$$U_{eff} = \frac{-MG}{r} + \frac{L^2}{2r^2}. \tag{3.62}$$

When a planet is orbiting, the system energy conservation says:

$$\Delta\left(\frac{-mMG}{r}\right) = -\Delta\left(\frac{1}{2}mV^2\right).$$

Having $V^2 = V_\omega^2 + V_r^2$, and $V_\omega^2 = \frac{L^2}{r^2}$, we deduce:

$$\Delta\left(\frac{1}{2}mV_r^2\right) = -\Delta\left(\frac{-mMG}{r}\right) - \Delta\left(\frac{mL^2}{2r^2}\right) = -\Delta U_{eff}\ \blacksquare \tag{3.63}$$

This relation leads to object trajectories that are either ellipses, circles or hyperbolas. Also, when a planet approaches the Sun, its orbital potential energy increases and offsets gravity, so that the planet does not fall into the Sun. These considerations will be impacted by General Relativity, as we will further see in Section 5.2.4.

3.6.6.3 The Gravitational Time Dilation Effect in General Relativity

We saw in Volume I Section 6.2.2.2 that a photon acquires more energy when it falls toward a great mass, and so its frequency increases. Alternatively, when struggling upward against a gravity field, a photon loses energy in the form of a reduction of its frequency.

We can now relate the work of the gravitational force incurred by the photon with the variation of gravitational potential, denoted by **Φ**, in a way that is common to classical physics and Relativity: $\mathbf{d\Phi} = \overrightarrow{grad}(\Phi)\cdot\overrightarrow{dM} = \vec{g}\cdot\overrightarrow{dM}$.

Hence, relation (6.2) of Volume I: $\frac{f_A - f_B}{f_B} = \frac{gH}{c^2}$ can be written: $\frac{df}{f} = -\frac{\mathbf{d\Phi}}{c^2}$.

Let's integrate this relation: $\log(f_A) - \log(f_B) = -\frac{\Phi_A - \Phi_B}{c^2}$, and then finally: $\frac{f_A}{f_B} = \exp\left(\frac{\Phi_B - \Phi_A}{c^2}\right)$, which can also be written:

$$\frac{f_B}{f_A} = \exp\left(\frac{\Phi_A - \Phi_B}{c^2}\right) \tag{3.64}$$ ■

These relations are also valid using the gravitational potential of General Relativity. Besides, we saw in Volume I that the variation of the photon frequency from a point A to a point B at a different altitude, is identical to the variation of the perception by an observer in B of the pace of a clock in A. Hence, if A is at a higher altitude than B, the time of a clock in B is seen dilated by an observer in A by the ratio: $\frac{f_B}{f_A}$.

3.7 Questions and Problems

Problem 3.1) The speed of a Neutrino with an energy of 10 MeV

The neutrino is the lightest particle, initially thought to be massless, but its rest mass is here taken to be 25 eV (in natural units).

Q1) What is the difference between c and the speed of a neutrino whose energy is 1.0 MeV?

Q2) How much more time does it take for this neutrino to go across our galaxy, than for a photon? Our galaxy is about 10^5 light-years across.

Problem 3.2) The Muon energy

The muon rest mass is: 106 MeV/c^2 (which is 207 times that of the electron). Its speed is: 0.992 c. Compare its kinetic energy with its total energy and its energy at rest.

Problem 3.3) Cosmic Rays: Extremely Energetic Protons

Protons are the main cosmic rays, and the most energetic ones. They have an energy of 5×10^{19}eV, as measured from the Earth frame K.

Q1) What are their speeds relative to K? You may use the approximations for ultra-high-speed particles and the proton mass ≈ 1 GeV/c^2.

Q2) Imagine that this proton hits a particle of 1 mg, which is fixed in the high atmosphere, and that the collision is inelastic: the proton is absorbed by the particle: what will be the particle speed after the collision? Before, calculate the proton mass in kg. Then, the same question but the collision is done with one of the finest particles, whose mass is 10^{-15} kg.

Problem 3.4) Accelerated Electrons

An electron is accelerated by a constant electrical field E in a linear accelerator measuring 10 m. Initially, its speed is null, and at the end of the accelerator, it is: 0.7 c.

Q1) What is the electron inertia increase? What is the 3D work of the electrical force that was applied?

Q2) What is the value of the electrical field inside the accelerator?

Q3) What is the electron acceleration at the start and at the end?

Problem 3.5) Electrons in a Magnetic Field

What is the trajectory of an electron that is moving at 0.75 c in a horizontal plane K where there is a magnetic field perpendicularly to K, with a strength of 0.2 T? What is the curvature radius? Same question with the electron moving at 0.9999 c.

Problem 3.6) Cosmic Protons Deflected by the Terrestrial Magnetic Field

Let's assume a cosmic proton of 12 GeV approaches the Earth in the equator frame K, where it is exposed to the terrestrial magnetic field, which is parallel to the North–South axis, meaning perpendicular to K. What will be the deviation of this proton, assuming that the magnetic field in its area is 5.7×10^{-5} T? Then, the same question but with a less energetic proton of 5 GeV.

Problem 3.7) Electrons Entering a Transverse Electrical Field

Electrons are going along a straight line OX with a constant velocity V. Then suddenly, they enter a region where there is an electrical field in a transverse direction, parallel to OY. Show that the component of the electrons' velocity along OX will decrease. (After Purcell E. M. (1963), *Electricity and Magnetism*, Mc Graw Hill, New York.)

Problem 3.8) The Fusion Helium-3 Neutron

A Helium-3 fuses with a Neutron while emitting a gamma ray: $^3He + n \rightarrow ^4He + \gamma$. Given the ^3He mass of 3.01493 u, and that of $^4He = 4.00151$ u, what is the energy of the gamma ray? You may use the masses of the nucleons given in Volume I Appendix 1.

Problem 3.9) Binding Energies in Uranium

Consider Uranium fission where:

$^{235}_{92}U + 1n \rightarrow ^{236}_{92}U \rightarrow ^{141}_{56}Ba + ^{92}_{36}Kr + 3n$

Q1) Express the binding energies of the atoms of Uranium, Barium and Krypton, knowing that their masses are, respectively: $^{235}_{92}U$: 235.0434 u ; 140.9144 u ; and 91.9262 u. Regarding the nucleons' masses, you may use the values given in Volume I Section 10. Then, express these binding energies as a percentage of the respective atomic mass, and comment on the stability of the atoms.

Q2) What is the sum of the kinetic energies of the atoms and neutrons after the reaction? We will make the calculation with the following approximation: we will neglect the kinetic energy of the neutron that initiates the reaction.

Problem 3.10) Creation of an Electron-Positron Pair

An electron-positron pair is created after an electron was struck by a gamma ray: $e^- + \gamma \rightarrow e^- + e^+ + e^-$.

What is the minimum energy the gamma ray must have? We assume that the resulting particles have the same Momenta. The electron mass is $m = 0.511\ \text{MeV}/c^2$.

Problem 3.11) Decay of a Lambda Particle

The Lambda baryon, denoted by $\Lambda^{0,}$ is a particle created by cosmic rays, and has a very short lifetime. It rapidly disintegrates, forming a proton and a negative pion: $\Lambda^0 \rightarrow p + \pi^-$. The masses of these particles, respectively, are: $m_{\Lambda^\circ} = 1.1157\text{GeV}/c^2$; $m_p = 0.9382\text{GeV}/c^2$; $m_{\pi^-} = 0.1396\text{GeV}/c^2$.

Q1) What are the energies of the resulting proton and the pion, considered in the rest frame of the Λ^0 particle?

Q2) What is the speed of the pion?

**

Topics related to the Potential Energy

Problem 3.12) The Speed of a Falling Object on the Earth

An object of mass 10 kg falls from the altitude of 100,000 km above sea level. What is its speed when reaching the Earth's surface? We assume that there is no friction with the air, and that the classical Newtonian gravitational law applies. Calculate using the classical kinetic energy, and then the relativistic kinetic energy. Use:

$MG \approx 4\times10^{14}\ m^3s^{-2}$ and the terrestrial radius $\approx 6{,}000$ km.

Problem 3.13) The Speed of a Falling Object on the Sun

The same as Problem 3.12), but the object of mass 10 kg falls from the altitude of 10 million km above the Sun's center. What is its speed when it reaches the distance of 100 km to the Sun's center, assuming again that there is no friction with any particle? Even if this assumption is not fully realistic, it will give the maximum theoretical speed that such object may acquire.

Q1) Calculate the speed using the relativistic kinetic energy. Take Sun mass $\approx 2\times10^{30}$ kg.

Q2) What is its acceleration, seen from its ITF? **Q3)** Seen from the frame of its departure?

Problem 3.14) Calculation of the Escape Velocity from Earth

An object is at the point A on the Earth's surface. We give an initial impulsion upward to this object, so that it has an initial speed V upward. It then continues moving upward and is submitted to the sole gravitational force of the Earth. The object has no engine and we neglect friction with the air. We want to know the speed V beyond which this object will never fall back on Earth, meaning that it will escape from terrestrial gravitation: this is the escape speed. Hint: posit that when the object altitude tends to the infinite, its speed tends to zero. Use the classical kinetic energy, and assess its approximation with the Relativistic one.

Problem 3.15) Theory – The Work of the Four-Force

Does the 3D work of the spatial part of the Four-Force give the energy variation in the context of constant mass? If not, what does it give?

Problem 3.16) Theory – The Newtonian Force

In classical physics, the fact that the Newtonian force was identical in all inertial frames supported the principle of relativity. In Relativity, the Newtonian-like Force is not identical in different inertial frames.

Q1) Does it mean that the Newtonian-like force does not respect the principle of Relativity? Explain the difference between both cases.

Q2) Can we obtain the Newtonian-like Force in K′ by considering the Newtonian-like Force in K, and converting the components from K to K′?

Q3) If not, what additional information is needed to do it? (Hint: use the Force of Minkowski.)

Notes

1 Cf. Volume I Section 4.2.2.2. The fundamental reason of the contravariance of $\overrightarrow{Q}$ Q is that all the terms, m, $\overrightarrow{dM}$ and the proper time are intrinsic.

2 This assumption is consistent with the fact that: $\Delta(m\cdot\gamma_v\cdot c^2)\approx\Delta\left(\frac{1}{2}mv^2\right)$ when $v \ll c$ (cf. Volume I Section 5.1.1).

3 This cycle is also called the Bethe cycle, from the name of its discoverer, H. A. Bethe, in 1939.

4 Cf. Section 2.8.2.

5 This notation is specific to this book and should be abandoned after the learning phase. It is only introduced with the aim of avoiding confusion with the usual Euclidean scalar product.

6 The gradient operator is also denoted by $\vec{\nabla}(\Phi)$, and is further explained in Section 5.1.6.2.

7 (V^{*T}) *means the transpose of* (V^*).

4

Introduction to General Relativity: Tensors, Affine Connection, Geodesic Equation

Introduction

The first part of this chapter is an in-depth presentation of the tensors of ranks 1 and 2. These mathematical tools are essential for General Relativity and useful but not mandatory for Special Relativity. This presentation uses a very detailed approach, enabling readers who are not familiar with these notions to understand them well. The relations between the tensors and the metric are carefully explained in the general case of curved surfaces. Then, the main remarkable rank 2 tensors of Relativity will be analyzed, in particular their frame-changing rules.

The second part presents the key concepts of the pseudo-Riemannian manifold by which the curved space–time universe is modeled. In particular, the Affine Connection, covariant derivative, parallel transport, affine parameter and geodesic are carefully presented, with a very incremental approach enabling the reader to visualize these rather abstract notions. A 2D illustration with the spherical case will be given, showing precisely how these notions function.

4.1 Tensors

A tensor is a set of values representing an intrinsic[1] *entity, a typical example being a space-time separation vector. The tensor theory has been developed in order to facilitate linear calculations in a vectorial space, and as many physical relations are linear, especially locally, it is logical to take advantage of this theory. We already used tensors in Special Relativity without noticing it, since only the simplest relations were involved. We will first see the tensors of rank 1, and then of rank 2.*

DOI: 10.1201/9781003201359-4

4.1.1 Tensors of Rank 1

Being in a 4D vectorial space, denoted by K, a tensor of rank 1 is a set of four values representing an intrinsic physical reality. There are two categories of tensors that we will successively examine: contravariant and covariant ones.

4.1.1.1 Contravariant Tensors of Rank 1

Contravariant definition: a set of four components in K is a contravariant rank 1 tensor if when changing the frame from K to K′, these components change the same way as if they were the four coordinates of an event (for the sake of simplicity, we assume that K and K′ share a common origin-event).

Hence, the event is a typical rank 1 contravariant tensor, and also the space–time separation vector between two events.

Let's explain the name "contravariant": For the sake of clarity and simplicity, we will take the example of a 2D Euclidean space, knowing that all will be replicable in 4D and even in non-Euclidean spaces. We take two frames, K and K′, whose unit basis vectors are: $\vec{i}$ and $\vec{j}$ in K, and $\vec{i'}$ and $\vec{j'}$ in K′.

The two-unit basis vectors of K′ can be expressed as a function of those in K, as follows:

$$\vec{i'} = a\vec{i} + b\vec{j} \quad \text{and} \quad \vec{j'} = c\vec{i} + d\vec{j}.$$

This expression can be represented by a matrix, denoted by M:

$$M = \begin{pmatrix} a & b \\ c & d \end{pmatrix} \tag{4.1}$$

For the sake of simplicity, we will skip the arrows on vectors, and these will be generally marked with bold characters as it is commonly done.

It is a well-known mathematical result that the coordinates of any vector **V** in K, denoted by $\mathbf{V}^K$, are transformed into K′ by applying the inverse of the matrix M, meaning:

$$\mathbf{V}^{K'} = (M^{-1})\mathbf{V}^K. \tag{4.2}$$

Hence, the name contravariant since (M^{-1}) is the inverse of matrix (M).

Remark 4.1: We should bear in mind that $\mathbf{V}^K$ and $\mathbf{V}^{K'}$ correspond to the same vector, meaning the same physical entity but seen in K and K′.

This contravariance property is essential; hence, a demonstration is presented hereafter as a reminder:

4.1.1.1.1 Contravariance Demonstration

For the sake of simplicity and clarity, we will still work in 2 dimensions knowing that the demonstration is valid for any dimension. Any vector **V** can be expressed in a frame having (**i**, **j**) as unit basis vectors as follows: $\mathbf{V} = x\mathbf{i} + y\mathbf{j}$. With the matrix format, this expression is written: $\mathbf{V} = (x, y)\begin{pmatrix}\mathbf{i}\\ \mathbf{j}\end{pmatrix}$.

When changing the frame, the new unit basis vectors are given by: $\begin{pmatrix}\mathbf{i'}\\ \mathbf{j'}\end{pmatrix} = \begin{pmatrix} a & b \\ c & d\end{pmatrix}\begin{pmatrix}\mathbf{i}\\ \mathbf{j}\end{pmatrix} = (M)\begin{pmatrix}\mathbf{i}\\ \mathbf{j}\end{pmatrix}$.

We then have: $\mathbf{V} = (x, y)\begin{pmatrix}\mathbf{i}\\ \mathbf{j}\end{pmatrix} = (x, y)(M)^{-1}\begin{pmatrix}\mathbf{i'}\\ \mathbf{j'}\end{pmatrix}$.

The coordinates (x′, y′) of the same vector **V** in K′ are: $\mathbf{V}^{K'} = (x', y')\begin{pmatrix}\mathbf{i'}\\ \mathbf{j'}\end{pmatrix}$.

Consequently, we have: $(x', y') = (x, y)(M)^{-1}$ ■

Thus, the coordinates (x, y) of the vector **V** are changed in K′ by applying the inverse of the matrix (M).

**

4.1.1.1.2 Contravariant Tensors of Rank 1 in Minkowski Space

Let's now consider the case of a space–time separation 4-vector, **A**, in the frame of Minkowski. We mentioned that **A** is a contravariant tensor of rank 1. We saw that when changing the frame from K to K′, it is the Lorentz transformation which gives the coordinates of **A′** in K′ knowing those in K. Using the Lorentz matrix, we have: $\mathbf{A}^{K'} = (\Lambda)\mathbf{A}^{K}$, meaning:

$$\begin{pmatrix} t' \\ x' \\ y' \\ z' \end{pmatrix} = \begin{pmatrix} \gamma & -\gamma\beta & 0 & 0 \\ -\gamma\beta & \gamma & 0 & 0 \\ 0 & 0 & 1 & 0 \\ 0 & 0 & 0 & 1 \end{pmatrix}\begin{pmatrix} t \\ x \\ y \\ z \end{pmatrix}.$$

This implies that the Lorentz matrix is the inverse of the 4D matrix (M) giving the unit basis vectors of K′ as a function of the unit basis vectors of K.

In a 4D frame, the 4 unit basis vectors can be denoted by $(\vec{t}, \vec{i}, \vec{j}, \vec{k})$, or in a more general way $(\vec{e_0}, \vec{e_1}, \vec{e_2}, \vec{e_3})$, or even $(\mathbf{e}_0, \mathbf{e}_1, \mathbf{e}_2, \mathbf{e}_3)$ when there is no ambiguity as to their vectorial status.

With Einstein's notation, the components of a contravariant tensor are denoted with their indices on the top: $\mathbf{A}(a^0, a^1, a^2, a^3)$ or simply: $\mathbf{A}(a^i)$. We thus have: $\mathbf{A} = \sum_0^3 a^i \vec{e_i}$, or in a more compact way: $\mathbf{A} = a^i\mathbf{e}_i$. Indeed, the same index on the bottom and on the top is enough to mean that we make the summation on this index. This index is called a 'dummy index' (or mute) as it is only used for this calculation, and it can be denoted by any letter. This notation innovation is called the 'summation convention'.

4.1.1.1.3 Variable Contravariant Tensors of Rank 1

Some vectors have their components which are a function of time, e.g., the Momentum of a moving object. A variable vector is a contravariant tensor of rank 1 if, when changing the frame, its four coordinates always change as if they were the coordinates of a fixed space-time separation vector, meaning by applying the Lorentz transformation. This is the case of the Momentum, as we saw in Volume I Section 4.2.2.2.

Typical examples of rank 1 contravariant tensors are: the Proper Speed: $\mathbf{S}=\mathbf{dM}/d\tau$; the Force of Minkowski' $\mathbf{\Phi}=\mathbf{dP}/d\tau$ (also called the Four-Force). Note that the name Four-X is dedicated to contravariant tensors of rank 1.

4.1.1.2 Covariant Tensors of Rank 1

A covariant tensor of rank 1 is 4D-vector whose components change from K to K′ like the unit basis vectors of K, meaning using the matrix M of relation (4.1).

We will see that covariant tensors of rank 1 are functions, denoted by **F**, which give to any vector **V** of K as input, a real number r as output: $r=\mathbf{F}(\mathbf{V})$. These functions **F** must have two properties: they must be linear, and they must be intrinsic:

To be intrinsic means that in any other frame K′, the function **F** gives the same number r to the same vector (tensor) **V**, despite the fact that **V** has different components in K and in K′.

These linear functions are also called "covectors," or "linear forms," or even "linear applications." We will see that these function can be characterized by a 4D vector in the 4D vectorial space, called the Dual space, and that this vector is covariant:

Lets' consider a random linear form **F**: we successively apply it to the four unit frame vectors $\mathbf{e}_i$ of K, and we obtain four numbers f_i:

$$\mathbf{F}(\mathbf{e}_i) = f_i \, . \tag{4.3}$$

Besides, we will consider the four linear forms, denoted by $\mathbf{e}^i$, which give the number 1 to the vector $\mathbf{e}_i$, and 0 to all other unit basis vectors $\mathbf{e}_j$ when $j \neq i$. Using the Kronecker delta symbol, we have:

$$\mathbf{e}^i(\mathbf{e}_j) = \delta_i^j \, . \tag{4.3a}$$

We will show the following important relation, where the compact summation notation is used:

$$\mathbf{F} = \sum_{0}^{3} f_i\, \mathbf{e}^i = f_i\, \mathbf{e}^i. \tag{4.4}$$

Let's apply the linear form **F** to a random vector $\mathbf{A}(a^i)$:

$$\mathbf{F(A)} = \mathbf{F}\left(\sum_{i=0}^{3} a^i \mathbf{e}_i\right) = \sum_{i=0}^{3} a^i \mathbf{F}(\mathbf{e}_i) = \sum_{i=0}^{3} a^i f_i = a^i f_i \tag{4.5}$$

Next, apply $f_i\, \mathbf{e}^i$ to the same vector $\mathbf{A}(a^i)$:

$$f_i\, \mathbf{e}^i(\mathbf{A}) = \sum_{i=0}^{3} f_i\, \mathbf{e}^i \left(\sum_{j=0}^{3} a^j \mathbf{e}_j\right) = \sum_{i=0}^{3} f_i \left(\sum_{j=0}^{3} a^j \delta_i^j\right) = \sum_{0}^{3} f_i\, a^i = a^i f_i \; \blacksquare$$

We have thus shown that: $\mathbf{F} = f_i \mathbf{e}^i$, which means that any linear form **F** is a linear combination of the four covectors $\mathbf{e}^i$. Mathematically, this means that all linear forms constitute a 4D vectorial space, called the Dual space of K, and denoted by K*. The four covectors $\mathbf{e}^i$ constitute the Dual basis of K* that corresponds to the basis of K, which is composed of the four $\mathbf{e}_i$.

Thus, any linear form such as **F** is a vector of K* having for components f_i when using the basis composed of the four covectors $\mathbf{e}^i$. Relation (4.5) shows that **F(A)** is obtained by the simple "dot product" between $\mathbf{V}(a^i)$ and $\mathbf{F}(f_i)$, meaning: $\mathbf{F(A)} = a^i f_i$. With the matrix format, this relation is written: $\mathbf{F(A)} = (\mathbf{F})^T(\mathbf{A})$, and means:

$$\mathbf{F(A)} = (\mathbf{F})^T(\mathbf{A}) = (f_0, f_1, f_2, f_3)\begin{pmatrix} a^0 \\ a^1 \\ a^2 \\ a^3 \end{pmatrix} = f_0 a^0 + f_1 a^1 + f_2 a^2 + f_3 a^3. \tag{4.6}$$

The notation $(\mathbf{X})^T$ means transpose of $(\mathbf{X})$. This notation concerns vectors as well as matrices.

We mentioned that **F** must be *intrinsic*: this imposes a condition when changing the frame on the f_i, and we will now see this condition:

We thus have in K: $\mathbf{F(A)} = a^i f_i = r$. In K′, we must have: $\mathbf{F'(A')} = a'^i f'_i = r$. (r *being the same as in K).*

in K: $(\mathbf{A'}) = (\Lambda)(\mathbf{A})$, where (Λ) is the Lorentz matrix. We thus have:

in K: $r = \mathbf{F}(\mathbf{A}) = (\mathbf{F})^T(\mathbf{A})$, and in K′: $r = \mathbf{F'}(\mathbf{A'}) = (\mathbf{F'})^T(\mathbf{A'}) = (\mathbf{F})^T(\Lambda)(\mathbf{A})$.

For this equality to be true for any vector **A**, we must have: $(\mathbf{F})^T = (\mathbf{F}')^T(\Lambda)$. Hence:

$$(\mathbf{F}')^T = (\mathbf{F})^T(\Lambda^{-1}). \tag{4.7}$$

We mentioned that the Lorentz matrix (Λ) was the inverse of the matrix giving the basis vectors of K′ as a function of those of K (hence the name contravariant). Consequently, when using (Λ^{-1}) for the transformation of the f_i, the name "covariant" is appropriate: the components of **F** change using the same matrix as the one giving the unit vectors of K′ from those of K.

To mark the difference with contravariant tensors, the components of covariant tensors are written with the index on the bottom, such as f_i, and this permits compact summation notations such as:

$$\mathbf{F}(\mathbf{A}) = a^i\, f_i = r. \tag{4.8}$$

4.1.1.3 Relation with the Lorentzian Distance

The Lorentzian norm², like any norm², is a scalar product between a vector and itself. The scalar product is taken with its general definition, the Euclidean scalar product being a particularly simple case, even if it is very commonly used and performed by a relatively simple operation: the "dot product" (also called "sumproduct") between the components of both vectors. Hence, we will use the notation **A•B** for the general scalar product, and **A · B** (or **A.B**) for the Euclidean one, to mark the fact that **A•B** is generally not obtained by such a simple "dot product" operation (cf. Volume I Section 6.3).

*When the reader is familiar with the general definition of scalar product, he or she should use the notation **A · B**, which is commonly used for any type of scalar product, and not just the Euclidean one.*

A scalar product gives a way to associate to any vector **A** a covector **A***, which is the linear form that associates to any vector **B** a number r which is:

$$\mathbf{A}^*(\mathbf{B}) = \mathbf{A} \bullet \mathbf{B} = r.$$

A* is linear and intrinsic because the scalar product has these properties. As any linear form, **A*** belongs to the Dual space, K*. Let's then find the components of **A*** in the Dual basis composed of the four $\mathbf{e}^i$:

The components of the vector **A** are: $\left(a^0, a^2, a^2, a^3\right)$ or simply: a^i, and the same for **B**: b^j. Using the compact summation notation, we have: $\mathbf{B} = b^i\mathbf{e}_i$.

Then, thanks to the linearity of the scalar product, we have:

$$\mathbf{A} \bullet \mathbf{B} = a^i\mathbf{e}_i \bullet b^j\mathbf{e}_j = a^i b^j \mathbf{e}_i \bullet \mathbf{e}_j. \tag{4.9}$$

Let's consider the Lorentzian scalar product which is used in Minkowski frames. The scalar product between the unit basis vector $\mathbf{e}_j$ and the unit basis vector $\mathbf{e}_j$ is a number classically denoted by g_{ij} :

$$\mathbf{e}_i \bullet \mathbf{e}_j = g_{ij}. \tag{4.9a}$$

These g_{ij} coefficients are classically gathered in a 4×4 matrix, where each g_{ij} is placed at the intersection of the line i and the column j. This matrix is called the "metric" matrix,[2] denoted by G:

$$G = \begin{pmatrix} 1 & 0 & 0 & 0 \\ 0 & -1 & 0 & 0 \\ 0 & 0 & -1 & 0 \\ 0 & 0 & 0 & -1 \end{pmatrix}.$$

Applying these coefficients, relations (4.9) and (4.9a) yield:

$$\mathbf{A} \bullet \mathbf{B} = a^i b^j \ \mathbf{e}_i \bullet \mathbf{e}_j = a^i b^j \ g_{ij} = a^0 b^0 - a^1 b^1 - a^2 b^2 - a^3 b^3,$$

which indeed corresponds to the Lorentzian scalar product.

Besides, we saw with relations (4.4) and (4.6) that **A•B**=**A*·B**, where **A*** is the linear form associated with **A** , and "·" is the usual "dot product". Consequently, the components a_i of **A*** are:

$$\mathbf{A}* \ \left(a_0 = a^0; a_1 = -a^1; a_2 = -a^2; a_3 = -a^3\right).$$

With the matrix format, this result comes from the following operation:

$$\left(\mathbf{A}^*\right)^T = (\mathbf{A})^T (G) \tag{4.10}.$$

The Lorentzian norm of any vector **A** can then be computed using equation 4.10:

$$ds^2(\mathbf{A}) = \mathbf{A}^2 = \mathbf{A} \bullet \mathbf{A} = \left(\mathbf{A}^*\right)^T (\mathbf{A}) = (\mathbf{A})^T (G)(\mathbf{A}) \tag{4.10a}$$

Thus, $ds^2(\mathbf{A}) = (a^0)^2 - (a^1)^2 - (a^2)^2 - (a^3)^2$.

Relations (4.10) and (4.10a) are particularly important.

4.1.1.4 Generalization to a Curved Surface

We saw (cf. Volume I Section 6.3.3) that in a curved surface, the scalar product between two vectors **A** and **B** in the tangent space K to an event M is: $\mathbf{A} \bullet \mathbf{B} = g_{\mu\nu} a^\mu b^\nu$ where the different $g_{\mu\nu}$ coefficients are also given by the relation: $\mathbf{e}_\mu \bullet \mathbf{e}_\nu = g_{\mu\nu}$, where $\mathbf{e}_\mu$ and $\mathbf{e}_\nu$ are the unit basis vectors of K, We again place the $g_{\mu\nu}$ coefficients in a 4×4 matrix denoted by (G), and we have:

$$\mathbf{A} \bullet \mathbf{B} = (\mathbf{A})^T (G)(\mathbf{B}). \tag{4.10b}$$

We recall that by convention, the number $g_{\mu\nu}$ is at the intersection of the line μ with the column ν. Now the $g_{\mu\nu}$ coefficients that are not in the diagonal of (G) can be not null. The scalar product being symmetric ($g_{\mu\nu} = g_{\nu\mu}$), the metric matrix (G) is symmetric, and there are 10 different $g_{\mu\nu}$ coefficients instead of 16.

From this scalar product •, we can again define a linear form **A*** such that:

$$\mathbf{A}^*(\mathbf{B}) = \mathbf{A} \bullet \mathbf{B} = g_{\mu\nu} a^\mu b^\nu. \tag{4.11}$$

By this method, we associate to any vector **A** a covector **A***, which is a covariant tensor of rank 1. From relation 4.10b, we have: $\left(\mathbf{A}^*\right)^T = (\mathbf{A})^T(G)$, meaning that:

$$a_\nu = a^\mu\, g_{\mu\nu}. \tag{4.12}$$

For instance, the component 2 of A* is: $a_2 = a^0 g_{02} + a^1 g_{12} + a^2 g_{22} + a^3 g_{32}$.

Besides, we have: $\mathbf{A}^* = a_\nu\, \mathbf{e}^\nu$, where the four $\mathbf{e}^\nu$ are the unit basis vector of the Dual, K*, and we recall that $\mathbf{e}^i(\mathbf{e}_j) = \delta_i^j$. Hence, equation 4.12 yields:

$$\mathbf{A}^* = a^\mu\, g_{\mu\nu}\, \mathbf{e}^\nu. \tag{4.12a}$$

Regarding the Lorentzian norm, we again have:

$$\mathbf{ds}^2(\mathbf{A}) = \mathbf{A}^2 = \mathbf{A} \bullet \mathbf{A} = \mathbf{A}^*(\mathbf{A}) = (\mathbf{A})^T(G)(\mathbf{A}) = g_{\mu\nu} a^\mu b^\nu.$$

which again explains why (G) is called the metric matrix (tensor).

Note that there are as many different scalar products as there are different g_{ij} while respecting $g_{ji} = g_{ij}$.

4.1.2 Tensors of Rank 2 and Remarkable Tensors

4.1.2.1 Rank 2 Tensors Definition and Main Properties

A rank 2 tensor is a linear and intrinsic function that, from a rank 1 tensor as input, gives another rank 1 tensor as output. A rank 2 tensor can be represented with a 4×4 matrix (with 16 coefficients in total).

Rank 2 tensors can be built by making the tensorial product of two tensors of rank 1. The tensorial product, denoted by $C = \mathbf{A} \otimes \mathbf{B}$, is defined as follows: its generic term is: $c^{ij} = a^i\, b^j$, where a^i and b^j, respectively, are the generic terms of the rank 1 tensors A and B.

In our example, the notation shows that both A and B are contravariant tensors, but this is not an obligation: the tensorial product can be made between covariant tensors, or even a mix of a covariant and a contravariant tensor.

The tensorial product enables us to make tensors of higher ranks than 2. However, we will focus on rank 2 tensors.

The matrix format facilitates representations: the above product $C = \mathbf{A} \otimes \mathbf{B}$ can be made by the matrix product: $(C) = (\mathbf{A})(\mathbf{B})^T$, which means:

$$(C) = \begin{pmatrix} a^1 \\ a^2 \\ a^3 \\ a^4 \end{pmatrix} (b^1, b^2, b^3, b^4) = \begin{pmatrix} a^1 b^1 & a^1 b^2 & a^1 b^3 & a^1 b^4 \\ a^2 b^1 & a^2 b^2 & a^2 b^3 & a^2 b^4 \\ a^3 b^1 & a^3 b^2 & a^3 b^3 & a^3 b^4 \\ a^4 b^1 & a^4 b^2 & a^4 b^3 & a^4 b^4 \end{pmatrix}$$

We will use this matrix representation to find how this rank 2 tensor (C) changes when changing the frame, and this will also show what the intrinsic character of this tensor means:

In another frame K′, we have: $(C') = (\mathbf{A}')(\mathbf{B}')^T$, where $\mathbf{A}'$ and $\mathbf{B}'$ are the tensors $\mathbf{A}$ and $\mathbf{B}$ expressed in K′. Having assumed that $\mathbf{A}$ and $\mathbf{B}$ are contravariant, we then have:

$(C') = (\mathbf{A}')(\mathbf{B}')^T = (\Lambda)(\mathbf{A})[(\Lambda)(\mathbf{B})]^T = (\Lambda)(\mathbf{A})(\mathbf{B})^T(\Lambda)^T = (\Lambda)(C)(\Lambda)^T$.

When using natural units, the Lorentz matrix is symmetric; hence:

$$(C') = (\Lambda)(C)(\Lambda) \blacksquare \tag{4.13}$$

Furthermore, it can be shown that any twice contravariant tensor is a linear combination of tensors that are obtained by tensorial product of two contravariant tensors of rank 1. Consequently, any rank 2 tensor which is twice contravariant follows relation (4.13) when changing the frame.

**

We will now present remarkable rank 2 tensor:

4.1.2.2 The Metric Tensor

The metric matrix (G): $g_{\mu\nu}$ is a rank 2 tensor according to the very definition of a rank 2 tensor: indeed, the relation: $\mathbf{A}^* = (\mathbf{A})^T(G)$ shows that (G) gives as output a rank 1 covariant tensor, $\mathbf{A}^*$, from a rank 1 contravariant tensor, $\mathbf{A}$. This also shows that the metric tensor is twice covariant.

This can also be seen by the relation $\mathbf{A} \bullet \mathbf{B} = g_{\mu\nu} a^\mu b^\nu = r$, which gives a scalar number after making the scalar product of two contravariant tensors.

4.1.2.3 The Lorentz Matrix

The Lorentz matrix is a rank 2 tensor because it is a linear function that gives a contravariant tensor of rank 1 (a space–time separation vector) from another contravariant tensor of rank 1 (a space–time separation vector). Using natural (homogeneous) units and orthonormal frames, the Lorentz matrix is the following:

$$\begin{pmatrix} t' \\ x' \\ y' \\ z' \end{pmatrix} = \begin{pmatrix} \gamma & -\gamma\beta & 0 & 0 \\ -\gamma\beta & \gamma & 0 & 0 \\ 0 & 0 & 1 & 0 \\ 0 & 0 & 0 & 1 \end{pmatrix} \begin{pmatrix} t \\ x \\ y \\ z \end{pmatrix}.$$

Using Einstein's compact notation, the generic term of a contravariant vector **V** is written V^{ν}. The generic term of the Lorentz tensor is written Λ_{μ}^{ν}, and we have: $V'^{\nu} = \Lambda_{\mu}^{\nu} V^{\mu}$. The Lorentz tensor is thus mixed covariant–contravariant.

4.1.2.4 The Electromagnetic Tensor

We show in Section 5.1.2 that the electromagnetic matrix (EM) verifies the relation 5.4: $\mathbf{\Phi}$=q (EM)(**S***) This relation shows that (EM) is a rank 2 tensor since it gives a rank 1 contravariant tensor (the Four-Force) from a covariant rank 1 tensor (the linear form **S*** associated to the proper speed **S**). Hence, (EM) is a twice contravariant tensor.

4.1.2.5 The Stress–Energy–Tensor

We will see that the logic of tensors help us find the Stress-Energy tensor:

We saw in Volume I Section 6.4.1 that the causes of the space–time deformations on a small volume are the energy and impulse of the particles inside this volume, and that these causes form a rank 2 tensor, the stress-energy tensor T.

Energy and impulse are part of the Momentum, **P** = m**S**, with **S** being the proper velocity of the particle, but this expression refers to a point object whereas the deformations caused by T concern a small volume. There are indeed stresses, such as shear forces, which affect a volume, but not a point; and these are taken into account by the rank 2 stress-energy tensor T.

Consider a small unit volume in an inertial frame K, and for the sake of clarity and simplicity, we assume that it contains n identical particles of mass m , going all at the same speed $\vec{u}$.

Thus in K, the mass density is: ρ=nm, and the energy density is: $e_K = \rho\, c^2 \gamma_u$.

In the frame K° where these particles are at rest, the mass density is ρ_0 and the energy density is $e_{K^\circ} = \rho_0\, c^2$. In another inertial frame K, this unit of volume of K° is seen contracted in the direction of the motion $\vec{u}$. Hence: $\rho = \rho_0\, \gamma_u$, and consequently: $e_K = \rho_0\, c^2\, \gamma_u^2 = e_{K^0}\, \gamma_u^2$.

This frame-changing rule induced us to consider e_K as a component of a twice contravariant tensor, which is: $T = \rho_0\, \mathbf{S} \otimes \mathbf{S}$, where **S** is the proper velocity of the particles. We recall that **S** is in K°: $\mathbf{S}^{K^\circ}(c,0,0,0)$, and in K: $\mathbf{S}^{K}(c,\, \gamma_u u_x,\, \gamma_u u_y,\, \gamma_u u_z)$.

We thus have in K°: $(T^\circ) = \rho_0\, \mathbf{S} \otimes \mathbf{S} = \rho_0 \begin{pmatrix} c \\ 0 \\ 0 \\ 0 \end{pmatrix} (c,0,0,0) = \begin{pmatrix} \rho_0 c^2 & 0 & 0 & 0 \\ 0 & 0 & 0 & 0 \\ 0 & 0 & 0 & 0 \\ 0 & 0 & 0 & 0 \end{pmatrix}$.

And in K: $T = \rho_0\, \mathbf{S} \otimes \mathbf{S} = \rho_0 \begin{pmatrix} c\gamma_u \\ u_x\gamma_u \\ u_y\gamma_u \\ u_z\gamma_u \end{pmatrix} (c\gamma_u, u_x\gamma_u, u_y\gamma_u, u_z\gamma_u)$

$$= \rho_0\gamma_u^2 \begin{pmatrix} c^2 & cu_x & cu_y & cu_z \\ cu_x & u_x^2 & u_y u_x & u_z u_x \\ cu_y & u_x u_y & u_y^2 & u_y u_z \\ cu_z & u_x u_z & u_y u_z & u_z^2 \end{pmatrix} \blacksquare$$

The term T^{00} represents the energy density as desired, $e_k = \rho_0 \gamma_u^2 c^2$, and the other terms represent:

- The flux of energy in the three spatial directions,
- the impulsion density in the three spatial directions
- the flux of component i of the impulsion in the direction j.

4.1.3 The Dual of the Dual Space, and the Possibility to Raise or to Lower the Indices

4.1.3.1 The Dual of the Metric Tensor

Consider a scalar product in a vectorial space K, defined a metric tensor (G):

$$\mathbf{A} \bullet \mathbf{B} = (\mathbf{A})^T (G)(\mathbf{B}) = r.$$

We also saw that this scalar product can associate to any vector **A** a linear form **A*** belonging to the Dual of K, denoted by K*. Thus: $\mathbf{A}^*(\mathbf{B}) = \mathbf{A} \bullet \mathbf{B} = r$, which shows that: $\mathbf{A}^* = (\mathbf{A})^T (G)$.

Let's now consider the Dual space K*: we can define a scalar product, denoted by ⊙, between any couple of covectors of K*, and which is such that: $\mathbf{A}^* \odot \mathbf{B}^* = r$, with r being the same number as above. This scalar product is defined by the metric tensor G′. We will show how G′ is related with G:

Remark 4.4: The notation ⊙ is used here to clearly mark the difference between this scalar product and the previous one defined by the metric G. However, when the reader is familiar with these notions, he should use the same notation for all types of scalar products, since it is commonly done this way.

The same reasoning tells us that this scalar product $\odot$ can be considered as a linear form belonging to the Dual of K*, denoted by K**. We then have: $r = \mathbf{A}^* \odot \mathbf{B}^* = (\mathbf{A}^{**})^T (\mathbf{B}^*)$ with $(\mathbf{A}^{**})^T = (\mathbf{A}^*)^T (G')$.

So: $r = (\mathbf{A}^*)^T (G')(\mathbf{B}^*)$, and also: $(\mathbf{A}^*)^T = (\mathbf{A})^T (G)$; hence:

$$r = (\mathbf{A})^T (G)(G')(\mathbf{B}^*). \tag{4.14}$$

While from A•B=r, we also have: $r = (\mathbf{A})^T (\mathbf{B}^*)$ hence, with equation 4.14, we finally obtain:

$$(G') = (G)^{-1} \ \blacksquare$$

This is an important result, which in particular shows that the Dual of the Dual K**, is isomorphic to the initial space K. We indeed have: $(\mathbf{B}^{**})^T = (\mathbf{B}^*)^T (G') = (\mathbf{B}^*)^T (G)^{-1} = (\mathbf{B})^T (G)(G)^{-1} = (\mathbf{B})^T$.

The coefficients of the tensor $(G') = (G)^{-1}$ are denoted with their two indices on the top, g^{ij}, since it is twice contravariant. This can indeed be seen by the relation $r = (\mathbf{B}^*)^T (G') (\mathbf{A}^*)$, which shows that a scalar number is obtained by applying two covariant tensors to (G′). Besides, we again have: $g^{ij} = \mathbf{e}^i \odot \mathbf{e}^j$.

Remark 4.5: The scalar product $\mathbf{e}^i \odot \mathbf{e}^j$ is different from $e_i \bullet e_j$ (cf. Problem 4.2).

4.1.3.2 Possibility to Raise or Lower the Indices

The scalar product between two vectors **A** and **B** of K, **A•B**=r, can be obtained in different ways:

- The direct way in K: $r = a^i \, b^j \, g_{ij}$.
- By applying the linear form **A*** to the vector **B**: $r = a_i \, b^i$.
- Similarly, the linear form of **B*** applied to **A**: $r = b_i \, a^i$.
- The scalar product between their associated linear forms **A*** and **B*** in K*: $r = \mathbf{A}^* \odot \mathbf{B}^* = a_i \, b_j \, g^{ij}$.

The fact that the relation $a^i \, b^j \, g_{ij} = b_i \, a^i$ is valid whatever **A** implies that: $b_i = b^i g_{ij}$ ■

Similarly, we have: $a^i = a_j g^{ij}$ ■

4.2 The Affine Connection, the Covariant Derivative and the Geodesic

Our main objective is to characterize the geodesic in a non-Euclidean space. We recall that it is the trajectory of a free-falling object, including the photon (light), and that this trajectory is dictated by the constancy of the Momentum. We will find a way to express the Momentum constancy in a 4D curved space, and this will lead us to see the Affine Connection and the Covariant Derivative. We will assume that our universe is a pseudo-Riemannian manifold having the Lorentzian distance; hence, let's first characterize this manifold.

4.2.1 The Riemannian Manifold and the Momentum Constancy Relation

4.2.1.1 The Pseudo-Riemannian Manifold

The first basic postulate is the existence of a system of four independent coordinates enabling us to characterize any event. These coordinates are denoted by (t, x, y, z) or in a more general way $\left(x^0, x^1, x^2, x^3\right)$, or even simply by x^μ, with μ taking any value among (0, 1, 2 and 3). There can be different systems of coordinates, which are curvilinear in the general case. This manifold has the following properties:

- It must be differentiable, meaning that for any point $M\left(x^0, x^1, x^2, x^3\right)$ and whatever the small values (dx^0, dx^1, dx^2, $d\,x^3$), there must exist a point $N\left(x^0 + dx^0,\ x^1 + dx^1,\ x^2 + dx^2,\ x^3 + dx^3\right)$.
- It must have a pseudo-Riemannian metric, meaning that for any point $M\left(x^0, x^1, x^2, x^3\right)$, the "distance" between M and any point N in its vicinity can be approximated by a quadratic form of the coordinates of **MN**, which is classically written using Einstein's notation:

$$ds^2(\mathbf{MN}) = g_{\mu\nu} dx^\mu dx^\nu. \tag{4.15}$$

We saw in Volume I Section 6.3.3 that the $g_{\mu\nu}$ coefficients and their variations reflect the chrono-geometry of the space. Besides, the metric is assumed to be is differentiable.

This quadratic form actually is the general expression of a distance in a Euclidean space, as seen in Volume I Section 6.2. This means that in the vicinity of M, we can make the approximation that the events are in a Euclidean space, which is called the **Euclidean tangent space** at the event M. The size of the vicinity of the event M, where this approximation is valid depends on the level of accuracy required, and on the space curvature around M. The Riemannian manifold is assumed to be differentiable; it is always possible to find a size for this vicinity that satisfies the level of approximation required.

This distance being an intrinsic notion, the distance ds^2(**MN**) is the same independently of the system of coordinates. Thus, in a different frame, there exists another set of coefficients, denoted by g'_{ij}, such that:

$$g'_{ij}\, dx^{i'} dxi^{j'} = g_{ij}\, dx^i\, dx^j .$$

In particular, we can use an orthonormal frame (i.e., a frame of Minkowski), where the Lorentzian distance is expressed more simply by: $ds^2(\mathbf{dM}) = c^2dt^2 - dx^2 - dy^2 - dz^2$. However, another frame is generally more useful: the one that is directly derived from the system of curvilinear coordinates chosen, as we will see.

In order for the metric to be compatible with our space–time universe, it must be the Lorentzian one, which in particular discriminates between time-like and space-like intervals. The fact that this metric can give negative values motivates calling this manifold *pseudo*-Riemannian. If $ds^2(\mathbf{MN})$ is positive, then **MN** is a time-like interval and the proper time follows the relation with natural units:

$$\tau_{MN} = \sqrt{ds^2(MN)} = \sqrt{g_{ij}dx^i dx^j} \quad (4.16).$$

4.2.1.1.1 The Natural Coordinate Tangent Frame

Consider an event N, which is close to an event M, and such that three coordinates of N are identical to those of M, and the other one, denoted by μ, differs from the one of M by a small quantity: dx^μ. When dx^μ tends to zero, MN can be considered as being in the Euclidean tangent space at M, and we can define the following vector:

$$\overrightarrow{e_\mu} = \lim_{dx^\mu \to 0} \frac{\overrightarrow{MN}}{dx^\mu}. \quad (4.17)$$

This vector is along the tangent in M to the curvilinear line of the coordinate μ, which is made of all points having the same coordinates as M except the coordinate μ.

This operation can be repeated to the three other coordinates, yielding four vectors that constitute the **natural coordinate tangent frame** at the event M.

Let's denote the coefficients of the metric in the natural coordinate tangent frame at M by $g_{\mu\nu}$. Any metric also defines a scalar product; hence, the $g_{\mu\nu}$ define the following scalar product: $\overrightarrow{e_\mu} \bullet \overrightarrow{e_\nu} = g_{\mu\nu}$ (cf. Volume I Section 6.3.3).

4.2.1.2 The Scheme for Characterizing the Momentum Constancy

Consider an inertial (free-falling) object passing by a point (event) M. Its trajectory is determined by the law of inertia expressing the constancy of its Momentum: $\vec{P}(M) = m\,\vec{S}(M) = \overrightarrow{cst}$, with $\vec{S}(M) = \overrightarrow{dM}/d\tau$.

At another event N on the object trajectory, we must also have: $\vec{P}(N) = m\,\vec{S}(N) = \overrightarrow{cst}$. However, we cannot compare $\vec{P}(N)$ and $\vec{P}(M)$ because these two vectors are in two different Euclidean tangent spaces, the one at the event M and the other one at the event N. However, we will see that if N is in the vicinity of M, a solution exists thanks to the Affine Connection that was formulated by T. Levi-Civita.

Visualization in the simpler context of a 2D space:

We will first use the simpler case of the 2D surface of a sphere, and we will then generalize the findings to our 4D universe: we regularly draw meridians and latitudes on a sphere, which makes a system of curvilinear coordinates. Any point M on the sphere can be characterized as being on the meridian X_M and the latitude Y_M; thus, the coordinates of M are: (X_M, Y_M) (Figure 4.1).

We will use the object's proper time τ as the parameter for identifying its position in its trajectory. At the time τ, the object is at the event $M(\tau)$, and then at the time $\tau + d\tau$, it is at the event $N = M(\tau + d\tau)$. The space–time separation vector $\overrightarrow{MN}$ is denoted by $\overrightarrow{dM(\tau)}$.

There is a Euclidean tangent space in M, and we can use the natural coordinate tangent frame in this space, with its two basis vectors denoted by $\vec{i}$ and $\vec{j}$. These are, respectively, tangent to the latitude and the meridian passing by M. The object's proper velocity at M, $\vec{S}(\tau) = \overrightarrow{dM}(\tau)/d\tau$, can be expressed in this basis:

$$\vec{S}(\tau) = a\,\vec{i} + b\,\vec{j}. \tag{4.18}$$

At the time $\tau + d\tau$, the object is at N; its proper velocity, $\vec{S}(\tau + d\tau)$, can be expressed in the natural coordinate tangent frame in N, with $\vec{i}'$ and $\vec{j}'$ being its two basis vectors:

$$\vec{S}(\tau + d\tau) = (a + da)\vec{i}' + (b + db)\vec{j}'. \tag{4.19}$$

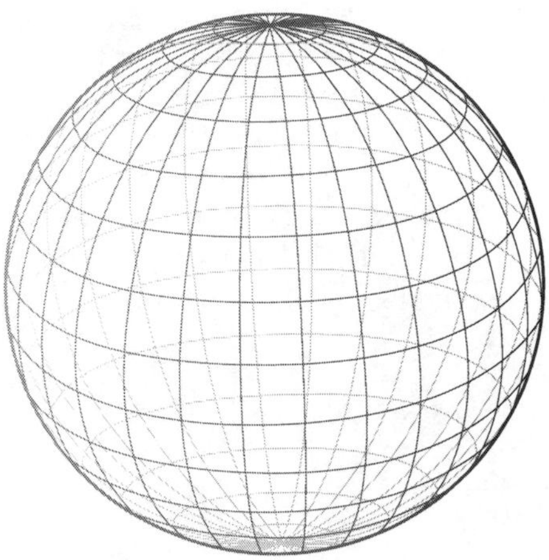

FIGURE 4.1
Sphere with its longitudes and latitudes forming a curvilinear system of coordinates (image from Shutterstock).

We will see that the "Affine Connection" will enable us to make a valid approximation of $\vec{i}'$ and $\vec{j}'$ in the coordinate tangent frame in M. This will enable us to express $\vec{S}(\tau + d\tau)$ in the frame $(\vec{i}, \vec{j})$, which will be referred to as the "covariant derivative."

4.2.2 The Affine Connection

The general principle of the Affine Connection of Levi-Civita is that each basis vector in the coordinate tangent space at an event N in the vicinity of the event M can be approximated by its orthogonal projection in the Euclidean tangent space at M.

Performing this projection assumes that our space–time universe can be plunged into a Euclidean space of dimension n+1. Note that it is not possible to make such an isometric embedding with any space, in particular surfaces that have a torsion. This orthogonal projection extremizes the distance between the point (event) on the curved surface and the tangent plane; hence, it provides the best approximation.

Besides, the pseudo-Riemannian metric must be the same everywhere, meaning that if we place a real small segment measuring 1 mm at one point, the metric will give the same value for the length of this segment independently of the location of this point. This also applies to the time: the metric will give the same value for 1 second of a perfect clock, independently of the location of this clock. This complies with the universal homogeneity postulate and ensures the proper time and distance universality. Consequently, when comparing $\overrightarrow{e_\mu}$ and the projection of $\overrightarrow{e'_\mu}$ on the coordinate tangent frame at M, we will be sure that the difference stems from the surface curvature and not from a change in the metric for the same proper distance or proper time.

An illustration with a 2D curved space is given in section 4.3.1.

We will again consider a 2D curved surface for the sake of clarity and simplicity, knowing that the reasoning will apply to any dimension: let's first consider the case when N tends to M along the direction given by the coordinate x. The point N only differs from the point M by the value dx of the coordinate x. At the point N, the basis vectors of the natural coordinate tangent frame at N are denoted by $\vec{i}'$ and $\vec{j}'$. Their projections in the coordinate tangent frame at M are denoted by $\overrightarrow{i'_m}$ and $\overrightarrow{j'_m}$. The small variation of the vector $\vec{i}$ along the coordinate x is: $\overrightarrow{di_x} = \overrightarrow{i'_m} - \vec{i}$.

When dx tends to zero the limit of $\dfrac{\overrightarrow{di_x}}{dx}$ is then by definition the partial derivative: $\dfrac{\partial \vec{i}}{\partial x} = \lim\limits_{dx \to 0} \dfrac{\overrightarrow{di_x}}{dx}$.

Remark: The word "partial" refers to the fact that the variations of $\vec{i}$ are along the x coordinate, the other coordinates being fixed.

In the general case, $\overrightarrow{i'_m}$ has its components along $\vec{i}$ and $\vec{j}$ which are not null.

Hence, $\frac{\partial \vec{i}}{\partial x} = u\,\vec{i} + y\,\vec{j}$.

Then, if we know the values of u and v, we are able for any point that only differs from M by dx on the coordinate x, to give an approximation of its corresponding $\overrightarrow{di_x}$ by making a linear approximation:

$$\overrightarrow{di_x} = \frac{\partial \vec{i}}{\partial x} \cdot dx. \tag{4.20}$$

In the same fashion, when N is in the vicinity of M but along the y coordinate, meaning: MN=(0, dy), the variation of the vector $\vec{i}$, denoted by $\overrightarrow{di_y}$ is equal to: $\overrightarrow{di_y} = \frac{\partial \vec{i}}{\partial y} \cdot dy$.

The same reasoning also applies to the vector $\vec{j}$, leading to: $\overrightarrow{dj_x} = \frac{\partial \vec{j}}{\partial x} \cdot dx$ and $\overrightarrow{dj_y} = \frac{\partial \vec{j}}{\partial y} \cdot dy$.

Then, when the point N is in any direction but still in the vicinity of M, with $\overrightarrow{MN} = (dx, dy)$, the variations of $\vec{i}$ and $\vec{j}$ from M to N are obtained by summing the variations due to the displacement dx along the x coordinate and to the displacement dy along the y coordinate. This is mathematically called the "total derivative":

$$d\vec{i} = dx\frac{\partial \vec{i}}{\partial x} + dy\frac{\partial \vec{i}}{\partial y} \quad \text{and} \quad d\vec{j} = dx\frac{\partial \vec{j}}{\partial x} + dy\frac{\partial \vec{j}}{\partial y}. \tag{4.21}$$

We can thus see that the knowledge of $\frac{\partial \vec{i}}{\partial x}, \frac{\partial \vec{i}}{\partial y}, \frac{\partial \vec{j}}{\partial x}, \frac{\partial \vec{j}}{\partial y}$ enables us to find the variations of $\vec{i}$ and $\vec{j}$ anywhere in the vicinity of M. These four partial derivatives are expressed in the frame $(\vec{i}, \vec{j})$ with two coefficients each; hence, there are eight coefficients in total. We then need a notation convention to handle this complexity, and we will use the Christoffel symbols, which give in the generic case:

$$\frac{\partial \vec{i}}{\partial y} = \Gamma^{i}_{iy}\vec{i} + \Gamma^{j}_{iy}\vec{j}. \tag{4.22}$$

Remark 4.7: In the Christoffel symbols, the indices i and x are equivalent; and the same for j and y.

Relation (4.21) thus becomes:

$$d\vec{i} = dx\left(\Gamma^{i}_{ix}\vec{i} + \Gamma^{j}_{ix}\vec{j}\right) + dy\left(\Gamma^{i}_{iy}\vec{i} + \Gamma^{j}_{iy}\vec{j}\right). \tag{4.23}$$

The same applies to $\overrightarrow{dj}$. As an example, the calculation of the Christoffel coefficients is presented in the case of a 2D sphere in Section 4.3.1.

In a space with n dimensions, the above reasoning still applies; hence, there are n^3 different Christoffel coefficients. This generalization is presented in Section 4.2.4, and we will see in particular that these coefficients are functions of the different $g_{\mu\nu}$ coefficients of the metric in M and its vicinity.

4.2.2.1.1 General Properties of the Connection

If the surface is without torsion, we have: $\Gamma^{a}_{bc} = \Gamma^{a}_{cb}$.
If the tangent frame at M is orthonormal, then the Christoffel coefficients in M are null.

These properties will be shown in the case of a sphere in Section 4.3.1.

4.2.3 The Covariant Derivative and the Parallel Transport

Let's return to the Covariant Derivative, which was introduced with relation (4.19). The object proper velocity at the time τ can be expressed in the natural coordinate tangent frame at M:

$$\vec{S}(\tau) = \frac{\overrightarrow{dM}(\tau)}{d\tau} = a(\tau)\vec{i} + b(\tau)\vec{j}. \tag{4.24}$$

We now know how to express the variation of $\vec{i}$ and $\vec{j}$ in the vicinity of M; hence, we can express the variation of $\vec{S}(\tau)$: we will use the same strategy as previously, that is to first examine the variation when the object moves by dx along the x-axis, and then by dy along the y-axis; finally, when it moves in any direction, i.e., by dx along the x-axis and dy along the y-axis, we will sum up the results found for dx and dy separately. We also assume the differentiability of the proper speed $\vec{S}$.

Thus, when the object moves by dx along the x coordinate, the variation of its proper velocity $\vec{S}$ is:

$$\frac{\partial\vec{S}}{\partial x} = \frac{\partial a}{\partial x}\vec{i} + a\frac{\partial\vec{i}}{\partial x} + \frac{\partial b}{\partial x}\vec{j} + b\frac{\partial\vec{j}}{\partial x}. \tag{4.25}$$

Then using the Affine Connection and the Christoffel symbols, we obtain:

$$\frac{\partial\vec{S}}{\partial x} = \frac{\partial a}{\partial x}\vec{i} + a\left(\Gamma^{i}_{ix}\vec{i} + \Gamma^{j}_{ix}\vec{j}\right) + \frac{\partial b}{\partial x}\vec{j} + b\left(\Gamma^{i}_{jx}\vec{i} + \Gamma^{j}_{jx}\vec{j}\right) = \nabla_x\vec{S}. \tag{4.26}$$

This is the Covariant Derivative of the vector $\vec{S}$ along the coordinate x, denoted by: $\nabla_x\vec{S}$.[3]

The word "covariant" used here has nothing to do with the word "covariant" used in the context of tensors. The meaning here is that a covariant expression has the same form in any system of coordinates while giving the same result for the same intrinsic object. The object's trajectory and the proper velocity vectors are intrinsic geometrical objects, which are independently of the system of coordinates chosen to represent them. The Covariant Derivative is the way to express the variation of a vector (here the proper velocity) while offsetting the bias due to the particular system of coordinates chosen. More generally, Einstein issued the General Covariance Principle, stating that physical laws are independent of the system of coordinates chosen, which reinforces the importance of covariant expressions.

The same reasoning applies if the object moves by dy along the y-axis. Then, in the general case where the object moves in any direction, we have:

$$\vec{S}(\tau+d\tau)=dx\,\frac{\partial\vec{S}}{\partial x}+dy\,\frac{\partial\vec{S}}{\partial y}. \tag{4.27}$$

This can be expressed as a function of the proper time, then using the Covariant Derivative symbol:

$$\frac{d\vec{S}}{d\tau}=\frac{dx}{d\tau}\cdot\frac{d\vec{S}}{dx}+\frac{dy}{d\tau}\cdot\frac{d\vec{S}}{dy}=\frac{dx}{d\tau}\nabla_x\vec{S}+\frac{dy}{d\tau}\nabla_y\vec{S}. \tag{4.28}$$

We show in Section 4.3.1 that in the case of a 2D sphere, the Covariant Derivative explains why the trajectory of an inertial (free-falling) object is the great circle, which thus is a geodesic.

Relation (4.25) can be reorganized by regrouping the terms along $\vec{i}$ and $\vec{j}$, so that:

$$\frac{\partial\vec{S}}{\partial x}=\vec{i}\left(\frac{\partial a}{\partial x}+a\,\Gamma^{i}_{ix}\vec{i}\;+b\,\Gamma^{i}_{jx}\right)+\vec{j}\left(\frac{\partial b}{\partial x}+a\,\Gamma^{j}_{ix}+b\,\Gamma^{j}_{jx}\right). \tag{4.29}$$

Note that the Covariant Derivative also applies to any differentiable field of vectors, and not just the proper velocity vector. For instance, the wind can be represented by a field of vectors.

4.2.3.1 The Parallel Transport

A vector $\vec{v}$ undergoes a parallel transport along a given curve if its Covariant Derivative is null at all points of the curve. In particular, we saw that the proper velocity vector $\vec{S}$ of an inertial object must remain constant, meaning that its Covariant Derivative is always null along its trajectory, which reads with the previous notation: $\nabla_S \vec{S} = 0$. (The index S indicates the direction along which the derivative is performed).

In the vicinity of a point M, the parallel transport corresponds to the classical definition of parallelism, and even to the notion of equality between vectors (parallel and with the same length). This can be seen by the fact that we can choose in the tangent space at M an orthonormal frame, and in this frame, we mentioned that the Christoffel coefficients are null, implying that the Covariant Derivative is identical to the classical one. Hence locally, we are left with the classical derivative $\overrightarrow{dv} = 0$, which in a Euclidean space means that $\vec{v}$ remains constant and always parallel to the same direction. This also means that the object trajectory is a geodesic *locally*, since a geodesic is a trajectory where the tangent to any point has the same direction.

However, this parallelism is valid only locally and this can be seen by the following example where the space is a 2D sphere: an inertial ball on this sphere pursues a trajectory that is a great circle (as shown in Section 4.3.1). We now plunge this sphere in a Euclidean 3D space, and we can see that the succession of proper velocity vectors of the ball along its trajectory are not parallel in our Euclidean 3D space since they are always tangent to the sphere.

Another important difference with Euclidean frames is the following: the vector at a point B, which results from the parallel transport from a given vector at the point A, is dependent on the curve that was chosen from A to B to perform this parallel transport.

Let's indeed take again the example of the 2S sphere: We have a vector $\overrightarrow{V_A}$ at a point A on the equator, and this $\overrightarrow{V_A}$ is parallel to the meridian passing at A. We then make a first parallel transport of this vector along the meridian passing by A until the North Pole N. The resulting vector in N, denoted by $\overrightarrow{V_{AN}}$, is parallel to the meridian passing by A.

Next, we make another parallel transport of $\overrightarrow{V_A}$ along the equator from the point A until a point B: the resulting vector in B, denoted by $\overrightarrow{V_{AB}}$, is parallel to the meridian passing by B because the succession of vectors undergoing the parallel transport along the equator is perpendicular to the equator. Then, if we make a last parallel transport of $\overrightarrow{V_{AB}}$ from B to the North Pole, we will have a vector, denoted by $\overrightarrow{V_{ABN}}$, which is parallel to the meridian passing by B. We thus have this strange situation where both $\overrightarrow{V_{ABN}}$ and $\overrightarrow{V_{AN}}$ result from $\overrightarrow{V_A}$ by parallel transport to B, but are not identical.

This illustrates the impossibility to compare vectors that are in distant tangent spaces, which is a major restriction brought by General Relativity.

If we plunge the 2D curved surface in a 3D Euclidean space, we can make a real Euclidean parallel transport of a vector $\overrightarrow{V_A}$ at point A to any point B, and the resulting vector, denoted by $\overrightarrow{V_B}$, is equal to $\overrightarrow{V_A}$: parallel and same length. However, $\overrightarrow{V_B}$ generally is not in the tangent plane to the curved surface at B. Nevertheless, if the point B is in the vicinity of A, the projection of the vector $\overrightarrow{V_B}$ on the tangent plane to the curved surface at B is identical to $\overrightarrow{V_{AB}}$, which is the vector obtained from $\overrightarrow{V_A}$ by parallel transport from A to B on the curved surface.

The projection of $\overrightarrow{V_B}$ on B's tangent plane will generally be of a different magnitude from $\overrightarrow{V_{AB}}$, because orthogonal projection doesn't preserve lengths, and orthonormal projections don't exist (the reason why maps are so hard to draw).

4.2.4 Generalization to Our 4D Space–time Universe

All that was presented in a space with two dimensions applies to pseudo-Riemannian spaces with n dimensions, in particular our 4D space–time universe. In order to handle the complexity induced by these additional dimensions, specific notations have been defined, as presented hereafter:

The unit frame vectors in the natural tangent frame in M are denoted by $\mathbf{e}_\mu$. The bold character indicates that it is a vector, and not a scalar. In our 4D universe, the unit vector along the time dimension is generally denoted by $\mathbf{e}_0$, whereas the space ones are denoted by $\mathbf{e}_1$, $\mathbf{e}_2$, $\mathbf{e}_3$,. Thus in our previous 2D case, $\vec{i}$ would be denoted by $\mathbf{e}_1$ and $\vec{j}$ by $\mathbf{e}_2$. There are now 64 (= 4^3) Christoffel coefficients.

The components of a vector $\mathbf{v}$ in the natural tangent frame are x^a, the index being on the top in order to show that it is a contravariant tensor (rank 1). We thus have: $\mathbf{v} = \sum_{a=0}^{3} x^a \mathbf{e}_a$; using Einstein's compact summation notation: $\mathbf{v} = x^a \mathbf{e}_a$.

4.2.4.1 *The Affine Connection and Its Relation with the Space–Time Curvature*

4.2.4.1.1 *The Affine Connection*

Consider an event N, which only differs from the event M by the value dx^j of its coordinate j, we then have: $\frac{\partial \mathbf{e}_i}{\partial x^j} = \sum_{a=0}^{3} \Gamma^a_{ij} \mathbf{e}_a$ with Γ^a_{ij} being the Christoffel coefficient as presented with relation (4.22).

Then using the notation: $\partial_j \mathbf{e}_i = \frac{\partial \mathbf{e}_i}{\partial x^j}$, and Einstein's compact summation, we have:

$$\partial_j \mathbf{e}_i = \Gamma^\mu_{ij}\, \mathbf{e}_\mu \,. \tag{4.30}$$

4.2.4.1.2 *Relation between the Christoffel Coefficients and the Metric (or the Space Curvature)*

The existence of a link between the chrono-geometry and the metric was already seen in Volume I Section 6.3.3. The following will confirm and specify this link.

We saw that: $g_{ab} = \mathbf{e}_a \bullet \mathbf{e}_b$. Let's then make the partial derivative along the coordinate c, and we obtain:

$$\partial_c g_{ab} = (\partial_c \mathbf{e}_a) \bullet \mathbf{e}_b + \mathbf{e}_a \bullet (\partial_c \mathbf{e}_b) = \Gamma^\mu_{ac}\, \mathbf{e}_\mu \bullet \mathbf{e}_b + \mathbf{e}_a \bullet \Gamma^\nu_{bc} \mathbf{e}_\nu = \Gamma^\mu_{ac}\, g_{\mu b} + \Gamma^\nu_{bc}\, g_{a\nu} \,.$$

These n^3 relations enable us to find the n^3 Christoffel symbols and express them as functions of the different g_{ab} and their partial derivatives; and we saw that these $g_{\mu\nu}$ coefficients characterize the space curvature, as well as the distance (metric). Note that the result is symmetric in the two lower indices, so there are 24 independent values. The calculation yields:

$$\Gamma^{\mu}_{cb} = \frac{1}{2}\left(\partial_c g_{ab} + \partial_b g_{ca} - \partial_a g_{bc}\right) g^{a\mu} . \tag{4.31}$$

The demonstration is given in Section 4.3.2.

4.2.4.2 The Covariant Derivative and Parallel Transport

4.2.4.2.1 The Covariant Derivative

In the general case of a field of vectors, the vector **v** (M) at the event M can be expressed in the natural coordinate tangent frame at M by: $\mathbf{v} = v^a\ \mathbf{e_a}$.

The partial derivative of **v** along the coordinate b is:

$\partial_b \mathbf{v} = (\partial_b v^a)\mathbf{e_a} + v^a\, \partial_b \mathbf{e_a} = \left[\partial_b v^a + v^a\, \partial_b\right]\mathbf{e_a}$.

Thanks to the Affine Connection (4.30), we have: $\partial_b \mathbf{v} = \left(\partial_b v^a\right)\mathbf{e_a} + v^a \Gamma^{\mu}_{ab}\mathbf{e_\mu}$.

This relation gives the Covariant Derivative along a given coordinate (b in our example). It is possible to further simplify this relation since in the second term, the indices a and μ are mute; hence, we can exchange them and obtain: $\partial_b \mathbf{v} = \left(\partial_b v^a\right)\mathbf{e_a} + v^{\mu}\Gamma^{a}_{\mu b}\mathbf{e_a} = \left[\partial_b v^a + v^{\mu}\Gamma^{a}_{\mu b}\right]\mathbf{e_a}$.

We then define the Covariant Derivative of the component of a vector:

$$\nabla_b v^a = \partial_b v^a + v^{\mu}\Gamma^{a}_{\mu b} . \tag{4.32}$$

And so relation (4.31) becomes compact:

$$\partial_b \mathbf{v} = \nabla_b v^a \mathbf{e}_a . \tag{4.33}$$

Remark 4.9: In the particular case where the frame is orthonormal (frame of Minkowski), the Covariant Derivative is identical to the classical derivative since the Christoffel coefficients are null (cf. Section 4.2.2.1.1).

Then in the general case, where v evolves in any direction, we must make the total derivative as with equation 4.21 and we will see an example with the parallel transport.

4.2.4.2.2 The Parallel Transport

A case is of special interest: when the variation of a vector $\mathbf{v} = v^a\, \mathbf{e_a}$ along a given curve is null, meaning that its Covariant Derivative is null, as it is for the proper velocity of an inertial object. This vector thus incurs a parallel transport along a curve. Any curve can be parameterized with one parameter, denoted by u, and we want to express $\frac{d\mathbf{v}}{du} = 0$. We then have:

$$\frac{d\mathbf{v}}{du} = \frac{dv^a}{du}\mathbf{e_a} + v^a\frac{d\mathbf{e_a}}{du} = \frac{dv^a}{du}\mathbf{e_a} + v^a\left(\frac{\partial \mathbf{e_a}}{\partial x^c}\frac{dx^c}{du}\right) = \frac{dv^a}{du}\mathbf{e_a} + v^a\partial_c\mathbf{e_a}\frac{dx^c}{du}$$

$$= \frac{dv^a}{du}\mathbf{e_a} + v^a\frac{dx^c}{du}\Gamma^i_{ac}\mathbf{e_i}. \tag{4.34}$$

We can factorize with $\mathbf{e_i}$ after replacing the index a with i in the first term since a is mute, and we obtain:

$$\frac{d\mathbf{v}}{du} = \mathbf{e_i}\left(\frac{dv^i}{du} + v^a\frac{dx^c}{du}\Gamma^i_{ac}\right). \tag{4.35}$$

which is also denoted by: $\frac{d\mathbf{v}}{du} = \frac{Dv^i}{Du}\mathbf{e_i}$.

The term $\frac{Dv^i}{Du} = \frac{dv^i}{du} + v^a\frac{dx^c}{du}\Gamma^i_{ac}$ is called the absolute derivative (intrinsic derivative).

In the case of an inertial object, its proper velocity has all its absolute derivatives null.

4.2.5 The Proper Time Is the Privileged Parameter for Indexing an Object Trajectory

We mentioned that any curve can be parameterized with one parameter, denoted by u. There can be various parameters, but we will show that there is a privileged one, the proper time:

We will assume that this parameter u is differentiable and monotone, which means the following: let's have an event M indexed by the value u_M of the parameter u, and an event N indexed by u_N, with N being in the vicinity of M. The limit of $(u_N - u_M)$ tends to zero when the Lorentzian distance MN tends to zero. Likewise, the greater the Lorentzian distance MN, the greater $u_N - u_M$.

The vector **MN** can be expressed in the tangent frame in M by: $\mathbf{MN} = dx^a\mathbf{e_a}$ (using Einstein's compact notation). The norm^2 of **MN** is $ds^2(\mathbf{MN}) = g_{\mu\nu}dx^\mu dx^\nu$ with $g_{\mu\nu}$ being the coefficients of the metric in M. Similarly, we recall that the distance $ds^2(\mathbf{MN})$ is the proper time^2 between M and N, provided that $ds^2(\mathbf{MN})$ is positive (and that natural units are used; otherwise, we should divide by c^2).

Let's then consider the vector **T**, which is tangent to the curve at M: $\mathbf{T} = \lim_{du \to 0} \frac{\mathbf{MN}}{du} = \lim_{du \to 0} \frac{dx^a}{du} \mathbf{e_a}$.

The norm of the vector **T** is: $ds^2(\mathbf{T}) = \lim_{du \to 0} g_{\mu\nu} \frac{dx^\mu}{du} \frac{dx^\nu}{du}$. We then conclude:

$$ds^2(\mathbf{T}) = \lim \frac{ds^2(\mathbf{MN})}{du^2}.$$

Consequently, if the parameter u is equal to the proper time, then the norm of the vector **T** is equal to 1 all along the curve. More generally, if the parameter u is an affine relation of the proper time, meaning $u = a\tau + b$, then the norm of **T** is constant all along the curve. Such parameter is called an **affine parameter**, and this constancy property of **T** is an additional reason to privilege affine parameters among all possible parameters.

This conclusion has an important consequence regarding the axioms relating to the relativistic Momentum: the homogeneity postulate indeed requires that the rate of variation of the events characterizing an inertial object must be constant, since the universe properties are the same everywhere and at all times, meaning that $\frac{\mathbf{dM}}{du} = \text{cst}$.

Then in order to satisfy this condition, the above demonstration shows that the parameter u, which is used to characterize the rate of variation of the events, must necessarily be the object proper time (or an affine parameter). This provides an additional demonstration why the relativistic Momentum must use the proper time (cf. more in Section 7.2).

4.2.6 The Geodesic Equation

The definition of the geodesic is an extension of the definition of the straight line in a Euclidean space, which was based on the two following characteristic properties:

A. The tangent to any point of the straight line is parallel to the same direction.

B. The distance between any couple of points is minimum.

4.2.6.1 Expression of the Property (A) in a Riemannian Space

Consider a curve: any of its points (events) can be referred to by a parameter, and we just saw the existence of a privileged parameter, the proper time τ (or any other affine parameter). The coordinates of a point M(τ) are denoted by $x^i(\tau)$.

The tangent vector **T** to the point M has for components: $\mathbf{T}(\tau) = t^a(\tau)\, \mathbf{e_a}$, with $t^a(\tau) = \lim_{d\tau \to 0} \frac{dx^a}{d\tau}$. This limit will be denoted by the derivative: $\frac{dx^a}{d\tau}$.

We saw that using this affine parameter, **T** always has the same norm all along the curve, which is a condition expressed by the proper speed constancy of an inertial object. Then, in order to satisfy the condition (A), we want **T** to undergo a parallel transport along the curve, and this implies that the Covariant Derivative of **T** is always null. Let's then apply the Covariant Derivative of **T** along the curve:

Relation (4.35) yields: $\frac{d\mathbf{T}}{d\tau}=0=\frac{Dv^a}{Du}\mathbf{e_a}=\left(\frac{dt^a}{d\tau}+t^b\frac{dx^c}{d\tau}\Gamma^a_{bc}\right)\mathbf{e_a}$.

Having: $t^a(\tau)=\frac{dx^a}{d\tau}$, we finally obtain:

$$\frac{d^2x^a}{d\tau^2}+\frac{dx^b}{d\tau}\frac{dx^c}{d\tau}\Gamma^a_{bc}=0 \qquad (4.36)\ \blacksquare$$

These four relations constitute the equation of the geodesic, which is the trajectory of a free-falling particle. It is one of the most important relations of General Relativity.

4.2.6.2 Expression of the Property (B) in a Riemannian Space

The distance between any couple of points (events) A and B is the curvilinear integrand along the curve from A to B; we thus have: $L=\int_A^B\sqrt{ds^2(dM)}$ with $ds^2(\mathbf{MN})=g_{\mu\nu}dx^\mu dx^\nu$.

Let the curve be parameterized with the parameter u; we then have: $L=\int_A^B\sqrt{g_{\mu\nu}(u)\frac{dx^\mu}{du}\cdot\frac{dx^\nu}{du}}\cdot du$.

We want this distance to be extremized when there are small variations of the curve, meaning that dL=0 when there are small variations of x^ν and x^μ. This is a typical case of variational calculation, which can be solved with the equations of Euler-Lagrange (cf. Section 3.5.3), and leads to the same equations as equation 4.36. This confirms that properties (A) and (B) are equivalent.

4.3 Supplements

4.3.1 Case Study of the 2D Sphere: Christoffel Coefficients and the Geodesic

We will consider the 2D surface of a sphere of radius R. A regular network of n meridians and n latitudes is drawn on the sphere, defining a curvilinear system of coordinates. We will first determine the Christoffel Coefficients, and then the trajectory of an inertial object of this sphere, i.e. the Geodesic.

We will first determine the Christoffel Coefficients at a point M on the equator, and then at any location on the sphere. In a second part, these results will be used to determine the geodesic.

We will denote by P_M^t the natural coordinate tangent plane at a point M on the sphere, with its two basis vectors $\vec{i}$ and $\vec{j}$ being, respectively, tangent to the latitude and to the meridian at the point M.

The network of meridians and latitudes is regular, meaning that the angle from the center of the sphere and between two consecutive meridians is always the same, and it is also the same as the angle between two consecutive latitudes. This angle is $2\pi/n$ since there are n meridians and n latitudes.

For example, if M is at the intersection of the meridian number X_M and the latitude number Y_M, then the coordinates of M are (X_M, Y_M).

4.3.1.1 Determination of the Christoffel Coefficients on the Equator

Figure 4.2 represents the equator and the evolution of the natural basis vector $\vec{i}$ from a point M on the equator to a point N in its vicinity and also on the equator. The coordinate basis vector in N, denoted by $\vec{i}'$, is on the tangent frame in N, and it has a projection on P_M^t, which is denoted by $\overrightarrow{i'_m}$.

The small angle $\widehat{MON}$ is denoted by $d\theta$, with O being the center of the sphere (the above picture exaggerates this angle for the sake of clarity). We can see that the magnitude of $\overrightarrow{i'_m}$ is: $i'_m = i\cos(d\theta)$.

The corresponding Affine Connection coefficients are given by: $\frac{\partial\vec{i}}{\partial x} = \lim_{dx\to 0}\frac{\overrightarrow{i'_m} - \vec{i}}{dx} = \Gamma^i_{ix}\,\vec{i} + \Gamma^j_{ix}\,\vec{j}$.

For symmetry reason on the equator, the component along $\vec{j}$ is null; hence: $\Gamma^j_{ix} = 0$.

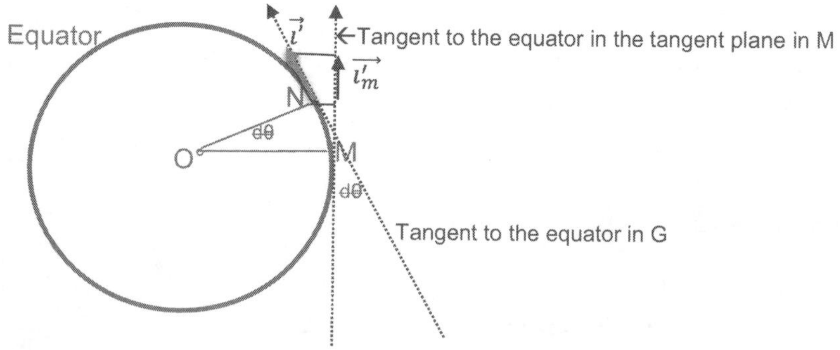

FIGURE 4.2
Evolution of the vector $\vec{i}$ along the equator.

The component along $\vec{i}$ is: $\Gamma^{i}_{ix} = \lim_{dx \to 0} \frac{[\cos(d\theta) - 1]}{dx}$.

Let's see the relation between dx and dθ: when dx is small, dx tends to be proportional to tg(dθ), and the latter tends to be equal to dθ. The proportionality coefficient is such that when $\theta = 2\pi/n$, then $x = 1$. Hence, $dx = n\, d\theta/2\pi$. We then have:

$$\Gamma^{i}_{ix} = \lim_{dx \to 0} \frac{2\pi[\cos(d\theta) - 1]}{n\, d\theta} = \lim_{dx \to 0} \frac{2\pi[\cos(0 + d\theta) - \cos(0)]}{n\, d\theta} = \frac{2\pi[-\sin(0)]}{n\, d\theta} = 0.$$

We indeed noticed that this limit matches with the definition of the derivate of the cos(x) function at $x = 0$.

This result is consistent with the fact that on the equator, the system of coordinates $(\vec{i}, \vec{j})$ is Cartesian (orthonormal), implying that the Christoffel coefficients are null.

4.3.1.2 Determination of the Christoffel Coefficients at any Latitude

Consider a point M on the North latitude corresponding to the latitude angle α. The meridian of M is again denoted by X_M and its latitude Y_M. The tangent plane in M is also again denoted by P^t_M.

We denote by E, G and F the points on the sphere having, respectively, coordinates: $E(X_M, Y_{M+1})$, $G(X_{M+1}, Y_M)$ and $F(X_{M+1}, Y_{M+1})$. We denote by E′, G′, F′ the projections of E, G, F on P^t_M, as shown in Figure 4.3.

4.3.1.2.1 *The Variation of $\vec{i}$ along the y-Axis (Toward the North)*

When $\vec{i}$ moves to the North, its length shrinks as a consequence of the fact that two consecutive meridians are not parallel since they coincide in the North Pole (the length of $\vec{i}$ even becomes null at the Pole). As a result, the latitude circle in E is shorter than the one in M; hence, when moving to the North, the length of $\vec{i}$ shrinks.

Determination of the Shrinking of $\vec{i}$ when M moves to the North

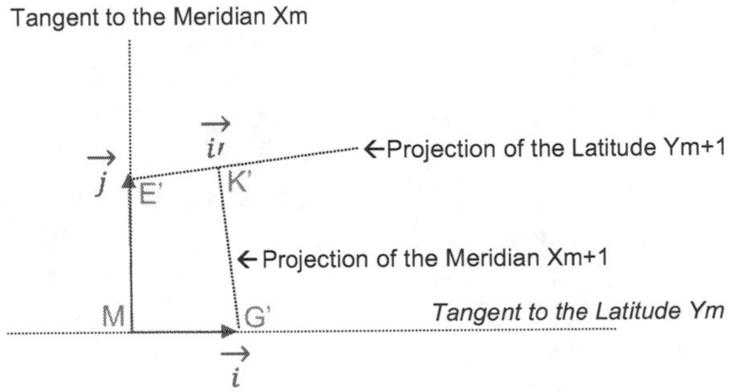

FIGURE 4.3
Projection of the meridians and the latitudes on the tangent plane at M.

The length of $\vec{i}$, denoted by i, is by definition equal to lim $\frac{MG'}{dx}$. We have: MG=R dθ cos (α), and dx =dθ n/2π (as previously). Thus: $i = 2\pi R \cos(\alpha/n)$. Then, when $\vec{i}$ moves to the North and becomes $\vec{i}'$, α increases by dα. We thus have: $di = i' - i = 2\ \pi R\left[\cos(\alpha + \ d\alpha) - \cos(\alpha)\right]/n$.

Our target actually is to calculate: $\lim \frac{di}{dy}$, where dy is the change in the ordinate of M at the latitude (α+dα). The relation between dy and dα is linear: dy=n dα/2π (we have dy=1 when $da = 2\pi/n(\alpha + d\alpha)$.

We then have: $\lim \frac{di}{dy} = \frac{R4\pi^2\left[\cos(\alpha + d\alpha) - \cos(\alpha)\right]}{n^2\, d\alpha} = \frac{-R4\pi^2 \sin(\alpha)}{n^2}$.

We can see that when M is on the equator, α=0, meaning that there is no shrinking of $\vec{i}$, which confirms our previous result. However, when α is not null, there is a shrinking effect that is all the more important as we are close to the Pole.

Regarding the Christoffel coefficients:

$$\Gamma^i_{iy} = \frac{-R4\pi^2 \sin\alpha}{n^2} / 2\pi R\cos\alpha/n = \frac{-\ 2\pi\ tg\alpha}{n}.$$

Regarding the evolution of the ordinate of the vector $\vec{i}$ when M moves to the North: When moving to the North, the vector $\vec{i}$ remains parallel to itself and to the plane P^t_M. Consequently, there is no variation of the ordinate of its projection on P^t_M. Thus: $\Gamma^j_{iy} = 0$. So finally: $\frac{\partial \vec{i}}{\partial y} = \frac{-\ 2\pi\ tg\alpha}{n}\vec{i}$ ■

4.3.1.2.2 Evolution of $\vec{i}$ When M Moves to the East

When M moves to the East, we face the problem that the projection on P^t_M of the latitude circle of M is a curve, which is oriented to the North (when M is in the North hemisphere). Let's first calculate this effect.

4.3.1.2.2.1 Calculation of the Projection on P^t_M of the Latitude Circle of M Consider a point N on the same latitude as M, and the meridian of N differs from the one of M by the angle θ. The X-axis of P^t_M is denoted by Mx: it is tangent to the latitude of M at the point M. The projection of N on P^t_M is denoted by Z. The projection of N on Mx is denoted by H. We want to calculate HZ (Figure 4.4).

To facilitate the calculation, we consider the point N′, which is the projection of N on the line QM. NN′ and MH are parallel and equal; hence, the projection of N′ on P^t_M, denoted by Z′, is such that Z′M=ZH. We then need to calculate Z′M, which indeed is simpler than ZH directly:

We have: Z′M=N′M sinα. Besides, N′M=QM − N′Q=QM − QN cosθ. We then apply: QM=QN=Rcosα, and finally obtain:

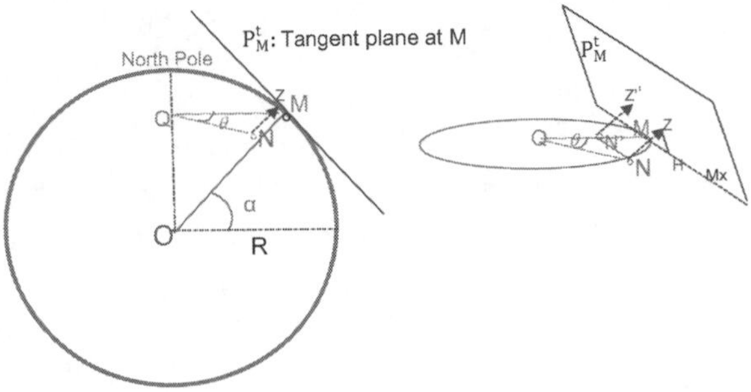

FIGURE 4.4
Projection of the latitude of M on the tangent plane at M.

$$ZH = Z'M = N'M \sin\alpha = (QM - QN\cos\theta)\sin\alpha = R\sin\alpha\cos\alpha\,(1 - \cos\theta).$$

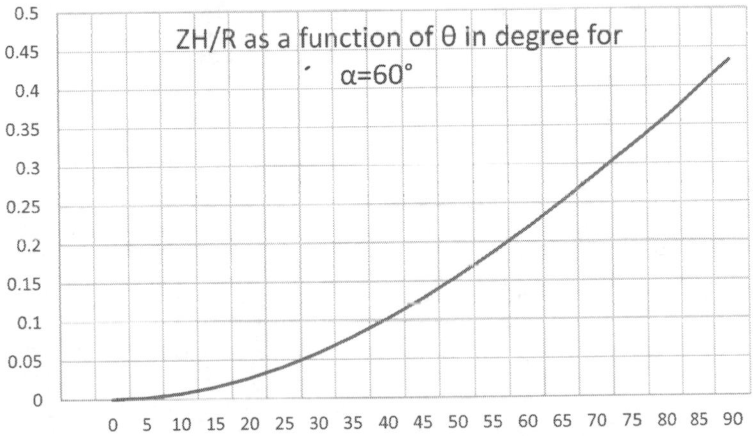

FIGURE 4.5
Length ZH for $\alpha = 60°$.

The graph in Figure 4.5 gives the projection of the latitude of M on the Tangent Plane at M. The latitude chosen is 60°, and we can see for instance that when $\theta = 45°$, we have: $ZH \approx 0{,}13\,R$.

When θ tends to zero, we have:

$\lim \dfrac{ZH}{dx} = \dfrac{R\sin\alpha\cos\alpha\left(1-\cos\theta\right)}{d\theta\, n/2\pi} = \dfrac{2\pi R\sin\alpha\cos\alpha\sin 0}{n} = 0$, which is consistent with the fact that the tangent to the above curve is flat at $\theta = 0$.

**

We are now prepared for the calculation of the variation of $\vec{i}$ when M moves to the East.

4.3.1.2.2.2 Variation of $\vec{i}$ When M Moves to the East When M moves to the East and becomes N, we denote by $\vec{i'}$ the unit frame vector along the X direction in the tangent plane in N, and we want to calculate the projection of $\vec{i'}$ on P^t_M (the tangent plane in M).

Both extremities of $\vec{i'}$ are in the tangent plane in N, meaning that $\vec{i'}$ is tangent to the latitude circle in N. The vector $\vec{i'}$ is represented by the vector $\overrightarrow{NA}$ in Figure 4.6. Its length is equal to: $i' = NA = R \cos\alpha\, 2\pi/n$. The projection of A on P^t_M is denoted by A′, and the projection of A on (Mx) is denoted by B. We want to calculate A′B, which is the ordinate of the projection of $\vec{i'}$ on P^t_M.

We have: $A'B = AB \sin\alpha$. Besides, we have: $AB \approx NA \sin\theta$ (this is a first-order approximation that is legitimate because MN is tangent to Mx). We then have: $A'B = NA\sin\theta = \frac{2\pi}{n} R \cdot \sin\alpha\cos\alpha \cdot \sin\theta$.

We actually want to calculate: $\lim \frac{A'B}{dx}$ with $dx = d\theta \frac{n}{2\pi}$.

We thus have:

$$\lim \frac{A'B}{dx} = \frac{\frac{2\pi}{n} R \sin\alpha\cos\alpha\sin(d\theta)}{d\theta\, n/2\pi} = \frac{4\pi^2}{n^2} R \sin\alpha\cos 0 = \frac{4\pi^2}{n^2} R \sin\alpha\cos\alpha.$$

Regarding the Christoffel coefficients, we have: $\frac{\partial \vec{i}}{\partial x} = \Gamma^i_{ix}\vec{i} + \Gamma^j_{ix}\vec{j}$; hence, $\Gamma^j_{ix} = \frac{4\pi^2}{n^2} R \sin\alpha\cos\alpha / j$, and as $j = R\frac{2\pi}{n}$, we finally obtain:

$$\Gamma^j_{ix} = \frac{2\pi}{n}\sin\alpha\cos\alpha \;\blacksquare$$

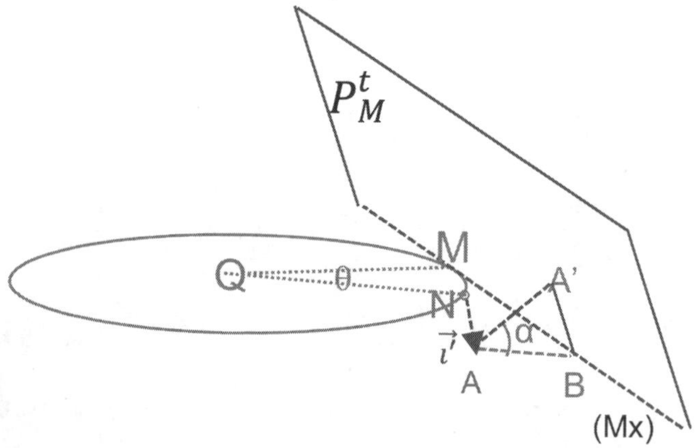

FIGURE 4.6
Projection of the vector $\vec{i'}$ on the tangent plane at M.

**

Regarding Γ^{i}_{ix}, the abscissa of $\vec{i'_m}$ is equal to HB. When θ tends to zero, we are in a similar situation as on the equator except that the length of R becomes R cos α, and we saw in Section 4.3.1.1 that the limit $\frac{HB - i}{dx} = 0$, meaning that: $\Gamma^{i}_{ix} = 0$. Hence, we finally have: $\frac{\partial \vec{i}}{\partial x} = \frac{2\pi}{n} \sin\alpha \cos\alpha \, \vec{j}$ ■

4.3.1.2.3 The Variations of the Vector $\vec{j}$

4.3.1.2.3.1 The Variation of the Vector $\vec{j}$ When M Goes to the North In the first order, there is no variation of the vector $\vec{j}$ when M goes to the North for the same reason as there is no variation on the equator. Indeed, the meridian of M is a great circle, and the distance ME is the same whatever the latitude of M. We thus have: $\Gamma^{j}_{jy} = 0$.

4.3.1.2.3.2 The Variations of the Vector $\vec{j}$ When M Goes to the East When M goes to the East, we must take into account the effects resulting from the fact that two consecutive meridians meet in the North Pole: the upper extremity of $\vec{j}$ moves by a smaller distance than its bottom extremity. This effect was seen in Section 4.3.1.2.1 and resulted in: $\Gamma^{i}_{iy} = \frac{-2\pi \, tg\alpha}{n}$. We deduce that the abscissa of $\frac{\partial \vec{j}}{\partial x}$ is also equal to this value; hence, we have: $\Gamma^{i}_{jx} = \frac{-2\pi \, tg\alpha}{n}$ ■

Remark 4.10: We mentioned that if the surface is without torsion, then we have: $\Gamma^{a}_{bc} = \Gamma^{a}_{cb}$. We can see that this is confirmed in the Sphere case.

Regarding the variation of the ordinate of $\vec{j}$ when M goes to the East, it is null because $\vec{j}$ remains parallel to itself and to P^{t}_{M}. We thus have: $\Gamma^{j}_{jx} = 0$.

**

4.3.1.3 Application: The Geodesic on a 2D Sphere

Consider a sphere like the Earth, with a ball at the point M at a random North latitude (the reasoning being the same in the South). We give the ball an impulse in the direction of the latitude at the point M, so that it starts moving with its speed $\vec{v}$ along this latitude.

Being submitted to no force, the ball pursues its trajectory on the sphere. We will show that its trajectory is not the latitude at the point M, but the great circle passing by M, which is then the geodesic. We will reason with the absurd: let's assume the ball moves among the latitude at the point M, and we will show that it is impossible.

If at the point M the ball moves along the latitude of M, meaning along $\vec{i}$, then the ball being inertial, the Covariant Derivative of the ball's proper velocity at M along $\vec{i}$ must be null. We thus have: $\nabla_i \vec{S} = 0$.

Also, $\vec{S} = a\,\vec{i}$; hence: $\nabla_i\vec{S} = 0 = \frac{\partial a}{\partial x}\vec{i} + a\,\frac{\partial \vec{i}}{\partial x}$. We saw that: $\frac{\partial \vec{i}}{\partial x} = \frac{2\pi}{n}(\sin\alpha\cos\alpha)\vec{j}$ (cf. Section 4.3.1.2.2.2); hence: $\nabla_i\vec{S} = \frac{\partial a}{\partial x}\vec{i} + \frac{2\pi}{n}(\sin\alpha\cos\alpha)\vec{j}$. This shows that $\nabla_i\vec{S}$ cannot be null since neither $\frac{\partial a}{\partial x}$ nor $\sin\alpha\cos\alpha$ are null, with the only exception when $\alpha=0$; and this is the case when the latitude is the equator, which is the great circle, and when the speed is constant.■

Remark 4.11: If we thought the object trajectory was the latitude Ym, it was because we were misled by the particular choice of coordinates. There are indeed many circles passing by M and tangent to the same latitude, among which the great circle is the solution also for symmetry reason: there is no reason that the ball moves to the left rather than to the right of the line of symmetry, which is the great circle.

4.3.2 Demonstration of the Relationship between the Connection Coefficients and the Metric

We saw that $g_{ab} = e_a \bullet e_b$. Let's make the partial derivative along the coordinate c, and we obtain:

$$\partial_c g_{ab} = (\partial_c \mathbf{e_a})\bullet \mathbf{e_b} + \mathbf{e_a}\bullet(\partial_c \mathbf{e_b}) = \Gamma^{\mu}_{ac}\mathbf{e_\mu}\bullet\mathbf{e_b} + \mathbf{e_a}\bullet\Gamma^{\nu}_{bc}\mathbf{e_\nu} = \Gamma^{\mu}_{ac}\,g_{\mu b} + \Gamma^{\nu}_{bc}\,g_{a\nu}\,.(4.37)$$

These are n^3 relations between the Christoffel symbols and the $g_{\mu\nu}$ coefficients and their derivatives. Let's express the relations obtained by permuting the indices a, b, and c in the relation (4.37), and we thus have:

$$\partial_b g_{ca} = \Gamma^{\mu}_{cb}\,g_{\mu a} + \Gamma^{\nu}_{ab}\,g_{c\nu} \tag{4.38}$$

and

$$\partial_a g_{bc} = \Gamma^{\mu}_{ba}\,g_{\mu c} + \Gamma^{\nu}_{ca}\,g_{b\nu}. \tag{4.39}$$

Next, make the following linear combination: 4.37+4.38−4.39, and we obtain:

$$\partial_c g_{ab} + \partial_b g_{ca} - \partial_a g_{bc} = \Gamma^{\mu}_{ac}\,g_{\mu b} + \Gamma^{\nu}_{bc}\,g_{a\nu} + \Gamma^{\mu}_{cb}\,g_{\mu a} + \Gamma^{\nu}_{ab}\,g_{c\nu} - \Gamma^{\mu}_{ba}\,g_{\mu c} - \Gamma^{\nu}_{ca}\,g_{b\nu}.$$

Let's consider: $\Gamma^{\mu}_{ac}\,g_{\mu b} - \Gamma^{\nu}_{ca}\,g_{b\nu}$. We saw that $\Gamma^{\mu}_{ac} = \Gamma^{\mu}_{ca}$, and that $g_{\mu b} = g_{b\mu}$. Also, the indices μ and ν are mute; hence, $\Gamma^{\mu}_{ac}\,g_{\mu b} = \Gamma^{\nu}_{ca}\,g_{b\nu}$; the same applies for: $\Gamma^{\nu}_{bc}\,g_{a\nu} = \Gamma^{\mu}_{cb}\,g_{\mu a}$.

Consequently, we have:

$$\left(\partial_c g_{ab} + \partial_b g_{ca} - \partial_a g_{bc}\right) = 2\Gamma^{\mu}_{cb}\,g_{\mu a}\,. \tag{4.40}$$

To bring clarity and simplicity, let's write this expression with the matrix format, and the left part will be denoted by: $2k^a_{cb}$, meaning that:

$$k^a_{cb} = 1/2\left(\partial_c g_{ab} + \partial_b g_{ca} - \partial_a g_{bc}\right)$$

Equation 4.40 thus becomes:

$$k^a_{cb} = \left(\Gamma^0_{cb}, \Gamma^1_{cb}, \Gamma^2_{cb}, \Gamma^3_{cb}\right) \cdot \begin{pmatrix} g_{0a} \\ g_{1a} \\ g_{2a} \\ g_{3a} \end{pmatrix}. \tag{4.41}$$

We will use the inverse matrix G^{-1}, which is the metric of the Dual space, as seen in Section 4.1.3. We saw that the generic term of G^{-1} is $g^{\mu\nu}$, with $g^{\mu\nu} = e^\mu \bullet e^\nu$, i.e., the scalar product of the unit basis vectors $\mathbf{e}^\mu$ and $\mathbf{e}^\nu$ of the Dual basis. Let's then multiply both terms of this relation (4.41) by $\left(g^{a0}, g^{a1}, g^{a2}, g^{a3}\right)$, and we obtain:

$$k^a_{cb}\left(g^{a0}, g^{a1}, g^{a2}, g^{a3}\right) = \left(\Gamma^0_{cb}, \Gamma^1_{cb}, \Gamma^2_{cb}, \Gamma^3_{cb}\right) \cdot \begin{pmatrix} g_{0a} \\ g_{1a} \\ g_{2a} \\ g_{3a} \end{pmatrix} \cdot \left(g^{a0}, g^{a1}, g^{a2}, g^{a3}\right),$$

$$= \left(\Gamma^0_{cb}, \Gamma^1_{cb}, \Gamma^2_{cb}, \Gamma^3_{cb}\right)$$

which means: $\Gamma^\mu_{cb} = k^a_{cb} g^{a\mu} = \dfrac{1}{2}\left(\partial_c g_{ab} + \partial_b g_{ca} - \partial_a g_{bc}\right) g^{a\mu}$ ■

4.4 Questions and Problems

Question 1) The Proper Speed and Proper Acceleration

Is the proper speed, denoted by **S**, a tensor? Of what type? Same question for the proper acceleration, denoted by $\mathbf{A} = \dfrac{d\mathbf{S}}{d\tau}$.

Question 2) Show That **S** is Perpendicular to **A**, Meaning That $\mathbf{S} \bullet \mathbf{A} = 0$.

Question 3) Is the Force of Minkowski a tensor, and why? Is it contravariant or covariant?

Question 4) Same question for the vector $\mathbf{V} = \dfrac{d\mathbf{M}}{dt}$.

Question 5) The same question for the Newtonian-like Force.

Question 6) Express the linear form associated with the proper velocity **S** of an object in the context of the absence of chrono-geometrical deformations.

Question 7) In the context of chrono-geometrical deformations, we perform the parallel transport of a 4-Vector **V** located in the tangent space at a point A to the tangent space at a point B. Can we determine the resulting vector? Can we determine its norm?

Question 8) Is the scalar product between two vectors conserved after a parallel transport?

**

Problem 4.1) Tensors of rank 1: Difference between $\mathbf{e}^1$ and $\mathbf{e}_1^*$.

The space K has chrono-geometrical deformations. We consider the two following vectors:

- $\mathbf{e}^1$ which is one vector of the Dual basis of K* that corresponds to the basis of K
- $\mathbf{e}_1^*$ which is the linear form associated with $\mathbf{e}_1$

Q1) Are they tensors? If yes, covariant or contravariant. Are they identical? If there is not chrono-geometrical deformations, are they identical?

Q2) Express $\mathbf{e}_i^*$ in the basis formed by the $\mathbf{e}^j$.

Problem 4.2) Tensors of rank 1: Difference between the scalar products $\mathbf{e}^i \bullet \mathbf{e}^j$ and $\mathbf{e}_i \bullet \mathbf{e}_j$.

The space K has chrono-geometrical deformations. Is $\mathbf{e}^i \bullet \mathbf{e}^j$ different from $\mathbf{e}_i \bullet \mathbf{e}_j$? Why?

Problem 4.3) Show that all metrics in the space–time satisfy the relation: $g^{ij}g_{ji} = 4$.

Problem 4.4) Show that the Trace operator is commutative.

The trace is the sum of the elements which are in the diagonal of a square matrix. Consider two square matrices A and B. Show that Trace[(A)(B)] = Trace[(B)(A)]

Problem 4.5) How does a mixed covariant–contravariant rank 2 tensor C_j^i change when changing the frame?

Problem 4.6) The Covariant rank 2 tensor associated to a Contravariant rank 2 tensor.

Consider a twice contravariant rank 2 tensor $F^{\alpha\beta}$. Show that we can associate to this tensor a twice covariant rank 2 tensor, $F_{\mu\nu}$, such that $F_{\mu\nu} = g_{\mu\alpha}g_{\nu\beta}F^{\alpha\beta}$.

Problem 4.7) The Christoffel coefficients of the Surface of a Sphere.

We want to determine the Christoffel coefficients of the surface of a sphere by applying the relation (4.31). The point M is on a sphere of radius R at the latitude α and the longitude φ. The point N in the vicinity of M on the sphere, has its coordinates that differ from M by $d\alpha$ and $d\varphi$.

Q1) Express the Euclidean distance^2: MN^2

Q2) Deduce the g^{ij} coefficient of the metric associated to the coordinate tangent frame at M, the two coordinates being α and φ.

Q3) We will use equation 4.31 to determine the Christoffel coefficients: $\Gamma^{\mu}_{cb} = 2(\partial_c g_{ab} + \partial_b g_{bc} - \partial_a g_{bc})g^{a\mu}$. But first, calculate the different g^{ij}. Hint: write the g_{ji} with the matrix form.

Q4A) Determine $\Gamma^{\alpha}_{\varphi\varphi}$. **Q4B)** Determine $\Gamma^{\varphi}_{\alpha\varphi}$.

Notes

1 Cf. definition and considerations on "intrinsic" concepts in Section 4.1.3.2.
2 The names G and g_{ij} for the metric tensor refer to "geometry", and are due to the Italian mathematicians Gregorio Ricci and Tullio Levi-Civita.
3 The ∇ symbol is called "nabla." An index can be added at the bottom to indicate the direction along which the derivative is done.

5

Important Consequences: Relativistic Maxwell's Laws, Schwarzschild's Solution

Introduction

This chapter presents two great victories of Relativity:

- The solution to the problems raised by Maxwell's laws in classical physics
- Schwarzschild's solution to Einstein's equation.

The first part presents how Relativity considerably simplified the axiomatic framework on which electromagnetism is grounded. The source of electromagnetism will be characterized according to the principles of Relativity, leading to the Current Density Tensor. Then, the relation between this tensor and the electromagnetic tensor will be shown by combining two approaches: one based on an invariant of the electromagnetic tensor, the other by adapting to Relativity the classical electrical and magnetic potentials. The mathematical tools used in Maxwell's equations will be carefully presented.

The second part begins with a reminder of polar coordinates and their adaptation to 4D Minkowski frames. Then, Schwarzschild's coordinates will be precisely explained and two important applications will be derived: the possible existence of black holes, and the computation of the bending of light in the neighborhood of a great mass, which was a brilliant confirmation of General Relativity.

5.1 Relativistic Maxwell's Laws

This chapter requires specific mathematical knowledge of tensors and vector algebra. Tensors were seen in Chapter 4, and a comprehensive mathematical support on vector operators is provided in Section 5.1.6.

DOI: 10.1201/9781003201359-5

5.1.1 Introduction

We saw in Section 3.4 that the relativistic electromagnetic force experienced by a particle with a charge q evolving at the classical 3D speed $\vec{v}$ in an inertial Frame K follows the relation:

$$\mathbf{F} = q\,(EM)\,\mathbf{V}^{*}. \tag{5.1}$$

where F is the 4D Newtonian-like Force, (EM) is the electromagnetic matrix and V* the 4D vector:

$$\mathbf{V}^{*} = (c\,,\ -dx/dt,\ -dy/dt, -dz/dt) = (\,c, -\vec{v}\,).$$

We saw that if the object mass remains constant, the Relativistic Force is always perpendicular to the object trajectory (meaning the Lorentzian scalar product $\mathbf{F} \bullet \mathbf{V} = 0$). This implies that the (EM) matrix must be antisymmetric, as shown in Section 5.1.6.6. An antisymmetric matrix only has 6 degrees of freedom, and we saw that three corresponds to the magnetic field B and three to the electrical one $\vec{E}$. We indeed saw in Section 3.4 that when these components are placed in the (EM) matrix as shown below, the spatial part of the relation (5.1) exactly matches with the classical electromagnetic force, named the Lorentz force, which is:

$$\vec{f} = q\,\vec{E} + \vec{B} \times v \tag{5.2}$$

Equation 5.1 indeed reads in matrix format:

$$\begin{pmatrix} F_t \\ F_x \\ F_y \\ F_z \end{pmatrix} = q \cdot \begin{pmatrix} 0 & -E_x/c & -E_y/c & -E_z/c \\ E_x/c & 0 & -B_z & B_y \\ E_y/c & B_z & 0 & -B_x \\ E_z/c & -B_y & B_x & 0 \end{pmatrix} \begin{pmatrix} c \\ -v_x \\ -v_y \\ -v_z \end{pmatrix} \tag{5.3}$$

We also saw that the time part of this relation yields the power of Relativistic Force, which matches the classical power of the Lorentz force. However, no further physical explanation was given as to the origin of this (EM) matrix, and this will be addressed now. We will also see how this matrix changes when changing frames, which was the initial problem that led to elaborating Relativity.

A general principle is to find the intrinsic elements from which are derived those which are measured in a given frame. Hence, we will follow this principle.

5.1.2 The Electromagnetic Force of Minkowski and Faraday's Tensor

The above electromagnetic force is a Newtonian-like Force (F=dP/dt) considered in an inertial frame K with its time t; hence, this Force is not intrinsic, as we previously saw. However, we also saw the relation between the intrinsic Force of Minkowski, Φ, and the Newtonian-like one: $\mathbf{\Phi} = \mathbf{F}\,\gamma$. Hence, from equation 5.1, we have: $\mathbf{\Phi} = \gamma\, q\,(EM)\,\mathbf{V}^*$.

Consider the term $\gamma\,\mathbf{V}^*$:

$$\gamma\,\mathbf{V}^* = (\gamma\, c\, dt/dt,\ -\gamma\, dx/dt,\ -\gamma\, dy/dt, -\gamma\, dz/dt)$$
$$= (\gamma\, c,\ -dx/d\tau,\ -dy/d\tau, -dz/d\tau).$$

This expression matches with the covector, denoted by **S***, associated with the object's proper velocity **S** in the absence of space-time deformations. We indeed have:

$$\mathbf{S}^* = (\gamma\, c, -\gamma\, \vec{v}) = \left(\gamma\, c, -\gamma\frac{dx}{dt}, -\gamma\frac{dy}{dt}, -\gamma\frac{dz}{dt}\right). \tag{5.4}$$

In the rest of this document, the electromagnetic tensor will be denoted by (F) in memory of M. Faraday instead of (EM). However, beware of possible confusion with the Newtonian-like Force. We thus have:

$$\Phi = q(F)\mathbf{S}^*. \tag{5.5}$$

It is then logical to assume that equation 5.5 is the root from which the classical force of Lorentz (equation 5.2) was derived as an approximation for usual speeds.

Equation 5.5 shows that the Faraday matrix is a tensor of rank 2, from the very tensor definition: it is a linear expression giving as result a rank 1 tensor, Φ, from another rank 1 tensor, **S***. Moreover, this reasoning shows that (F) is a twice contravariant since **S*** is covariant and Φ contravariant.

We will now see the frame-changing rules for this Faraday's tensor, which will indicate the way the magnetic and electrical fields change when changing inertial frames, and we will compare them with the classical rules.

5.1.2.1 The Frame-Changing Rule for Faraday's Tensor

We just mentioned that Faraday's tensor is twice contravariant, and we saw in Section 4.1.2 the frame-changing rule for such a tensor; hence:

$$(F') = (\Lambda)(F)(\Lambda) \tag{5.6}$$

with (Λ) being the Lorentz matrix. Thus, in an inertial frame K′ moving at the speed v relative to K, the electromagnetic four-force is:

$$\Phi' = q\,(F')\,\mathbf{S}^{*\prime} = q\,(\Lambda)(F)(\Lambda)\,\mathbf{S}^{*\prime}.$$

Application: The Frame-Changing Rules of the Electrical and Magnetic Fields

We set the axis of K and K′ so that the speed v of K′ relative to K is along their Z′ and Z axes; we then have:

$$\begin{bmatrix} 0 & -E'_x/c & -E'_y/c & -E'_z/c \\ E'_x/c & 0 & -B'_z & B'_y \\ E'_y/c & B'_z & 0 & -B'_x \\ E'_z/c & -B'_y & B'_x & 0 \end{bmatrix} =$$

$$\begin{bmatrix} \gamma & 0 & 0 & -\gamma_v/c \\ 0 & 1 & 0 & 0 \\ 0 & 0 & 1 & 0 \\ -\gamma_v/c & 0 & 0 & \gamma \end{bmatrix} \begin{bmatrix} 0 & -E_x/c & -E_y/c & -E_z/c \\ E_x/c & 0 & -B_z & B_y \\ E_y/c & B_z & 0 & -B_z \\ E_z/c & -B_y & B_x & 0 \end{bmatrix} \begin{bmatrix} \gamma & 0 & 0 & -\gamma_v/c \\ 0 & 1 & 0 & 0 \\ 0 & 0 & 1 & 0 \\ -\gamma_v/c & 0 & 0 & \gamma \end{bmatrix}$$

After expanding, we obtain: $E'_z = E_z$ and $B'_z = B_z$ which match with the classical physics results. We also have: $E'_x = \gamma(E_x - v\,B_y)$; $E'_y = \gamma(E_y - v\,B_x)$; $B'_x = \gamma\left(B_x + \frac{v}{c^2} E_y\right)$; $B'_y = \gamma\left(B_y + \frac{v}{c^2} E_x\right)$.

Note that the classical relations indeed were: $\vec{B'} = \vec{B}$, and: $\vec{E'} = \left(\vec{E} + \vec{v} \times \vec{B}\right)$

Besides, we recall that the Newtonian-like Force in K′ is : Φ'/γ. Hence, we can see that the components of $\vec{E}$ and $\vec{B}$ along the transverse directions (X or Y) are different from those of classical physics, but slightly only for low speeds. Thus, the classical 3D Maxwell's equations were also good approximations, but for low speeds only.

The next step to finding the relativistic Maxwell's laws is to characterize the source of the Electromagnetic Field.

5.1.3 The Source of the Electromagnetic Field: The Current Density Tensor

The elementary source of the electromagnetic field is a set of charged particles located in a small volume, thus corresponding to the notion of charge density, denoted by ρ. In the inertial frame K° where these charges are fixed, the charge density is denoted by ρ_0. In another inertial frame K, these charges are seen moving, and their motion impacts the electromagnetic field they generate. We will assume that all charges inside the small volume move at the same velocity. However, we saw that in Relativity the classical 3D velocity is not intrinsic, hence we are induced to consider the proper velocity **S** of these charges. Hence, the Current Density Tensor **J** can be expressed knowing its value in K°:

$$\mathbf{J} = \rho_0\,\mathbf{S}. \tag{5.7}$$

We can see that like the proper speed **S**, the current density vector **J** is a contravariant rank 1 tensor, its invariant Lorentzian norm being $ds^2(\mathbf{J}) = c^2 . \rho_0^{\,2}$.

We have assumed that the value of the electrical charge of a particle is intrinsic (identical in all frames). Hence when changing frame, the charge density is only affected by the difference of the size of the volume hosting these charges, due to the distance contraction is the direction of the motion direction of K relative to K°. Hence, the charge density is greater in K than in K°, and we thus have:

$$\rho = \rho_0 \, \gamma_v \tag{5-8}$$

The components of the current vector in K are obtained by applying the Lorentz transformation to its expression in K°: We thus have: $\mathbf{J}^K = \rho_0 \left(c\,\gamma_v,\ v\,\gamma_v \right)$. Then with equation 5.8:

$$\mathbf{J}^K = \left(\rho\, c,\ \rho\, v \right). \tag{5.9}$$

5.1.3.1 *The Divergence of the Current Vector J Is Null If the Charge Density Is Constant*

In a vectorial field, the divergence is a useful mathematical operator, which gives a number to a vector in an intrinsic way, meaning the same number whatever the frame where the vector is considered.[1] Hence, the divergence is important from a physical perspective.

By definition, the divergence of a vector is the sum of the diagonal elements of its Jacobian matrix (cf. more in Section 5.1.6.3). The Jacobian is the matrix that contains all the partial derivatives of the components of the vector relative to the four dimensions. Let's then apply the divergence operator to the four-vector **J**:

The Jacobian matrix of **J** is:

$$\mathrm{Jac}(\mathbf{J}) = \begin{pmatrix} \frac{\partial J^t}{c.\partial t} & \frac{\partial J^X}{c.\partial t} & \frac{\partial J^y}{c.\partial t} & \frac{\partial J^z}{c.\partial t} \\ \frac{\partial J^t}{\partial x} & \frac{\partial J^X}{\partial x} & \frac{\partial J^y}{\partial x} & \frac{\partial J^z}{\partial x} \\ \frac{\partial J^t}{\partial y} & \frac{\partial J^y}{\partial y} & \frac{\partial J^y}{\partial y} & \frac{\partial J^y}{\partial y} \\ \frac{\partial J^t}{\partial z} & \frac{\partial J^X}{\partial z} & \frac{\partial J^y}{\partial z} & \frac{\partial J^z}{\partial z} \end{pmatrix} \tag{5.10}$$

We saw that the classical derivatives are identical to the covariant derivatives when Cartesian coordinates are used, which is assumed to be the case here. Hence, these derivatives are intrinsic. Besides, we use homogeneous units as we did for the definition of **J**; this is why, the time in seconds had to be multiplied by c. We then have:

$$\mathrm{Div}(\mathbf{J})=\frac{\partial J^t}{c\partial t}+\frac{\partial J^X}{\partial x}+\frac{\partial J^y}{\partial y}+\frac{\partial J^z}{\partial z}=\frac{d(\gamma\rho_0)}{dt}+\frac{\partial J^X}{\partial x}+\frac{\partial J^y}{\partial y}+\frac{\partial J^z}{\partial z}. \tag{5.11}$$

Remark 5.1: The divergence of a four-vector is denoted with a capital D, otherwise in 3D with a small d.

Being intrinsic, Div(J) is the same in all inertial frames. Let's then calculate it in K°, where $J^x = J^y = J^z = 0$. Besides, we assume that the charge is constant; hence: Div(**J**)=0. Then, in any other inertial frame, we have:

$$\mathrm{div}(\vec{j})+\frac{d\rho}{dt}=0. \tag{5.11a}$$

This relation involves the 3D divergence of the classical 3D current $\vec{j}$. It is similar to the 3D relation of classical physics stating the conservation of the charge, except that now we must consider the relativistic definition of ρ, which is $\rho = \rho_0\gamma_v$.

Green–Ostrogradsky's theorem illustrates the meaning of the divergence: the total flow of current passing through a closed surface is equal to the integrand of the divergence of $\vec{J}$ inside the small volumes contained in this closed surface. Thus, this total flow of current is null if the quantity of charges inside the volume remains constant.

Having expressed with intrinsic elements both the electromagnetic field and its source, we are now prepared to address the main issue: how these are related.

5.1.4 The Relativistic Maxwell's Laws

We will follow successively two approaches which we will then combine, and this will lead us to the relativistic Maxwell's laws.

5.1.4.1 The First Approach

Mathematically, it is possible to derive an intrinsic four-vector from the Faraday tensor F; hence, it is logical to assume that this four-vector is related to the current tensor **J**, since the latter is the cause of F. This method consists of making the "Quadri-Divergence[2]" of the Faraday tensor, denoted by ∇F: it is a four-vector whose component ∇F^ν is the Divergence of the four-vector formed by the four components of the column ν of F. This gives, using Einstein's summation:

$$\nabla F^\nu=\frac{\partial F^{\mu\nu}}{\partial x^\mu},\text{ which is also denoted by }\partial_\mu F^{\mu\nu}. \tag{5.12}$$

This approach is fruitful: Reality (experiments and observations) shows that this four-vector ∇F is proportional to the source vector **J**. The proportionality coefficient is with our international system of units: μ_0 (i.e., the magnetic vacuum permeability coefficient; and we have: $\mu_0 \varepsilon_0 = 1/c^2$).

We thus have: $\nabla\mathbf{F} = \mu_0\,\mathbf{J}$, meaning:

$$\partial_\mu F^{\mu\nu} = \mu_0 J^\nu. \tag{5.13}$$

The index μ of ∂_μ is on the bottom, meaning that the covariant components of the gradient are used (cf. Section 5.1.6.3), which is necessary to obtain a contravariant tensor, **J**, from a double contravariant tensor, F.

Noteworthy, this fundamental relation (5.13) works both ways: **J** is the source of the electromagnetic field, but conversely a change in the electromagnetic field has an impact on **J**.

However, the relation (5.13) cannot provide all the components of the Faraday tensor from the knowledge of four components of **J**. We saw that the Faraday tensor, being antisymmetric, has 6 components, but this is still two more than **J**. Hence, additional considerations are required to characterize F, which will be fulfilled by the second approach.

5.1.4.2 The Second Approach

In classical physics, the electric and magnetic fields could be derived from the gradients of two potentials:

- The 3D magnetic field vector $\vec{B}$ could indeed be derived from a 3D potential vector $\vec{A}$, and we had the relation:

$$\vec{B} = \overrightarrow{\text{curl}}\ \vec{A}. \tag{5.14}$$

- The 3D electrical field vector E was derived from a scalar potential, denoted by Φ, and also by: $V = \dfrac{\Phi}{c}$. We indeed had the relation:

$$\vec{E} = -\overrightarrow{\text{grad}}\,(\Phi) - \frac{d\vec{A}}{dt}. \tag{5.15}$$

Physically, relation (5.14) corresponds to the non-existence of a magnetic pole (as opposed to an electrical one), which implies that $\text{div}(\vec{B})=0$. We then deduced that $\vec{B}$ was the $\overrightarrow{\text{curl}}$ of a 3D potential vector since the divergence of a $\overrightarrow{\text{curl}}$ is always null. (cf. more in Section 5.1.6.5).

Besides, relation (5.15) is derived from the Maxwell-Faraday relation: $\overrightarrow{\text{curl}}\ \vec{E} = -\dfrac{d\vec{B}}{dt}$, knowing that the curl of a gradient is always null (cf. Section 5.1.6.5).

With Relativity, the electric and magnetic fields belong to a single entity: the Faraday's tensor F. Hence, it is logical to assume the existence of a common

4D potential, which is a four-vector denoted by **A**: $\left(A^0, A^1, A^2, A^3\right)$, where A^0 is the analog of the classical scalar potential V of the electric field, and the three other A^i are the analog of the magnetic potential vector $\vec{A}$, meaning that:

$$\mathbf{A}:\left(A^0, A^1, A^2, A^3\right)=\left(V, A^x, A^y, A^z\right)=\left(\frac{\Phi}{c}, \vec{A}\right). \quad (5.16)$$

We will first see that there exists a relation between **A** and (F). Then, as (F) is related to **J**, we will see the relation between **A** and **J**.

5.1.4.2.1 The Relation between the Potential A and the Faraday's Tensor F

Having assumed that the four-vector **A** is the potential from which $\vec{E}$ and $\vec{B}$ are derived, and as the components of these 3D vectors are regrouped within the Faraday's tensor F, it is logical to assume that the components of F involve the derivatives of **A**, meaning the gradients of the four components of **A** and denoted by A^i. Besides, the Faraday's tensor F being twice contravariant, we must use the contravariant components of these gradients, which are:

$$\partial^\mu\left(A^i\right)=\left(\frac{\partial\left(A^i\right)}{cdt},-\frac{\partial\left(A^i\right)}{dx},-\frac{\partial\left(A^i\right)}{dy},-\frac{\partial\left(A^i\right)}{dz}\right)$$ cf. more in Section 5.1.6.2.

These four gradients constitute a 4×4 matrix (tensor), which is the Jacobian of **A**, denoted by Jac(**A**). We saw that with Cartesian coordinates, these derivatives are intrinsic; hence, the Jacobian of **A** is intrinsic.

Then, in order to satisfy the requirement that the Faraday's tensor F is antisymmetric, a relatively simple way is to perform the following operation: $F = Jac(\mathbf{A}) - Jac(\mathbf{A})^T$, meaning with a tensorial notation:

$$F^{\mu\nu} = \partial^\mu A^\nu - \partial^\nu A^\mu . \quad (5.17)$$

The matrices Jac (**A**) and $Jac(A)-Jac(A)^T$ are shown in Section 5.1.6.7.

This operator $Jac()-Jac()^T$ has no name; it is a 4D extension of the curl operator, but the curl is a 3D operator. However, as we will refer to it, we will denote it by "Jacobian's Dissymmetry," since its values reflect the extent to which the Jacobian is not symmetric.

We will now check that this relatively simple way fulfills the condition of giving the same Faraday's tensor as the one previously deduced from the natural extension to 4D of the Lorentz force (cf. relation 5.3). Let's then expand the relation (5.17) and use the relations (5.14) and (5.15), which we assume to be valid also in Relativity:

$$F^{10}=-\frac{\partial A^0}{\partial x}-\frac{\partial A^1}{c.\partial t}=-\frac{\partial \Phi}{c.\partial x}-\frac{\partial A^x}{c.\partial t},$$ which matches with E^x/c since (5.15) is:

$\vec{E} = -\overrightarrow{Grad}(\Phi) - \dfrac{d\vec{A}}{dt}$. Thus: $F^{10} = E^x/c$.

$F^{21} = -\dfrac{\partial A^1}{\partial y} - \left(-\dfrac{\partial A^2}{\partial x}\right) = \dfrac{\partial A^2}{\partial x} - \dfrac{\partial A^1}{\partial y} = \dfrac{\partial A^y}{\partial x} - \dfrac{\partial A^x}{\partial y}$, which matches with B^z

since equation 5.14 is: $\vec{B} = \overrightarrow{Curl}\,\vec{A}$. Thus: $F^{21} = B^z$.

The same applies to all other elements of F. Hence, we finally obtain:

$$[F] = \begin{pmatrix} 0 & -E_x/c & -E_y/c & -E_z/c \\ E_x/c & 0 & -B_z & B_y \\ E_y/c & B_z & 0 & -B_x \\ E_z/c & -B_y & B_x & 0 \end{pmatrix}.$$

This tensor is the same as the previous Faraday tensor F of the relation (5.3), which was deduced from the natural extension to 4D of the Lorentz force. Faraday's tensor thus is the "Jacobian's Dissymmetry" of the potential four-vector **A**.

Noteworthy, the relation (5.17) applies the time distance equivalence principle when adding components of the electrical potential V (being the analog of the time) with the ones of A (being the analog of the space). It thus was a requirement to use homogenous units for time and distance ($t \rightarrow ct$).

We will now address the relation between the potential ***A*** *and the current vector* ***J****.*

5.1.4.2.2 *Relation between the Potential* ***A*** *and the Current Vector* ***J***

The combination of the two fundamental relations (5.13) and (5.17) enables us to derive the relation between the potential four-vector **A** and the current four-vector **J**:

$$\nabla\left[\mathrm{Jac}(\mathbf{A}) - \mathrm{Jac}(\mathbf{A})^T\right] = \mu_0 \mathbf{J}. \tag{5.18}$$

We will develop this relation using the tensorial notations that have the interest of being very compact. However, some readers may find it too abstract; hence, another demonstration is provided in Section 5.1.6.7.

The relation $\nabla\left[\mathrm{Jac}(\mathbf{A}) - \mathrm{Jac}(\mathbf{A})^T\right]$ means: $\partial_\mu F^{\mu\nu} = \mu_0 J^\nu$, and then: $\partial_\mu\left(\partial^\mu A^\nu - \partial^\nu A^\mu\right) = \mu_0 J^\nu$.

The term $\partial_\mu \partial^\mu A^\nu$ matches with the d'Alembertian operator, denoted by $\square$, and which is:

$$\square\left(A^\nu\right) = \partial_\mu \partial^\mu A^\nu = \frac{\partial^2\left(A^\nu\right)}{cdt^2} - \frac{\partial^2\left(A^\nu\right)}{dx^2} - \frac{\partial^2\left(A^\nu\right)}{dy^2} - \frac{\partial^2\left(A^\nu\right)}{dz^2}.$$

Regarding the last term $\partial_\mu \partial^\nu A^\mu$: We have $\partial_\mu \partial^\nu A^\mu = \partial^\nu \partial_\mu A^\mu$ due to the possibility to exchange the partial derivatives.[3] We then notice that $\partial_\mu A^\mu$ is the Divergence of **A**. Besides, we will see in Section 5.1.4.2.3 that using the Lorenz Gauge, the Divergence of **A** is null; hence, this second term is null.

Consequently, equation 5.18 becomes:

$$\square\left(A^\nu\right) = \frac{\partial^2\left(A^\nu\right)}{cdt^2} - \frac{\partial^2\left(A^\nu\right)}{dx^2} - \frac{\partial^2\left(A^\nu\right)}{dy^2} - \frac{\partial^2\left(A^\nu\right)}{dz^2} = \mu_0 J^\nu. \quad (5.19)$$

We can see that if there is no electrical charge, and thus no current, these equations 5.19 represent the propagation of waves at the speed c; and these are the same equations as in classical physics.

Remark 5.2: These wave propagation equations were first issued by the French scientist J. d'Alembert, hence his name for the $\square$ operator.

The relativistic relations (5.13), (5.17) and (5.19) enable us to retrieve the four famous classical Maxwell's equations, which are:

$$\mathrm{div}\left(\vec{E}\right) = \frac{\rho_0}{\varepsilon_0}, \quad \overrightarrow{\mathrm{curl}}\ \vec{E} = -\frac{d\vec{B}}{dt}, \quad \mathrm{div}\left(\vec{B}\right) = 0, \quad \overrightarrow{\mathrm{curl}}\ \vec{B} = \mu_0 \vec{j} + \varepsilon_0 \mu_0 \frac{d\vec{E}}{dt}.$$

The corresponding demonstrations are presented with the problems 5-3 to 5-6.

Conclusion: Relativity has considerably simplified the axiomatic framework on which electromagnetism is built: a current **J** generates a field of potential four vectors **A** such that:

- the "Jacobian's Dissymmetry" of **A** is the Faraday's tensor hosting the components of the electromagnetic field.
- The d'Alembertian of each component of **A** is the corresponding component of **J**.

An electrical charge q in an electromagnetic field experiences a Four-Force that is equal to its proper speed multiplied by Faraday's tensor and by q.

However, the above calculation involved the Lorenz Gauge; hence, we will now explain it:

5.1.4.2.3 The Lorenz Gauge

Looking at the relation (5.17), if we add to the potential four-vector **A** another rank 1 tensor denoted by **Z** having its Jacobian symmetric, then the resulting rank 1 tensor (**A**+**Z**) will yield the same Faraday tensor F. The additional term $\mathrm{Jac}(\mathbf{Z}) - \mathrm{Jac}(\mathbf{Z})^T$ indeed is null since Jac(**Z**) is symmetric. We thus have the liberty to add any such tensor **Z** to **A**; hence, we will do so in order to obtain Divergence (**A**+**Z**)=0.

The Divergence operator is linear. Notably, Jac(**Z**) is symmetric if **Z** is the gradient of a scalar potential, denoted by φ. We thus choose **Z** = grad (φ) with Div(grad(φ)) = −Div(**A**).

It is thus possible to find a 4D potential vector that satisfies equation 5.17 and with its Divergence null. For the sake of simplicity, this potential vector we still be denoted by **A**, and its zero Divergence means:

$$\mathrm{Div}(\mathbf{A}) = 0 = \partial_\mu A^\mu = \frac{\partial(A^0)}{cdt} + \frac{\partial(A^1)}{dx} + \frac{\partial(A^2)}{dy} + \frac{\partial(A^3)}{dz}, \tag{5.20}$$

which is the same Lorentz gauge as in classical physics, which was: $\frac{\partial(V)}{dt} + \mathrm{div}(\vec{A}) = 0.$

5.1.5 Further Topics: The Covariant Faraday's Tensor

We started our reasoning with the relation (5.5): $\mathbf{\Phi} = q(F)\,\mathbf{S}^*$. Actually, we made an implicit choice whereby the proper speed expression is covariant and the Four-Force contravariant. This stems from the initial relation stating the mass conservation $ds^2(\mathbf{P})$ =cst, from which we deduced that $\mathbf{P} \cdot \frac{d\mathbf{P}}{d\tau} = 0$; hence: $\mathbf{S} \bullet \mathbf{\Phi} = 0$. From this relation, we can as well derive (**S***) (**Φ**) = 0, as we did, or (**Φ***) (**S**) = 0 where the covariant components of the Four-Force are used, and the contravariant ones of the proper speed.

With this last choice, the relation (5.5) becomes **Φ*** = q (F**) **S** where F** is the Faraday tensor expressed as a twice covariant rank 2 tensor. To find its components, one can either make the same reasoning as we did to find F, but adapting with the change covariant-contravariant, or apply the general rule for changing any rank 2 twice contravariant into a twice covariant one, and which is[4]: $F_{\mu\nu} = g_{\mu\alpha}\, g_{\mu\beta}\, F^{\alpha\beta}$. The result is:

$$[F_{\mu\nu}] = \begin{pmatrix} 0 & E_x/c & E_y/c & E_z/c \\ -E_x/c & 0 & -B_z & B_y \\ -E_y/c & B_z & 0 & -B_x \\ -E_z/c & -B_y & B_x & 0 \end{pmatrix}$$

Consequently, an invariant combination of E and B can be derived from making a contraction of the indices of $F_{\mu\nu}$ and $F^{\mu\nu}$, which is:

$$-\frac{1}{2} F^{\mu\nu} F_{\mu\nu} = E^2/c^2 - B^2.$$

Thus, if B is null in one frame K, then in other frames K′, the classical norm of E′ in K′ will be greater than the norm of E in K.

In addition, another invariant exists[5]: the classical 3D scalar product $\vec{E}\cdot\vec{B}$. Thus, if $\vec{E}$ and $\vec{B}$ are perpendicular in one frame, they are perpendicular in all frames.

Remark 5.3: The covariant Faraday's tensor is naturally interpreted as a two-form, the exterior derivative of a covector **A**. The two-form version is naturally integrated over surfaces to measure flux, so the contravariant form is less 'natural'.

5.1.6 Mathematical and Physical Supplements

5.1.6.1 The Vector Product

In the classical 3D space, the vector product between two 3D vectors $\vec{U}$ and $\vec{V}$ is a vector $\vec{W} = \vec{U} \times \vec{V}$, which is perpendicular to both $\vec{U}$ and $\vec{V}$, and with its norm being the product of the norm of $\vec{U}$ and the norm of $\vec{V}$ multiplied by the sine of the angle $(\vec{U}, \vec{V})$. The product vectors are also denoted by Λ: thus $\vec{W} = \vec{U} \wedge \vec{V}$.

The vectors $\vec{U}$ and $\vec{V}$ define a plane, and there always exists one perpendicular line to this plane. There are two possible directions in this perpendicular line, and $\vec{W}$ is along the direct one, which is defined as follows: imagine a screw with a groove along U; the action of turning it toward V makes the screw move in the correct direction. Since this direction is a convention, $\vec{W}$ is termed a pseudo-vector (Figure 5.1).

The vector product is a 3D concept, and it is named so because the result is a vector. It is an intrinsic notion since the angles and the norms are intrinsic notions (invariant when changing frame). The norm of $\vec{W}$ is equal to the surface of the parallelogram formed by $\vec{U}$ and $\vec{V}$.

If the components of $\vec{U}$ and $\vec{V}$, respectively, are (u_x, u_y, u_z) and (v_x, v_y, v_z), then the components of $\vec{W}$ are: $\vec{W}\left(u_y v_z - u_z v_y,\; u_z v_x - u_x v_z,\; u_x v_y - u_y v_x\right)$

Note that the vector product is not intrinsic in 4D since the classical 3D norm is not intrinsic (e.g., if $\vec{U}$ and $\vec{V}$ are perpendicular to the motion of K′ relative to K, then $\vec{W} = \vec{U} \wedge \vec{V}$ will be seen contracted in K′ while $\vec{U}$ and $\vec{V}$ won't).

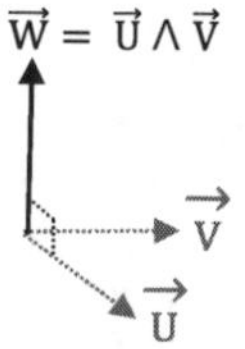

FIGURE 5.1
Vector product.

5.1.6.2 The Gradient

Consider a differentiable scalar field k(M), and two neighboring points M and N. In 3D, we have: $\overrightarrow{MN} = \overrightarrow{dM} = (dx, dy, dz)$. The gradient of this scalar field at the point M is the vector, also denoted by $\vec{\nabla}(k)$, such that:

$$\overrightarrow{grad}(k) = \left(\frac{\partial k}{\partial x}, \frac{\partial k}{\partial y}, \frac{\partial k}{\partial z}\right), \text{ and so: } \vec{\nabla}(k) = \frac{\partial k}{\partial x}\cdot\overrightarrow{e_x} + \frac{\partial k}{\partial y}\cdot\overrightarrow{e_y} + \frac{\partial k}{\partial z}\cdot\overrightarrow{e_z}.$$

This operator $\vec{\nabla}()$ is also called "nabla"; in 4D, it is "4-nabla."The Gradient is useful to assess the variation of k in the vicinity of M:

$$dk(M) = k(N) - k(M) = k(M + dM) - k(M) = \vec{\nabla}(\mathbf{k}) \cdot \overrightarrow{\mathbf{dM}}. \qquad (5.20a)$$

The Gradient actually concerns vectorial spaces of any dimension. If the space is 4D and curved, orthonormal Cartesian coordinates are assumed (Minkowski's frame) so as to have intrinsic (covariant) derivatives.

In 4D, we thus have with homogeneous units:

$$\overrightarrow{grad}(k) = \vec{\nabla}(k) = \left(\frac{\partial k}{c dt}, \frac{\partial k}{\partial x}, \frac{\partial k}{\partial y}, \frac{\partial k}{\partial z}\right).$$

However, relation 5.20a uses the classical Euclidean scalar product, but it is not intrinsic in 4D, whereas the Lorentzian scalar product is. Conversely, the value of $dk(M) = k(N) - k(M)$ is the same in all inertial frames, hence it is intrinsic. Consequently, the relation (5.20a) is not the expression of a scalar product in 4D, but it of the dot-product which performs the application of the linear form, the Gradient, to the contravariant 4-vector dM. The result is thus intrinsic. This shows that the above Gradient components are covariant: the 'natutal' expression of the Gradient is thus covariant.

Any covariant vector, in particular $\vec{\nabla}(k)$, has an associated contravariant vector, which is obtained by multiplying $\vec{\nabla}(k)$ by the inverse metric matrix: G^{-1} (cf. Section 4.1.3). Hence, in a frame with Cartesian coordinates (space of Minkowski), the contravariant components of the Gradient are $\left(\frac{\partial k}{c dt}, -\frac{\partial k}{\partial x}, -\frac{\partial k}{\partial y}, -\frac{\partial k}{\partial z}\right)$, the generic term being denoted by ∂^i ( ) with the index on the top.

With an orthonormal Minkowski's frame, we have $G^{-1} = G$. Also, the metric can be (–1, +1, +1, +1) since it is a matter of convention. In such a case, the contravariant gradient is: $\left(-\frac{\partial V}{c dt}, +\frac{\partial V}{\partial x}, +\frac{\partial V}{\partial y}, +\frac{\partial V}{\partial z}\right)$.

5.1.6.3 The Divergence

In a field of four vectors, denoted by **V**(M), the Divergence of the vector **V**(M) is a number which is the Trace of the Jacobian matrix associated to **V**(M). We recall that:

- the Trace of a square matrix is the sum of its elements on its diagonal;
- the Jacobian matrix is the square matrix containing all partial derivatives of **V**(M), as with equation 5.11.

The Divergence concerns vectorial spaces of any dimension. Each column of the Jacobian matrix is formed by the Gradient of each component of **V** at the point M.

We will show that when changing frame from K to K′, the value of the Divergence remains the same, meaning that the divergence is an intrinsic notion:

The relation $\mathbf{V}(\mathrm{M}+\mathrm{dM})=[\mathrm{Jac}(\mathbf{V})](\mathrm{dM})$ shows that Jac(**V**) is a rank two tensor, which is mixed covariant-contravariant. Hence, when changing frame from K to K′ [6]: $\mathrm{Jac}(\mathbf{V})'=(\Lambda)\mathrm{Jac}(\mathbf{V})(\Lambda^{-1})$.

Besides, we will use the property of the Trace to be commutative,[7] Hence, the Divergence in K′ is :

$$\mathrm{Div}(\mathbf{V})'=\mathrm{Trace}\left[\mathrm{Jac}(\mathbf{V})'\right]=\mathrm{Trace}\left[(\Lambda)\mathrm{Jac}(\mathbf{V})(\Lambda^{-1})\right]=\mathrm{Trace}\left[\mathrm{Jac}(\mathbf{V})(\Lambda^{-1})(\Lambda)\right]$$

$$=\mathrm{Trace}\left[\mathrm{Jac}(\mathbf{V})\right]=\mathrm{Div}(\mathbf{V}) \quad \blacksquare$$

We previously saw that in the 4D space-time of Minkowski, the Gradient has two expressions: a covariant one (which is the usual one), and a contravariant one. Hence, this is also the case for the Divergence.

The covariant Divergence is: $\mathrm{Div}(\mathbf{V})=\frac{\partial V^t}{c\,\partial t}+\frac{\partial V^X}{\partial x}+\frac{\partial V^y}{\partial y}+\frac{\partial V^z}{\partial z}$; the contravariant one: $\frac{\partial V^t}{c\,\partial t}-\frac{\partial V^X}{\partial x}-\frac{\partial V^y}{\partial y}-\frac{\partial V^z}{\partial z}$.

In 3D, the intrinsic character of the divergence can be seen by its following important property: consider a 3D vectorial field $\overrightarrow{\mathrm{A}(\mathrm{M})}$, and a small volume dv around a point M. The flow of the vectors $\vec{\mathrm{A}}$ going outside the small surface delimited by dv is the divergence of $\vec{\mathrm{A}}$ at the point M: $d\Phi=\mathrm{div}\left(\vec{\mathrm{A}}\right)\mathrm{dv}$.

This is generalized by Green–Ostrogradski's theorem: $\iint \vec{\mathrm{A}}\cdot\overrightarrow{\mathrm{ds}}=\iiint \mathrm{div}\left(\vec{\mathrm{A}}\right)\mathrm{dv}$. It shows that the outgoing flow is all the more important as its divergence is important. Alternatively, if the divergence is null, so is the outgoing flow.

The importance of the Divergence can be illustrated with the following application:

5.1.6.3.1 The Gauss–Maxwell Law: $\text{div}\left(\vec{E}\right) = \frac{\rho}{\varepsilon_\circ}$

Let's apply this relation to an electrical charge q placed in the center of a sphere of radius R, we have: $\iint \vec{E} \cdot \vec{ds} = \iiint \text{div}\left(\vec{E}\right) dv$, meaning: $E\,4\,\pi\,R^2 = \iiint \frac{\rho}{\varepsilon_\circ} dv = \frac{q}{\varepsilon_\circ}$, which is Coulomb's famous law showing that the electrical field generated by an electrical charge decreases with the square of the distance to this charge.

Similarly, the Maxwell–Thomson law: $\text{div}(\vec{B}) = 0$ expresses the non-existence of a magnetic charge.

5.1.6.4 The d'Alembertian Operator

The d'Alembertian Operator is defined as: $\square\left(A^\nu\right) = \partial_\mu \partial^\nu A^\nu$, which yields in an orthonormal frame of Minkowski: $\square\left(A^\nu\right) = \partial_\mu \partial^\mu A^\nu = \frac{\partial^2\left(A^\nu\right)}{cdt^2} - \frac{\partial^2\left(A^\nu\right)}{dx^2} - \frac{\partial^2\left(A^\nu\right)}{dy^2} - \frac{\partial^2\left(A^\nu\right)}{dz^2}$.

We also have the notation: $\partial_\mu \partial^\mu \mathbf{A} = \frac{\partial^2}{cdt^2}\mathbf{A} - \vec{\nabla}^2 \mathbf{A} = \mu_0\,\mathbf{J}$.

This operator is equivalent to the covariant Divergence of the contravariant Gradient; consequently, it is intrinsic.

5.1.6.5 The Curl Operator

The curl operator, also called rotational, only applies to 3D spaces and concerns fields of 3D vectors $\vec{A}$ (u, v, w). This operator associates the following 3D vector to the vector $\vec{A}$:

$$\overrightarrow{\text{curl}}\left(\vec{A}\right) = \left(\frac{\partial w}{\partial y} - \frac{\partial v}{\partial z}, \frac{\partial u}{\partial z} - \frac{\partial w}{\partial x}, \frac{\partial v}{\partial x} - \frac{\partial u}{\partial y}\right). \qquad (5\text{-}21)$$

The curl represents the degree to which a field of vectors is turning around itself.

The curl operator is also denoted by $\nabla\Lambda\left(\vec{A}\right)$ or $\nabla \times \left(\vec{A}\right)$ due to its similitude with the vector product.

The following mathematical results are useful:

- The $\overrightarrow{\text{curl}}$ of the $\overrightarrow{\text{grad}}$ of any scalar field φ is always null: $\overrightarrow{\text{curl}}\left(\overrightarrow{\text{grad}}\,(\varphi)\right) = 0$.
- The divergence of a $\overrightarrow{\text{curl}}$ is always null: $\text{div}(\overrightarrow{\text{curl}}) = 0$.

The corresponding demonstrations are proposed as Problems 5.1 and 5.2.

Illustration with the Maxwell–Thomson law: $\mathrm{div}(\vec{B}) = 0$ suggests that $\vec{B}$ is the $\overrightarrow{\mathrm{curl}}$ of a vector $(\vec{A})$.

5.1.6.6 The Electromagnetic Tensor Must Be Antisymmetric

We saw in section 3.4 the relation 3.26a : $(V^*)^T(EM)(V^*) = 0.$ (5.22)
We will first show that the elements of the diagonal of (EM) are all null:
The relation (5.22) is:

$$(c, -v_x, -v_y, -v_z)\begin{pmatrix} a^{11} & a^{12} & a^{13} & a^{14} \\ a^{21} & a^{22} & a^{23} & a^{24} \\ a^{31} & a^{32} & a^{33} & a^{34} \\ a^{41} & a^{42} & a^{43} & a^{44} \end{pmatrix}\begin{pmatrix} c \\ -v_x \\ -v_y \\ -v_z \end{pmatrix} = 0.$$

This relation is valid whatever the covector V*; besides, the a^{ij} coefficients are independent of the velocity of the particle since they characterize the electromagnetic field at the place where the particle is. Hence, let's choose a covector V* having all its components null except the one of rank i having the value 1. For instance, V*: (0, 0, 1, 0). The result of (EM)(V*) is the column i of (EM) as shown below (the third column on our example).

$$\begin{pmatrix} a^{11} & a^{12} & a^{13} & a^{14} \\ a^{21} & a^{22} & a^{23} & a^{24} \\ a^{31} & a^{32} & a^{33} & a^{34} \\ a^{41} & a^{42} & a^{43} & a^{44} \end{pmatrix}\begin{pmatrix} 0 \\ 0 \\ 1 \\ 0 \end{pmatrix} = \begin{pmatrix} a^{13} \\ a^{23} \\ a^{33} \\ a^{43} \end{pmatrix}.$$

Then, the product of $(V^*)^T$ with this column i gives the element of line i of this column i, meaning a^{ii} (in our example: a^{33}).

$$(V^*)^T(EM)(V^*) \text{yields:} (0,0,1,0)\begin{pmatrix} a^{13} \\ a^{23} \\ a^{33} \\ a^{43} \end{pmatrix} = a^{33}.$$

The relation (5.22) implies $a^{ii}=0$, meaning that the diagonal of (EM) is null.

This enables us to show that (EM) is antisymmetric:

Let's take the case of a vector V* such that only two elements in the positions i and j are equal to 1, the others being equal to zero. Then the result of (EM)(V*) is the sum of the columns i and j of (EM), as shown below, where we have chosen i=2 and j=4. Then the product of V* with this column gives the sum of the four elements of (EM): $a^{ii} + a^{ij} + a^{ji} + a^{jj}$.

$$\text{(EM)(V*) is: } \begin{pmatrix} a^{11} & a^{12} & a^{13} & a^{14} \\ a^{21} & a^{22} & a^{23} & a^{24} \\ a^{31} & a^{32} & a^{33} & a^{34} \\ a^{41} & a^{42} & a^{43} & a^{44} \end{pmatrix} \begin{pmatrix} 0 \\ 1 \\ 0 \\ 1 \end{pmatrix} = \begin{pmatrix} a^{12}+a^{14} \\ a^{22}+a^{24} \\ a^{32}+a^{34} \\ a^{42}+a^{44} \end{pmatrix}$$

$$\text{Then, } (V^*)^T\,(EM)\,(V^*) \text{yields: } (0,1,0,1) \begin{pmatrix} a^{12}+a^{14} \\ a^{22}+a^{24} \\ a^{32}+a^{34} \\ a^{42}+a^{44} \end{pmatrix} = a^{22}+a^{24}+a^{42}+a^{44}.$$

We can see that two elements a^{ii} and a^{jj} are in the diagonal of (EM), implying that they are null. Then only remains $a^{ij} + a^{ji}$, and the relation (5.22) implies that $a^{ij} + a^{ji} = 0$, so that $a^{ij} = -a^{ji}$, which means that the matrix (EM) is antisymmetric. ■

5.1.6.7 Demonstration That $\Box(A^{v}) = \mu_0 J^{v}$ Using Classical Notations

We want to develop the relation 5.18a: $\nabla F = \nabla\left[\mathrm{Jac}(\mathbf{A}) - \mathrm{Jac}(\mathbf{A})^T\right]\mu_0 \mathbf{J}$.

We saw that the appropriate Jacobian of **A** is the matrix (tensor) comprising the *contravariant* gradient of the components of **A**. We remind that the components of **A** are (V, A^x, A^y, A^z). Hence, Jac (**A**) is the following:

$$\mathrm{Jac}(A) = \begin{pmatrix} \dfrac{\partial V}{c.\partial t} & \dfrac{\partial A^x}{c.\partial t} & \dfrac{\partial A^y}{c.\partial t} & \dfrac{\partial A^z}{c.\partial t} \\ -\dfrac{\partial V}{\partial x} & -\dfrac{\partial A^x}{\partial x} & -\dfrac{\partial A^y}{\partial x} & -\dfrac{\partial A^z}{\partial x} \\ -\dfrac{\partial V}{\partial y} & -\dfrac{\partial A^x}{\partial y} & -\dfrac{\partial A^y}{\partial y} & -\dfrac{\partial A^z}{\partial y} \\ -\dfrac{\partial V}{\partial z} & -\dfrac{\partial A^x}{\partial z} & -\dfrac{\partial A^y}{\partial z} & -\dfrac{\partial A^z}{\partial z} \end{pmatrix}$$

We also saw that Faraday's tensor is equal to: $F=\text{Jac}(\mathbf{A})-\text{Jac}(\mathbf{A})^T$, meaning:

$$F=\begin{pmatrix} \frac{\partial V}{c.\partial t}-\frac{\partial V}{c.\partial t} & \frac{\partial A^x}{c.\partial t}+\frac{\partial V}{\partial x} & \frac{\partial A^y}{c.\partial t}+\frac{\partial V}{\partial y} & \frac{\partial A^z}{c.\partial t}+\frac{\partial V}{\partial z} \\ -\frac{\partial V}{\partial x}-\frac{\partial A^x}{c.\partial t} & -\frac{\partial A^x}{\partial x}+\frac{\partial A^x}{\partial x} & -\frac{\partial A^y}{\partial x}+\frac{\partial A^y}{\partial y} & -\frac{\partial A^z}{\partial x}+\frac{\partial A^x}{\partial z} \\ -\frac{\partial V}{\partial y}-\frac{\partial A^y}{c.\partial t} & -\frac{\partial A^x}{\partial y}+\frac{\partial A^y}{\partial x} & -\frac{\partial A^y}{\partial y}+\frac{\partial A^y}{\partial y} & -\frac{\partial A^z}{\partial y}+\frac{\partial A^y}{\partial z} \\ -\frac{\partial V}{\partial z}-\frac{\partial A^z}{c.\partial t} & -\frac{\partial A^x}{\partial z}+\frac{\partial A^z}{\partial x} & -\frac{\partial A^y}{\partial z}+\frac{\partial A^z}{\partial y} & -\frac{\partial A^z}{\partial z}+\frac{\partial A^z}{\partial z} \end{pmatrix}$$

Let's calculate the first component of ∇F: it is by definition the Divergence of the first column of F, denoted by F°, and it is equal to $\mu_0 J^0$ according to equation 5.18a.

We have:

$$\text{Div}(F^\circ)=\text{Div}\left[\frac{\partial(V)}{cdt}-\frac{\partial(V)}{cdt},-\frac{\partial(V)}{dx}-\frac{\partial(A^x)}{cdt},-\frac{\partial(V)}{dy}-\frac{\partial(A^y)}{cdt},\frac{\partial(V)}{dz}-\frac{\partial(A^z)}{cdt}\right].$$

Then, $$k.J^0=\frac{\partial}{cdt}\left[\frac{\partial(V)}{cdt}-\frac{\partial(V)}{cdt}\right]+\frac{\partial}{\partial x}\left[-\frac{\partial(V)}{dx}-\frac{\partial(A^x)}{cdt}\right]+\frac{\partial}{\partial y}\left[-\frac{\partial(V)}{dy}-\frac{\partial(A^y)}{cdt}\right]$$

$$+\frac{\partial}{\partial z}\left[-\frac{\partial(V)}{dz}-\frac{\partial(A^z)}{cdt}\right].$$

Then,

$$k.J^0=\frac{\partial^2(V)}{c^2dt^2}-\frac{\partial^2(V)}{dx^2}-\frac{\partial^2(V)}{dy^2}-\frac{\partial^2(V)}{dz^2}+\frac{\partial}{c\partial t}\left[-\frac{\partial(V)}{cdt}+\frac{\partial(A^x)}{dx}+\frac{\partial(A^y)}{dy}+\frac{\partial(A^z)}{dz}\right].$$

The second term is: $\frac{\partial}{c\partial t}$ [–Divergence (**A**)], and we saw that the Lorenz Gauge yields: Div(**A**)=0.

We then have: $\mu_0\, J^0=\frac{\partial^2(V)}{c^2dt^2}-\frac{\partial^2(V)}{dx^2}-\frac{\partial^2(V)}{dy^2}-\frac{\partial^2(V)}{dz^2}$. We recognize that this is the d'Alembertian operator. We thus have: $\square\, A^0=\mu_0\, J^0$.

The same reasoning applies to the three other components of ∇F, denoted by ∇F^i, leading to the same result when replacing the index 0 by the index i. Hence, we finally obtain: $\nabla F^i=\mu_0\, J^i=\frac{\partial^2(A^i)}{c^2dt^2}-\frac{\partial^2(A^i)}{dx^2}-\frac{\partial^2(A^i)}{dy^2}-\frac{\partial^2(A^i)}{dz^2}=\square A^i$ ■

5.2 Schwarzschild's Solution

Only 1 month after Einstein published his General Relativity theory, the mathematician and astronomer K. Schwarzschild found a solution to Einstein's equation in the context of a massive spherical body. This solution enabled us to calculate the deviation of light near the Sun, which could then be checked experimentally in 1919, providing a brilliant confirmation of Relativity. Moreover, this solution opened the door to the possibility of black holes.

The Schwarzschild's solution reads:

$$ds^2(dM) = \left(1 - \frac{r_S}{r}\right)c^2dt^2 - \left(1 - \frac{r_S}{r}\right)^{-1} dr^2 - r^2\left(d\theta^2 + \sin^2\theta d\varphi^2\right). \quad (5.23)$$

where r_s is the Schwarzschild radius, $r_s = \frac{2GM}{c^2}$, and $ds^2()$ refers to a small space–time interval in the vicinity of the event M whose polar coordinates are (t, r, θ, φ). These coordinates take advantage of the symmetry relative to the center of the sphere. However, Schwarzschild introduced some specificities, as we will see.

The Schwarzschild solution is valid provided that the mass is not turning and doesn't have an electrical charge. Besides, the radius of the spherical mass must be greater than the Schwarzschild radius. This condition is generally met since usually r_s is very small: for instance, it equals 8.9 mm for the Earth, and 3 km for the Sun.

5.2.1 Polar Coordinates: Reminder and Adaptation to the 4D Space-Time Universe

5.2.1.1 Classical Polar Coordinates in 3D

We will start presenting the polar coordinates in the classical 3D space, and then apply its principles and results to our space-time universe. In this part, we don't place any mass in this frame, so there is no distortion. In the next part, a massive object will be placed at the point O.

For the sake of clarity, we assume the plane XOY is horizontal, and OZ is vertical. Consider a random point M with coordinates in K being (x, y, z), and its projection on the horizontal plane XOY is denoted by H (Figure 5.2). The polar coordinates of M are defined by the three following parameters: r, θ and φ, where:

- r is the length OM.
- θ is the angle ZOM: this angle refers to the altitude of M, we indeed have: $z = MH = r\cos\theta$.
- φ is the angle YOH.

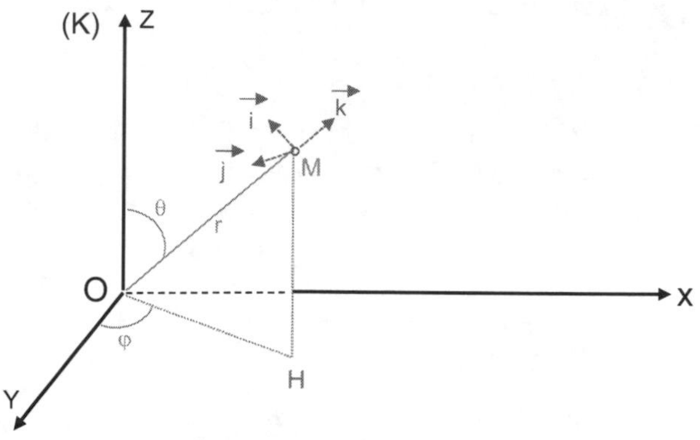

FIGURE 5.2
Classical polar coordinates.

The polar system of coordinates, also called spherical coordinates, constitutes a real frame: every point can be located in an unambiguous manner. Moreover, there is a relationship between the polar coordinates of any point M and its coordinates in K; we indeed have:

- $z = r \cos \theta$.
- $y = OH \cos \varphi$, and besides, $OH = r \sin \theta$, so: $y = r \sin \theta \cos \varphi$.
- $x = OH \sin \varphi$, and so: $x = r \sin \theta \sin \varphi$.

5.2.1.1.1 Classical Distance in the Vicinity of Point M

We will show that the polar system of coordinates enables us to define at any point M an orthonormal tangent frame, which in turn enables us to calculate the distance from M to any point N in its vicinity.

Let's imagine the sphere centered in O and having $r = OM$ as the radius. The angle θ represents the *latitude* of M; the angle φ represents its *longitude*. We can then draw two sets of lines on this sphere:

- One set of lines of latitude where points on the same line have the same latitude; two consecutive latitudes have an angle difference noted $\Delta\theta$.
- One set of lines having the same longitude: these are meridians; two consecutive meridians have an angle difference noted $\Delta\varphi$. We can choose $\Delta\theta = \Delta\varphi$ so to respect the symmetry.

We assume that two such lines cross at the point M. For an observer in M, or in its vicinity, these two lines are perpendicular. We can then define a local orthonormal frame, noted Km, with:

- the first unit vector is along the meridian passing by M: it is $\vec{i}$ in our picture;

- the second unit vector is along the latitude passing by M: $\vec{j}$ in our picture;
- the third unit vector is along the radius OM: it is $\vec{k}$ in our picture 5.2 (it is perpendicular to $\vec{i}$ and $\vec{j}$).

We choose the same length for these three unit basis vectors, and this length is the same as the one of the unit basis vectors of the frame K. These vectors $\vec{i}, \vec{j}, \vec{k}$ thus constitute an orthonormal tangent frame, denoted by K_M, enabling us to calculate the distance between M and any point N in its vicinity[8]:

Consider a point N in the vicinity of M, with its polar coordinates being: r+dr, θ+dθ, and φ+dφ. We want to calculate the distance MN, which is denoted by dM since its length is small. We first need to express N in the tangent frame K_M; the coordinates of N are:

- coordinate relative to $\vec{i}$ (direction of the latitude of M): $OH\, d\varphi = r \sin\theta\, d\varphi$.
- coordinate relative to $\vec{j}$ (direction of the meridian of M): $OM\, d\theta = r\, d\theta$.
- coordinate relative to $\vec{k}$ (along the radial line OM): dr.

Thus, the components of $\overrightarrow{MN}$ in the frame K_m are: $(r \sin\theta\, d\varphi, r\, d\theta, dr)$.

Then K_M being orthonormal:

$MN^2 = dM^2 = r^2 \sin^2\theta\, d\varphi^2 + r^2\, d\theta^2 + dr^2 = dr^2 + r^2 (\sin^2\theta\, d\varphi^2 + d\theta^2)$.

The term $(\sin^2\theta\ \ d\varphi^2 + d\theta^2)$ is usually denoted by $d\Omega^2$; hence, we have:

$$MN^2 = dM^2 = dr^2 + r^2 d\Omega^2. \qquad (5.24)$$

This polar representation can be extended to our 4D space-time universe, as we will see.

5.2.1.2 Polar Coordinates in the 4D Space-Time Universe

Being in a space-time universe, the point M must be replaced with the *event* M. Our previous orthonormal frame K becomes a 4D orthonormal frame, still denoted by K, by adding the time dimension. Using polar coordinates in K, an event M is defined by a set of four parameters M (t, r, θ, φ), where the time part is usually placed at the first position.

The distance OM becomes the Lorentzian norm of the space–time interval OM: $ds^2(OM) = ct^2 - r^2$.

Similarly, our previous 3D orthonormal tangent frame in M, K_M, becomes a 4D orthonormal frame, still denoted by K_M, with the addition of the time dimension and its associated fourth unit vector, denoted by $\vec{f}$. This unit vector $\vec{f}$ is perpendicular to the three existing vectors $\vec{i}, \vec{j}, \vec{k}$. Note that the perpendicularity means that the Lorentzian scalar product between any couple of these unit vectors is null. For the spatial ones, it matches with the classical scalar product (in the case of this frame where no vector has both a time and a spatial part).

Consider an event N in the vicinity of M, and we want to calculate the Lorentzian distance of MN:

The polar coordinates of N in K are: (t+dt, r+dr, θ+dθ, φ+dφ). In the tangent frame in M, K_M, the components of the small space–time vector MN, denoted by dM are with homogeneous units:

dM (c dt, r sinθ dφ, r dθ, dr).

The space–time interval MN is: $ds^2(dM) = dM \bullet dM$ *(with meaning the Lorentzian scalar product).*

$$ds^2(\mathbf{dM}) = c^2dt^2 - r^2\sin^2\theta\, d\varphi^2 - r^2d\theta^2 - dr^2, \text{ then :}$$

$$ds^2(\mathbf{dM}) = c^2dt^2 - dr^2 - r^2\left(\sin^2\theta \cdot d\varphi^2 + d\theta^2\right) = \mathbf{c^2dt^2 - dr^2 - r^2\ d\Omega^2}. \quad (5.25)$$

Remark 5.4: The lines of common polar coordinates are curved (meridians and latitudes), but this does not mean that the space is curved. These curved lines are just a convenient mathematical representation; the frames K and K_M are Minkowski frames. The universe is Euclidean as long as there is no great mass, which was the case until now. Let's then introduce a great mass at the point O and examine Schwarzschild's solution.

5.2.2 Meaning of Schwarzschild's Coordinates and the Possibility of Black Holes

Schwarzschild's coordinates are defined as follows.

The radial coordinate r is built the following way: Let's denote by L the length of the circumference of the sphere centered in O and passing by M. Then r is defined as: r=L / 2π. In case of Euclidean geometry, this leads to r=OM, but as the geometry is distorted, r is not equal to OM, as measured in the massive object frame. Still, r increases in a monotonic way with OM.

The time coordinate is built the following way: a perfect (atomic) clock is placed at the infinite, and special Schwarzschild's clocks are regularly placed along a radial line, giving the Schwarzschild's time coordinate where they are located. These special clocks are built as follows: let's consider two consecutive Schwarzschild's clocks, one in A and the other one in B, with A being closer to O.

O.·········.r_s·········.A·········.B

B emits a flash at each second of the Schwarzschild's clock in B, then at each flash received in A, the Schwarzschild's clock in A increments by 1 second, and simultaneously emits a flash giving the time to the next clock. We saw that gravity affects time: If the clocks in A and B were identical, marking the proper time in A and in B, the observer in A would see B's clock running faster than his or her own (similarly, B would see A's clock running slower than his or her own). Thus, Schwarzschild's time in A reflects how an observer in A sees the proper time at infinity. Schwarzschild's time at the point A runs faster than A's proper time, and the closer A is to r_s, the faster it is. Conversely, an observer at infinity sees the clock in A running slower than his or her own.

Let's compare Schwarzschild's metric (5.23) with Minkowski's metric (5.25): We first notice that when we are extremely far from the mass, both metrics are identical, which is normal since at the infinite the mass generates no distortions.

When the universe is distorted, we saw in Volume I Section 7.1.4 how to calculate $ds^2(dM)$ in the tangent frame at the event M, denoted K'_M. We recall that K'_M generally is not orthonormal; hence, we choose the following unit basis vectors of K′ to match with polar coordinates:

- f′ is the basis vector along the time axis;
- i′ is the basis vector along the radial axis OM;
- j′ is the basis vector tangent to the meridian passing by M;
- k′ is the basis vector tangent to the latitude passing by M.

We then have: $ds^2(dM) = dM \bullet dM$, meaning:

$$ds^2(dM) = \left(c.dt'f' + dx'i' + dy'j' + dz'k'\right) \bullet \left(c.dt'\vec{f'} + dx'\vec{i'} + dy'\vec{j'} + dz'\vec{k'}\right) \quad (5.26),$$

which yields after expanding and changing notation:

$$ds^2(dM) = \sum_{\mu\nu} g_{\mu\nu} dx^{\mu} dx^{\nu}, \tag{5.27}$$

where each $g_{\mu\nu}$ coefficient is given by:

$$g_{\mu\nu} = \vec{i'}_{\mu} \bullet \vec{i'}_{\nu}. \tag{5.28}$$

Bringing together relations (5.23), (5.25), (5.26) and (5.28), it appears that Schwarzschild metric has considerably simplified the metric relation: it has only four $g_{\mu\nu}$ coefficients, each one giving the norm of one basis vector, indeed:

- The coefficient of c^2dt^2 gives the norm² of the **time** basis vector:

$$ds^2\left(\vec{f'}\right) = \vec{f'} \bullet \vec{f'} = g_{00} = \left(1 - \frac{r_s}{r}\right). \tag{5.29}$$

- The coefficient of dr^2 gives the norm of the **radial** basis vector:

$$ds^2\left(\vec{i}'\right) = \vec{i}' \cdot \vec{i}' = g_{11} = \left(1 - \frac{r_s}{r}\right)^{-1}. \tag{5.30}$$

- The coefficient along the latitude and the longitude are the same in equations 5.23 and 5.25; hence, g_{22} and g_{33} are equal to 1. This is consistent with the fact that the transverse directions, i.e., the meridians and the latitudes, are not affected by the mass in O according to the universal isotropy postulate.

The fact that the $g_{\mu\nu}$ are null when $\mu \neq \nu$ means that each basis vector $\vec{f}', \vec{i}', \vec{j}', \vec{k}'$ is perpendicular to all others (the perpendicularity being with regard to the Lorentzian scalar product). Concerning $\vec{i}', \vec{j}', \vec{k}'$, this perpendicularity is due to the symmetry relative to the point O. The perpendicularity of the time basis vector, $\vec{f}'$, is a choice made by Schwarzschild to have an entire time unit. Besides, we notice that these $g_{\mu\nu}$ are not a function of the time, which is consistent with the universal homogeneity (invariance in time).

Let's further see the Time and the Radial Unit Basis Vectors.

5.2.2.1 The Radial Basis Vector $\vec{i}'$

The relation (5.30) shows that the norm of the Radial basis vector is a function of r, and it is greater than 1 when assuming $r > r_s$. However, when r is infinite, there is no norm increase, which is consistent with the absence of distortions. When r tends to r_s, the radial basis vector tends to the infinite, but this is not a physical singularity: it results from Schwarzschild's choice of the coordinates.

5.2.2.2 The Time Basis Vector $\vec{f}'$ and the Possibility of Black Holes

We can see from equation 5.29 that the norm of the time basis vector is a function of r, which is normal due to the way this time coordinate was defined. Since we consider cases where $r > r_s$, this relation shows that the norm of the time basis vector decreases with r, and that when r is infinite, we have: $g_{00} = 1$. This means that at the infinity, there is no norm reduction in the time basis vector, which is consistent with Schwarzschild's choice to set the time in the infinite at infinity as being the one of a perfect clock.

However, when r equals r_s, then: $g_{00}=0$, and this has important consequences: Consider an observer located at a point A whose radial coordinate is r_s. We mentioned that Schwarzschild's time in A reflects how an observer in A sees the proper time at infinity: 1 second at infinity thus represents zero second for him or her. Conversely, 1 second at A represents an infinite duration for observers at infinity. This is also true for observers between A and infinity, as we can show:

Consider an observer B with $r_B > r_s$. One second at the infinite is seen by B lasting $(1 - r_s / r_B)$ seconds. We mentioned that one second in A is seen at the infinite lasting an infinite duration, hence the latter is seen in B multiplied by $(1 - r_s / r_B)$, meaning that it is still an infinite duration.

Suitably, the sphere of circumference $2\pi r_s$ was called the "event horizon": the length of its circumference is 18.5 km multiplied by the black hole mass expressed in solar mass. When someone approaches the event horizon, his movements are seen from the outside as slower and slower; photons are received with larger and larger wavelengths; until no movement is seen, no photon is received. The interior of this sphere is a black hole (cf. Volume I Section 7.2).

The vanishing of g_{00} is connected to another important feature: at the event horizon, the nature of the radial vector changes from space-like to time-like. After crossing the event horizon, the radial geodesics are time-like. The gravitational force bringing objects to the "center" is time-like itself, and so the singularity is in the time direction. It is time itself that grags everything into the "center" where the laws of General Relativity don't apply: it is as if space-time were torn.

5.2.3 The Bending of Light Near a Great Mass

Light follows a geodesic of distance null, meaning $ds^2(dM) = 0$. Schwarzschild's metric then gives the equation of this geodesic:

$$ds^2(dM) = 0 = \left(1 - \frac{r_s}{r}\right)c^2dt^2 - \left(1 - \frac{r_s}{r}\right)^{-1} dr^2 - r^2 d\Omega^2 \tag{5.31}$$

with $r_s = 2GM/c^2$ and $d\Omega^2 = (\sin^2\theta \, d\varphi^2 + d\theta^2)$.

Consider a ray coming near the Sun O. We observe the scene from a frame K far from the Sun but fixed with it (Figure 5.3).

Figure 5.3 exaggerates this phenomenon for the sake of clarity. The OX axis is parallel to the direction of the ray when it comes from very far. Hence, the longitude angle φ is always null, and so equation 5.24 becomes:

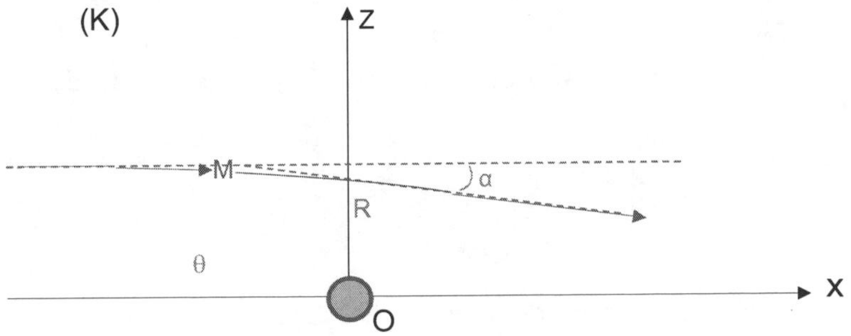

FIGURE 5.3
The bending of light near a great mass.

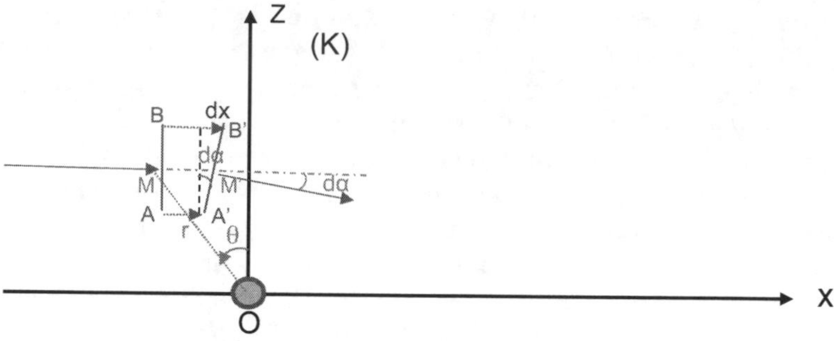

FIGURE 5.4
Zoom on the light wavefront.

$$0=\left(1-\frac{r_s}{r}\right)c^2dt^2-\left(1-\frac{r_s}{r}\right)^{-1}dr^2-r^2d\theta^2. \tag{5.32}$$

We already saw that light must bend in the vicinity of a great mass (cf. Volume I Section 6.2.1). To calculate the deviation angle α, let's make a zoom at the small area around a point M on the ray trajectory: the wavefront is always perpendicular to the ray motion; at the time t, it is represented by the segment AB, its lowest part being A and the highest one being B (Figure 5.4).

Then at the time t+dt, the wavefront is at the point M′, which is not on the tangent in M to the ray trajectory, but it makes a very small deviation angle dα with it. The wavefront in M′ is represented by A′B′. (We remind that the light speed is always c in free-falling frames, but this is not the case of the frame K where the ray is observed). The small segment A′B′ shows a deviation, BB′ being longer than AA′.

The small segment AB is denoted by dz, and the small variation (BB′ - AA′) by dx. Thus:

dz=AB and dx=BB′ - AA′.

We will make first-order approximations in $1/c^2$: we have: dx ≈ dz tg(dα), hence: dα ≈ dx/dz.

Besides: dx=[v(B)−v(A)] dt with v being the speed of the ray at the corresponding point.

Then, using the partial derivative, we have: $dx=\frac{\partial v}{\partial z}dz.dt$, so that: $d\alpha\approx\frac{dx}{dz}\approx\frac{\partial v}{\partial z}\cdot dt$. Hence:

$$\frac{d\alpha}{dx}=\frac{\partial v}{\partial z}\cdot\frac{dt}{dx}=\frac{1}{v}\cdot\frac{\partial v}{\partial z}. \tag{5.33}$$

We will obtain the complete deviation α by integrating:

$$\alpha=\int_{-\infty}^{+\infty}\frac{d\alpha}{dx}\cdot dx=\int_{-\infty}^{+\infty}\frac{1}{v}\cdot\frac{\partial v}{\partial z}\,dx. \tag{5.34}$$

To obtain $\frac{1}{v}\cdot\frac{\partial v}{\partial z}$, we will use Schwarzschild's relation (5.32), assuming $r \gg r_s$.

We first adapt it to the Cartesian coordinates of K, having:

$$r^2=z^2+x^2, \quad \cos\theta=\frac{z}{r}, \quad \sin\theta=-\frac{x}{r}. \tag{5.35}$$

We want to calculate: $\frac{\partial r}{\partial x}$ and $\frac{\partial\theta}{\partial x}$. From the first relation 5.35, we have: $\frac{\partial r}{\partial x}=\frac{x}{r}$.

From the 2nd relation 5.35 : $-d\theta.\sin\theta=-\frac{z}{r^2}\frac{\partial r}{\partial x}\,dx$.

Then: $d\theta\frac{x}{r}=-\frac{z}{r^2}\frac{x}{r}\,dx$, so: $\frac{\partial\theta}{\partial x}=-\frac{z}{r^2}$. Hence:

$$0=\left(1-\frac{r_s}{r}\right)c^2dt^2-\left(1-\frac{r_s}{r}\right)^{-1}\left(\frac{\partial r}{\partial x}\right)^2dx^2-r^2\left(\frac{\partial\theta}{\partial x}\right)^2dx^2$$

$$=\left(1-\frac{r_s}{r}\right)c^2dt^2-\left(1-\frac{r_s}{r}\right)^{-1}\frac{x^2}{r^2}dx^2-r^2\frac{z^2}{r^4}dx^2\text{, then:}$$

$$0\approx\left(1-\frac{r_s}{r}\right)c^2dt^2-\left[\left(1+\frac{r_s}{r}\right)\frac{x^2}{r^2}+\frac{z^2}{r^2}\right]dx^2=\left(1-\frac{r_s}{r}\right)c^2dt^2-\left[\frac{x^2}{r^2}+\frac{z^2}{r^2}+\frac{r_sx^2}{r^3}\right]dx^2$$

$$=\left(1-\frac{r_s}{r}\right)c^2dt^2-\left(1+\frac{r_sx^2}{r^3}\right)dx^2.$$

We then have the speed:

$$v^2\approx\frac{dx^2}{dt^2}=\frac{\left(1-\frac{r_s}{r}\right)c^2}{\left(1+\frac{r_sx^2}{r^3}\right)}\approx\left(1-\frac{r_s}{r}\right)c^2\cdot\left(1-\frac{r_sx^2}{r^3}\right)\approx c^2\left[1-\frac{r_s}{r}\left(1+\frac{x^2}{r^2}\right)\right].$$

So: $v\approx c\left[1-\frac{r_s}{2r}\left(1+\frac{x^2}{r^2}\right)\right]$ then:

$$\frac{1}{v}\frac{\partial v}{\partial z}\approx\frac{1}{2}r_s\left(3\frac{zx^2}{r^5}+\frac{z}{r^3}\right). \tag{5.38}$$

In order to use equation 5.34, we will resume using polar coordinates, and we denote by R the shortest distance from O to the ray. We have the following approximations:

$$z \approx R;\ r \approx R/\cos\theta;\ x \approx R\,\mathrm{tg}\theta.$$

Thus, from equations (5.34) and (5.36): $\alpha \approx \frac{r_s}{2R}\int_{-\pi/2}^{+\pi/2}(3\sin^2\theta + 1)\cos\theta\, d\theta.$

This integral is equal to:

$$\int_{-\pi/2}^{+\pi/2}\left(3\sin^2\theta\right)\cos\theta\, d\theta + \int_{-\pi/2}^{+\pi/2}\cos\theta\, d\theta = 3\left[\frac{1}{3}\sin^3\theta\right]_{-\pi/2}^{\pi/2} + \left[\sin\theta\right]_{-\pi/2}^{\pi/2} = 2 + 2.$$

So, finally:

$$\alpha \approx \frac{2r_s}{R} = \frac{4GM}{c^2R} \tag{5.39}$$ ■

Remark 5.5: At the time of Newton, light was assumed to be carried by very light massive particles. Hence, Newton could predict their deviation in the neighborhood of the Sun, which is half the deviation of Relativity. We now know that the photon is massless, but still in the Newtonian theory, the trajectory of an object is independent of its mass, like in Relativity. However, this Newtonian law has been established for massive particles, whereas General Relativity doesn't have this restriction.

5.2.3.1 Other Consequences

5.2.3.1.1 Gravitational Lenses

Figure 5.5 shows that light coming from a distant star and passing by a great mass before arriving at the observer's eye, appears to him or her with a greater angle than if there were no great mass in between. Moreover, this effect amplifies the flux received from the distant star.

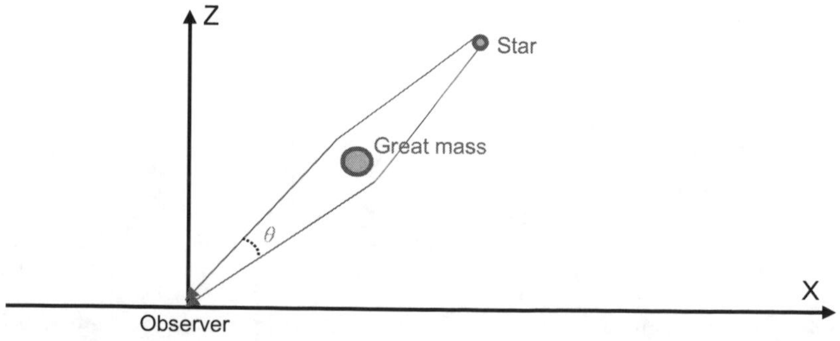

FIGURE 5.5
Gravitational lensing.

5.2.3.1.2 The Shapiro Effect

Let our observer in the point 0 on Figure 5.5 send a light beam toward the body (on the right), and then part of this beam is reflected by the surface of that body and comes back to our observer, like with a radar echo. The trajectories followed by the beam are longer than in the case where there is no deflection due to the presence of a great mass. Consequently, if the observer precisely knows the position of the distant body relative to him, the excess of time taken by the round trip of the beam compared to trajectories along straight lines enables us to calculate the great mass value.

5.2.4 Trajectories of Massive Objects and the ISCO Orbit

We saw in Section 3.6.6.2 that the movements of orbiting bodies are driven by relation (3.63):

$$\Delta\left(\frac{1}{2}mV_r^2\right) = -\Delta E_{eff} = -\Delta\left(\frac{-MG}{r} + \frac{L^2}{2r^2}\right).$$

This general principle remains valid, but one must replace the Newtonian potential with the one of General Relativity, which leads to adding the term $-\frac{r_s L^2}{2 r^3}$, where r_s is the Schwarzschild radius. Thus, the total effective potential becomes:

$$U_{eff} = \frac{-MG}{r} - \frac{r_s L^2}{2 r^3} + \frac{L^2}{2 r^2} = \frac{-c^2 r_s}{2 r} - \frac{r_s L^2}{2 r^3} + \frac{L^2}{2 r^2} \qquad (5.40) \blacksquare$$

The relativistic relation yields very close results to the classical one in the cases of moderate gravity such as the planets in the solar system. The additional term, $-\frac{r_s L^2}{2r^3}$, becomes important when the object is very close to the center of the great mass, acting as an increased gravitational attraction due to the increased inertia arising from the kinetic energy associated with the angular momentum. As a result in extreme cases, the centrifugal force becomes insufficient to offset gravity: orbiting objects fall in the great mass, which was impossible with Newton.

For such an event to occur, the object must be closer than the "innermost stable orbit radius," denoted by ISCO, and which is equal to $\mathbf{3r_s}$ using the Schwarzschild metric (and assuming the great mass is not rotating). Indeed, if an object comes closer than its ISCO orbit, its trajectory becomes unstable and likely to be an inward spiral toward the center of the great mass, which will ultimately absorb the object. The ISCO orbit plays an important role in the formation of accretion disks around black holes.

In the case of the Sun, the ISCO orbit would be at a distance of ~9 km from the center of the Sun, which is far less than its surface; hence, no orbiting object can fall into the Sun. However, they can in the case of black holes or even neutron stars.

An object orbiting at the ISCO radius has a speed of c/2, independently of the black hole mass, and it experiences a gravity that is twice the Newtonian result as shown in Problem 5.10.

5.3 Questions and Problems

Problem 5.1) Show that the $\overrightarrow{\text{curl}}$ of the $\overrightarrow{\text{grad}}$ of any scalar field φ is always null: $\overrightarrow{\text{curl}}\left(\overrightarrow{\text{grad}}\ (\varphi)\right)=0$.

Problem 5.2) Show that Divergence of a $\overrightarrow{\text{curl}}$ is always null: $\text{div}(\overrightarrow{\text{curl}})=0$.

Problem 5.3) Demonstrate the first classical Maxwell Equation from the Relativistic Maxwell Equations. This equation is: $\mathbf{div}(\vec{E})=\frac{\rho_0}{\varepsilon_0}$.

Problem 5.4) Same as 5.3) but for the second classical Maxwell Equation: $\overrightarrow{\text{curl}}\ \vec{E}=-\frac{dB}{dt}$.

Problem 5.5) Same as 5.3) but for the third Maxwell Equation: $\overrightarrow{\text{Curl}}\ \vec{B}=\mu_0 j+\varepsilon_0\mu_0\cdot\frac{dE}{dt}$.

Problem 5.6) Same as 5.3 but for the fourth Classical Maxwell Equation: $\overrightarrow{\text{curl}}\ \vec{E}=-\frac{dB}{dt}$.

Problem 5.7) The Relativistic potential when changing frame

Q1) Express the relativistic potential **A** in an inertial frame K′ going at the speed s along the OX axis of K, knowing that its components in K are $\left(\frac{\Phi}{c}, A^x, A^y, A^z\right)$. We assume both frames are orthonormal (frames of Minkowski) and that the space has no curvature.

Q2A) Show that we have: $\Phi^2-(A^x)^2-(A^y)^2-(A^z)^2=\text{cst}$ whatever the inertial frame.

Q2B) Is this cst also constant in time?

Problem 5.8) Schwarzschild and Black Hole

Q1) What is the radius of the Event Horizon of a black hole of 10 solar masses?

Q2) An observer at a point A is located at the radius: $r_s+10\%$. He or she has two clocks: the first one gives the Schwarzschild time at A and the second one gives the proper time at A (e.g., the atomic clock giving the official second). When the second clock beats 1 second, how many units of the Schwarzschild time have beat the first clock?

Q3) This observer emits photons with the frequency ν_A toward the infinite. What is the frequency received by an observer at the infinite?

Problem 5.9) Black Hole – Approaching the Danger Part 1

Consider again a black hole of 10 solar masses. *Note: the planet's mass being irrelevant, use quantities normalized by the mass: E/m, L/m, etc.*

Q1) What is the gravitational field at the event horizon, applying the classical Newtonian law? Does General Relativity yield a greater or a lower value?

Q2) What is the gradient of the gravitational field at the event horizon (still applying the Newtonian law)? What is the gravity difference between two points separated by 1 m in the radial direction? Does General Relativity yield a greater or a lower value? Please comment.

Q3A) What is the value of the ISCO radius?

Q3B) Suppose that a planet having the same angular momentum as the Earth was orbiting at the ISCO radius; how much time would it take to make one orbital rotation?

Q3C) What would be its speed, still using the classical laws?

Q3D) Adapt the expression of the angular momentum to relativity, and then calculate the planet's speed.

Problem 5.10) Black Hole – Approaching the Danger Part 2

Q1) What is the effective potential of an object having a circular trajectory at the ISCO radius?

Q2A) Deduce the angular momentum, and then the speed of such an object.

Q2B) What is the difference between the relativistic acceleration at the ISCO radius and the Newtonian result?

Q3) Can an object with the angular momentum of the Earth reach the ISCO? Can it fall into the black hole?

Notes

1 Cf. more in Section 5.1.6.3.

2 This operator doesn't have a standard name; it may be called differently in some books.

3 Clairaut's theorem indeed shows that $\frac{\partial^2 f}{\partial x \partial y} = \frac{\partial^2 f}{\partial y \partial x}$.

4 Cf. Problem 4.6.

5 The demonstration of this relation is out of the scope of this document.

6 Cf. Problem 4.5.

7 Cf. Problem 4.4.

8 Anywhere also, but those in the vicinity of M will benefit from this frame.

6

Introduction to Cosmological Models

Introduction

This chapter presents the principles on which model universes are built. The geometry of a homogeneous and isotropic universe is discussed in the general context of non-Euclidean 4D spaces, leading to the Friedmann–Lemaître–Robertson–Walker metric, and to the important relation between the redshift and the scale factor. Then, the fundamental Friedmann equation will be presented; its consequence regarding the relationship between the space curvature and the energy density will be shown, leading to the recognition that our universe must be flat or almost so. Then, by modeling the universe as a fluid, the important fluid and acceleration equations will be derived, expressing the relations between the scale factor, the energy density, the pressure and their evolutions. The introduction of the cosmological constant "lambda" will be discussed, enabling the theory to explain the observations of the acceleration of the universe expansion. Finally, four ideal scenarios will be analyzed, comprising the main phases of the universe: radiation only, matter only, empty universe, "lambda" only.

*

When General Relativity was conceived, the known universe was confined to the Milky Way. In the early 1920s, other galaxies were found, starting with our closest one, Andromeda. Since then, their number has always increased: the current estimated number is 2,000 billion!

On very large scales, meaning above 100 Mpc (1 Megaparsec ≈ 3.26 Million light-years), the universe appears to be homogeneous and isotropic. The same quantity of superclusters of galaxies and voids is observed in all directions and at all distances. According to the Copernician principle, the Earth is not in a privileged location; hence, the same repartition of the galaxies should be observed from any point in the universe. Any point can consider itself as the center of the world.

As more galaxies were discovered, E. Hubble noticed in 1924 that the more distant the galaxy, the greater its redshift. This was first interpreted by the Doppler effect, which implied that the galaxies were moving away from us at a speed that is proportional to their distance, leading to the relation:

DOI: 10.1201/9781003201359-6

$$V = H_0 r, \tag{6.1}$$

where r is the distance of a galaxy to us, V its speed and H_0 the Hubble constant, which was estimated to be $H_0 \approx 500$ (km/s)/Mpc after analyzing the results from 42 galaxies. Since then, different methodological errors were corrected, and more galaxies observed with more accurate instruments, leading to the same conclusion of the existence of a linear relation between V and r, but with a different Hubble constant, which is now: $H_0 \approx 70$ (km/s)/Mpc.

The redshift is characterized by the ratio: $z = \frac{\lambda_O - \lambda_e}{\lambda_e}$, with λ_e being the wavelength of the photons emitted by the galaxy, and λ_O their wavelength observed on Earth. An extreme example is the galaxy named " GN-11 ", its observed redshift is z=11.1, and its light was emitted 13.4 billion years ago (400 million years only after the Big-Bang).

However, this interpretation based to the galaxy speed was replaced soon after by the assertion that the universe is expanding: A. Friedmann realized in 1922 that General Relativity equations suggest that the world is expanding, otherwise their solutions are unstable. This was the first seeds of the Big Bang theory, although Friedman didn't accept such an extrapolation. After he passed away (1925), G. Lemaître pursued this idea and showed that the Hubble redshift was due to the expansion of the universe. Noteworthy, this vision also satisfies the Copernician principle, otherwise the Earth would be the center from which all galaxies move away. Lemaitre's 1931 proposal didn't receive much theoretical support until some of the other necessary ingredients, like stellar structure and nucleosynthesis, began to be developed.

Much later in 1965, the cosmic microwave background was discovered, providing an amazing confirmation of the extremely homogeneous and isotropic aspect of our universe (cf. Volume I Section 7.1).

6.1 The Friedmann–Lemaître–Robertson–Walker Metric

H.P. Robertson and A.G. Walker pursued Friedmann's and Lemaitre's work, independently of each other, and wondered: "What form can the metric of space-time assume if the universe is spatially homogeneous and isotropic at all times, and if distances are allowed to expand or contract as a function of the time?" This led to the Friedmann–Lemaître–Robertson–Walker (FLRW) metric that we will see now. We will first see the general characteristics of a homogeneous and isotropic space, which will enable us to derive the FLRW metric:

6.1.1 The General Shape of a Homogeneous and Isotropic Universe

Is the world finite or infinite? The question is still open. In case of a finite world, the representation of a homogeneous and isotropic universe is not straightforward: one cannot for instance represent its 3D spatial part by a 3D sphere (a ball) because it has an edge that breaks homogeneity and a center that beaks both homogeneity and isotropy (the direction from any point to the center being privileged).

However, we can have a spatial representation in 2D of a universe in expansion by supposing that it is a sphere which is inflated like a balloon.

One may think that homogeneity and isotropy imply that the space is Euclidean, but this is not necessarily the case. For instance, if we again imagine a 2D world such as the surface of a sphere, this world is isotropic and homogeneous, but not Euclidean.

One can prove that the surface of a sphere is not a Euclidean space by relatively simple means. Its North Pole, denoted by the point P, can be arbitrarily chosen. We draw a triangle on the sphere as follows: we mark a portion of the equator from a point A to a point B on the equator, then the portions of the meridians from A to P, and the meridian from B to P. Seen by an observer on the sphere, these three lines form a triangle, but two characteristics enable him to realize that he is not on a Euclidean space:

- The angle between the meridian AP and the equator is 90°; the same is true for the angle between the meridian BP and the equator. Hence, the observer on the sphere concludes that these two meridians are parallel, both being perpendicular to the line AB, but they meet at the pole P, contradicting a Euclidean postulate.
- The observer on the sphere notices that in the triangle formed by the points A, B and N, the sum of the three angles is greater than 180°, since the two sides AP and BP are perpendicular to AB.

This is actually a general way to determine if any space is Euclidean or not: we make a triangle formed by three geodesics (which is the general notion covering the straight line), then if the sum of the three angles is different from 180°, it means that the space is not Euclidean.

More specifically, let's denote the three angles by α, β and γ, and we will show that:

$$\alpha+\beta+\gamma = \pi + A/R^2 \tag{6.2}$$

with A being the surface of this triangle ABP and R the radius of the sphere.

In the above case, we have: $A = 4\pi R^2 \dfrac{\alpha}{2\pi}\dfrac{1}{2} = \alpha R^2$; hence: $A/R^2=\alpha$.

Consequently, equation 6.2 is satisfied as: $\alpha + \pi / 2 + \pi / 2 = \pi + \alpha$.

In the general case, non-Euclidean spaces are not always spherical, but the sum $(\alpha+\beta+\gamma)$ is different from π.

- The space is said positively curved if this sum is greater than π (e.g., the sphere): $\alpha + \beta + \gamma = \pi + A/R^2$.

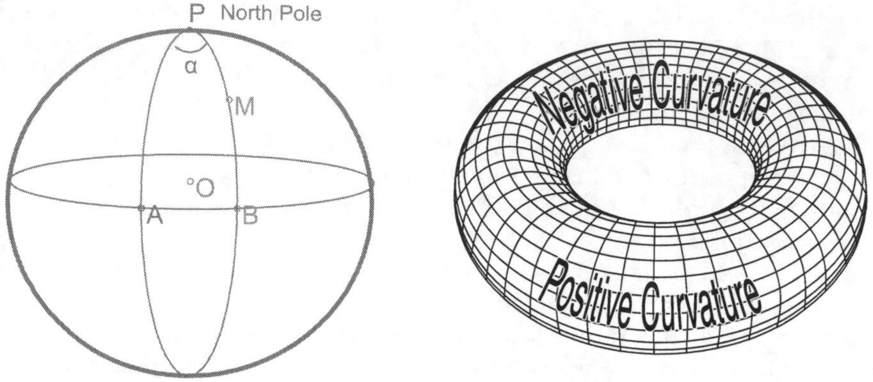

FIGURE 6.1
Surfaces with different curvature: sphere and torus.

- The space is said negatively curved if this sum is less than π : $\alpha + \beta + \gamma = \pi - A/R^2$.

A surface like a torus has both a negatively curved part (inner side) and a positive one (outer side), as shown in the right part of Figure 6.1.

6.1.2 The FLRW Metric

We will now examine how the distance can be assessed in the general case of a curved space; we will start with the relatively simple case of a 2D curved surface:

6.1.2.1 Classical Distance in a 2D Homogeneous and Isotropic Curved Space

Let's assess the distance between two neighboring points M and N on a sphere of radius R, based on the following system of coordinates:

- The curvilinear length from the Pole P to the point M is denoted by r.
- One meridian is arbitrarily chosen as being the reference one: for instance, let's take the meridian passing by the point A in our above figure.

The point M can then be characterized by two parameters: r and the angle α to the reference meridian. Likewise, N is characterized by: $(r+dr, \alpha+d\alpha)$. We will show that the distance MN is assessed by the relation:

$$MN^2 = dr^2 + R^2 \sin^2(r/R)\, d\alpha^2 .$$

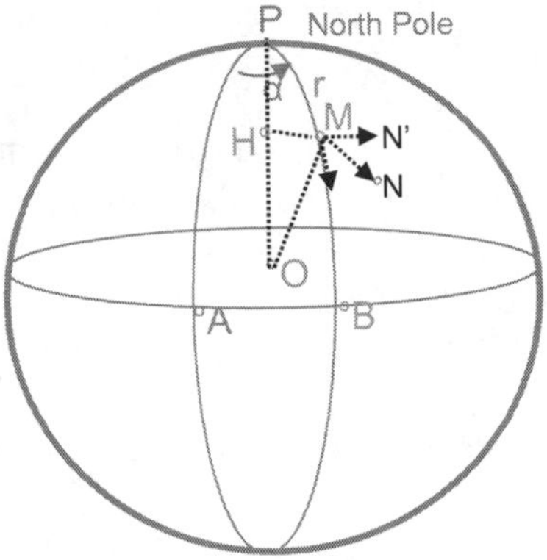

FIGURE 6.2
Distance on a positively curved surface.

Let's denote by N′ the point characterized by (r , α+dα). We have:

$$MN^2 = dr^2 + MN'^2.$$

The plane passing through M and parallel to the equator crosses OP at the point H. We have: MN′=MH.dα. Also: MH=R sin(OH, OM)=R.sin(r/R) ; hence:

$$MN' = R\sin(r/R)d\alpha\,. \tag{6.3}$$

So finally:

$$MN^2 = dr^2 + R^2 \sin^2(r/R)\, d\alpha^2 \;\blacksquare$$

Remark 6.1: The choice of the point P as the North Pole is arbitrary, but once it is taken, the South Pole and the Equator are automatically derived. Note that this choice induces artificial singularities at the Poles, which do not reflect the homogeneity-isotropy of the surface.

In the case of a uniform negative curvature, we have: MN′=R sinh(r/R) dα .

6.1.2.2 First Consequence: Limits in the Curvature of our Universe

In the case of a positively curved surface, we can see that MN′ is smaller than in the Euclidean case where we would have: MN′=r α . By contrast if the surface were negatively curved, we would have MN′>r α .

Let's return to positively curved surfaces, and let's imagine that an observer is at the pole P and looks at a distant galaxy whose diameter is D. The relation (6.3) tells us that he sees the galaxy with an angle:

$$d\alpha = \frac{D}{R\sin\left(\frac{r}{R}\right)}\,. \tag{6.3a}$$

This angle is greater than if the universe were not curved, as we would have had: $d\alpha=D/r$, and $\sin(x)$ is always inferior to x. Thus, a positively curved space acts as a magnifying lens.

This relation (6.3a) also shows that if the distance r is such that $r/R=\pi$, then $d\alpha$ is infinite: we would see the galaxy all around us! As this is not the case, and as the furthest galaxies observed are at a distance $r \approx c/H_0$, we can deduce that $c/H_0R \ll \pi$, so the current radius of curvature must be: $R \gg cH_0\pi$.

In the case of a negatively curved universe, the same reasoning shows that the universe curvature acts like a reducing lens: $d\alpha = \dfrac{D}{R\sinh\left(\dfrac{r}{R}\right)}$ (since $\sinh(x)$ is always greater than x).

In the case where $r \gg R$, we have: $\sinh(r/R) \approx \dfrac{e^{r/R}}{2}$; hence: $d\alpha \approx 2D\,\dfrac{e^{-r/R}}{R}$.

Thus, the angle $d\alpha$ exponentially decreases with the distance r, and since extremely distant galaxies can be observed with a distance that is in the range of c/H_0, the curvature radius must be greater than c/H_0. With the current estimate of H_0, this distance is $\approx$ 13.9 billion light-years.

6.1.2.3 *Application to a 3D Homogeneous and Isotropic Curved Space and the FLRW Metric*

We saw in Section 5.2.1.2 that if the space has no curvature, the spatial distance between two neighboring points, denoted by $dl^2(dM)$, is:

$$dl^2(\mathbf{dM}) = dr^2 + r^2\left(\sin^2\theta\, d\varphi^2 + d\theta^2\right) = \mathbf{dr}^2 + \mathbf{r}^2\mathbf{d\Omega}^2.$$

If the space has a uniform positive curvature, the previous reasoning in 2D can be extended to 3D, leading to: $\left(\sin^2\theta d\varphi^2 + d\theta^2\right) = \mathbf{dr}^2 + \mathbf{r}^2\mathbf{d\Omega}^2$, where R is the radius of curvature.

If the space has a uniform negative curvature, the distance is:

$$dl^2\,(dM) = dr^2 + R^2\sinh^2(r/R)\,d\Omega^2\,.$$

Thus in the general case of a homogeneous-isotropic space, the metric is classically denoted by:

$$dl^2(dM) = dr^2 + S_k(r)^2\,d\Omega^2 \tag{6.4}$$

where $S_k(r)$ is :

- If uniformly positive curvature: k=+1, and $S_k(r) = R\sin(r/R)$.
- If no curvature or flat (Euclidean) universe: k=0, and $S_k(r) = r$.
- If uniformly negative curvature: k=−1, and $S_k(r) = R\sinh(r/R)$.

We can see that if $r \ll R$, then $S_k(r) \approx r$. If the space is positively curved, $S_k(r)$ increases with r until: $r/R=\pi/2$, reaching $S_{max} = R$.

The FLRW metric takes into account that the universe is a 4D space-time with a Lorentzian metric, which is the metric of Minkowski in case of a flat universe (cf. Section 5.2.1.2):

$$ds^2(dM)=c^2\,dt^2-dr^2-r^2\,d\Omega^2.$$

Assuming that the universe is homogeneous, isotropic and uniformly curved, the distances are a function of the time, and they all change according to a same factor, called the scale factor: a(t). By convention, the present time is denoted by (t_0) and still by convention $a(t_0)=1$. More generally, the subscript "0" means the current state.

The FLRW metric then reads:

$$\mathbf{ds^2(dM) = -\,c^2dt^2 + a(t)^2\left[dr^2 + \ S_k(r)^2\,d\Omega^2\right].} \tag{6.5}$$

The time of the FLRW metric is called the cosmic time: it is the proper time of a fundamental observer who sees the universe expanding uniformly around him. A cosmological principle states that this time is the same for all fundamental observers: their clocks beat at the same pace (proper time universality), and furthermore, their times are synchronized, which makes an exception to the general rule of General Relativity: indeed, at the same cosmic time t, which is the same for all fundamental observers, they can see the universe at the same state of evolution, with the same density of matter and energy, the same expansion rate, and the same scale factor.

Remark 6.2: In cosmology, the fundamental observer is a passive (inertial) observer at rest relative to the cosmic fluid. The matter-energy-void in the universe is considered as a smooth, idealized fluid (also called substratum); its properties are described similarly as those of a gas, by macroscopic quantities, such as density, pressure, and temperature, without specifying the behavior of each individual component (atoms, molecules).

The spatial parameters are called the "comoving" coordinates (r, φ, θ). If the homogeneity and isotropy are perfect, then the comoving coordinates of a physical point, which is dragged by the universe expansion, are constant. A particle that is moving relative to such a physical point has a "peculiar" motion.

This metric is relatively simple and powerful as it describes the universe with very few parameters. However, one should bear in mind that the assumption of homogeneity and isotropy concerns huge distances only (>100 Mpc). For instance, at the scale of our galaxy (20,000 pc), the expansion rate is much less than the universal one, due to the gravitational forces between the numerous astral bodies of our galaxy.

The relation (6.5) can also be written with different systems of coordinates; we can choose in particular: $x = S_k(r)$, then we have:

$$ds^2(dM) = -c^2dt^2 + a(t)^2\left[\frac{dx^2}{1-kx^2/R^2} + dx\,^2d\Omega^2\right]. \tag{6.6}$$

6.1.3 Relation between the Redshift and the Scale Factor

6.1.3.1 The Proper Distance with the FLRW Metric, and the Apparent Recession Speed

The proper distance is given by measuring points (events) that are simultaneous: $dt=0$. Besides, for the sake of clarity and simplicity, let's consider points (events) that are along the dr line, meaning $d\theta=0=d\varphi$, so that $d\Omega=0$.

We then have from equation 6.5: $ds(t)=a(t)\,dr$, then for a galaxy located at a longer distance but still along the same comoving coordinate r, the proper distance is:

$$d = a(t)\int_0^r dr = a(t)\,r. \tag{6.7}$$

The time rate of change of this proper distance is: $\dot{d} = \dot{a}(t)r$. We indeed assume in the FLRW model that the comoving coordinate r of the galaxy is constant. Then, equation 6.6 yields:

$$\dot{d} = \frac{\dot{a}(t)}{a(t)}\,d. \tag{6.8}$$

This relation (6.8) shows that there is a linear relation between $\dot{d}$, which is the perceived recession speed of the galaxy, and its distance. This actually matches with Hubble's observations that led to equation 6.1. We denote by t_0 the current time of these observations, hence we have:

$$\frac{\dot{a}(t_0)}{a(t_0)} = H_0 \approx 70(\text{km/s})/\text{Mpc}. \tag{6.9}$$

Thus, the linear relation between the recession speed and the distance is:

$$\dot{d} = H_0\,d. \tag{6.10}$$

One consequence of equation 6.10 is worth noticing: $\dot{d}$ can be greater than c, which is the case of galaxies that are further than $\frac{c}{H_0}$, meaning 4,283 Mpc (with the current estimate of H_0). Note that this does not contradict the Causality principle nor Relativity since there is still no possibility of exchange of information at a greater speed than c.

6.1.3.2 Relation between the Redshift and the Scale Factor

At the cosmological time t_0, a fundamental observer sees a galaxy. He receives photons that were emitted at the cosmological time t_e, and notices a certain redshift $z = \frac{\lambda_0 - \lambda_e}{\lambda_e}$.

We will show that there is a relation between the redshift z and the ratio between the scale factor at the time of observation and emission:

The photon pursues a null geodesic (light-like) between the galaxy and the observer. We can choose the spatial origin to be the observer's location at t_0, and the r coordinate to be along this spatial geodesic. Thus: $d\theta = d\varphi = 0 = d\Omega$. We then have from equation 6.5:

$ds^2(dM) = 0 = -c^2dt^2 + a(t)^2[dr^2 + S_k(r^2)\, d\Omega^2]$. This gives: $c\, dt = a(t)\, dr$, hence:

$$dr = \frac{c\,dt}{a(t)}. \tag{6.11}$$

A wave crest of a photon emitted at t_e and received at t_0 satisfies:

$$\int_{t_e}^{t_0} \frac{c dt}{a(t)} = \int_0^r dr = r.$$

Let's consider the next wave crest: it is emitted at the time $t_e + \lambda_e / c$ and received at $t_0 + \lambda_0 / c$.

We then have: $\int_{t_e + \lambda_e/c}^{t_0 + \lambda_0/c} \frac{c\,dt}{a(t)} = r$ (the comoving coordinate r of the galaxy is constant).

We thus have: $\int_{t_e + \lambda_e/c}^{t_0 + \lambda_0/c} \frac{c\,dt}{a(t)} = \int_{t_e}^{t_0} \frac{c\,dt}{a(t)}$.

Let's subtract $\int_{t_e + \lambda_e/c}^{t_0} \frac{c\,dt}{a(t)}$ to both sides, and we obtain:

$$\int_{t_0}^{t_0 + \lambda_0/c} \frac{c\,dt}{a(t)} = \int_{t_e}^{t_e + \lambda_e/c} \frac{c\,dt}{a(t)}.$$

We can assume that during the very short time-lapse $[t_0, t_0 + \lambda_0/c]$ the universe expansion was negligible, and the same during the very short time-lapse $[t_e, t_e + \lambda_e / c]$; hence, we have: $\frac{c\left(\frac{\lambda_0}{c}\right)}{a(t_0)} = \frac{c\left(\frac{\lambda_e}{c}\right)}{a(t_e)}$, So finally:

$$\frac{\lambda_0}{a(t_0)} = \frac{\lambda_e}{a(t_e)} \text{ or : } \frac{\lambda_0}{\lambda_e} = z + 1 = \frac{a(t_0)}{a(t_e)}.$$

With the commonly used convention that $a(t_0) = 1$, we obtain:

$$\mathbf{z = \frac{1}{a(t_e)} - 1 \quad or: \quad a(t_e) = \frac{1}{Z+1}} \tag{6.12}$$ ■

For instance, if the observed redshift is 1.5, then the scale factor at emission was: $\frac{1}{2.5} = 0.40$.

Regarding the farthest galaxy observed to date, GN-11, the scale factor was: $1/(11.1+1) = 0.09$.

Noteworthy, the redshift is only a function of the scale factor at the time of emission and the one of observation. The way the universe has expanded in between has no influence.

6.2 The Friedmann Equation

We will first present the principles of the Friedmann equation in classical physics, and we will then adapt the result to General Relativity in a qualitative manner.

6.2.1 Friedmann Equation with Classical Physics

We assume the universe is a sphere (ball) whose radius R(t) may change with time, but its total mass M remains constant. We imagine a small object of mass m on the edge of the ball. This object experiences the Newtonian gravitational force: $F = -\frac{mMG}{R(t)^2}$. Then, the second Newtonian law gives its acceleration: $\ddot{R}(t) = -\frac{MG}{R(t)^2}$. We indeed assume that the small mass remains at the edge of the universe.

Multiplying both terms by $\dot{R}(t)$, then integrating yields: $\frac{1}{2} \cdot \dot{R}^2(t) = \frac{MG}{R(t)} + U$, with U being a constant. Let's express the mass M as a function of mass density $\rho(t)$ of the ball: $M = \rho(t)V(t) = \frac{4\pi}{3}\rho(t)R(t)^3$.

We saw that R(t) is a function of the scale factor, $R(t) = R_0\, a(t)$. Hence:

$$1/2\dot{a}(t)^2 R_0^2 - \frac{4\pi}{3}G\rho(t)a(t)^2 R_0^2 = U.$$

We divide both terms by ½ $a(t)^2R_0^2$, and obtain the Friedmann equation in the Newtonian form:

$$\left(\frac{\dot{a}(t)}{a(t)}\right)^2 = \frac{8\pi}{3}G\rho(t) + \frac{2U}{a(t)^2 R_0^2} \quad \blacksquare \tag{6.13}$$

This equation permits three different fates for the universe, depending on the sign of U:

- If U is positive, $\dot{a}$ is always positive, meaning that the universe is expanding forever.
- If U is negative, the right term will be null when $a(t) = \frac{GM}{UR_0}$, and so

  the universe stops expanding at that time; it will then contract since $\ddot{a}$ is negative.
- If U=0, the universe is always expanding; its expansion rate decreases, since $\rho(t)$ diminishes, and tends to zero at the infinite.

Remark 6.3: This situation is analogous to the one encountered when an object is launched upward from the surface of the Earth: either its initial speed is beyond the escape velocity and it will escape the gravity field, or it will return and fall back, or it will stop at the infinite.

6.2.2 Adaptation to General Relativity and First Consequences

The previous classical model raises some issues, some not only with regard to Relativity:

- it assumes the universe is spherical and Euclidean, but we saw that the spherical shape contradicts the homogeneity-isotropy postulate.
- Another difference is that General Relativity showed that gravitation is replaced by space-time deformations, and that their cause is the energy (i.e., the time part of the Momentum) and not only the mass. Hence, the mass density $\rho(t)$ needs to be replaced with the energy density: $\varepsilon(t)/c^2$. (The division by c^2 gives to the energy the dimension of the mass).

The Friedmann equation based on General Relativity thus becomes:

$$\left(\frac{\dot{a}(t)}{a(t)}\right)^2 = \frac{8\pi G}{3c^2}\varepsilon(t) - \frac{kc^2}{R_0^2}\frac{1}{a(t)^2}. \qquad (6.14)$$

We can see that $\frac{2U}{R_0^2}$ has become $-\frac{kc^2}{R_0^2}$, which means that the classical case

with a positive U corresponds to a negatively curved space, a negative U to a positive curved and a zero U to a flat (Euclidean) universe.

6.2.2.1 The Current Situation and the Relation with the Hubble Constant

Regarding the current state, we saw with the relation (6.9) that: $\left(\frac{\dot{a}(t_0)}{a(t_0)}\right) = H_0$.
Hence at the present time the Friedmann equation is:

$$H_0^2 = \frac{8\pi G}{3c^2}\varepsilon(t) - \frac{kc^2}{R_0^2}\frac{1}{a(t)_0^2} \approx 4{,}900\ (\text{km/s})^2/\text{Mpc}^2 \qquad \blacksquare$$

6.2.2.2 Relation between Space Curvature and Energy Density: The Critical Energy Density

In case of no space curvature, the Friedmann equation becomes:

$$H(t)^2 = \frac{8}{3c^2}G\pi\varepsilon(t).$$

The energy density $\varepsilon_c(t) = \frac{3c^2}{8G\pi}H(t)^2$ has then been called the **critical energy density**: it is the energy density required for having a flat universe. If the energy density is greater than this critical value, k is positive and the universe is positively curved. Conversely, if it is less, the universe is negatively curved.

With the current estimate of H_0, the current critical energy density is: 9.2×10^{-27} kg/m^3.

This is an extremely low density compared to our standards on Earth and even considering the Milky Way; but one should bear in mind that the universe is mainly composed of intergalactic voids. Amazingly, the current energy density estimate is extremely close to this critical value with an uncertainty of only 0.5%! This is another indication that the universe is either flat, or very close.

It is often useful to express the energy density $\varepsilon(t)$ by the ratio $\Omega(t) = \frac{\varepsilon(t)}{\varepsilon_c(t)}$.

The Friedmann equation (6.14) becomes: $1 - \Omega(t) = -\frac{kc^2}{R^2(t)\,H(t)^2}$.

We can see that the right-hand side always has the same sign; hence, if the energy density is lower than the critical one at any point of time, it means that it is always so. It also means that the universe curvature always remains with the same type.

6.2.3 The Fluid Equation

We may model the universe as a fluid (cf. Remark 6.2); hence, we are induced to apply the first principle of thermodynamics as for a gas: $dQ = dE + PdV$, with E being the fluid internal energy, P its pressure and V its volume. It is assumed that the universe doesn't exchange heat with its exterior (since there is no exterior); hence:

$$\dot{E} + P\,\dot{V} = 0\,. \tag{6.15}$$

The fluid internal Energy at the time t is: $E(t) = \varepsilon(t)\ V(t)$, where $\varepsilon(t)$ is the energy density that was previously used.

Regarding V(t), if we assume the universe is a sphere of comoving radius r_s , its radius at any time is a function of the scale factor: $R(t) = a(t)\ r_s$. Thus, the sphere volume is: $V(t) = \frac{4}{3}\pi r_s^3 a(t)^3$.

Its derivative is: $\dot{V}(t) = \frac{4}{3}\pi r_s^3\, 3a(t)^2\, \dot{a}(t)$, then:

$$\frac{\dot{V}(t)}{V(t)} = 3\,\frac{\dot{a}(t)}{a(t)}\,V(t)\,. \tag{6.16}$$

Then, having $E(t) = \varepsilon(t)V(t)$ we obtain:

$$\dot{E}(t) = \dot{\varepsilon}(t)\ V(t) + \varepsilon(t)\,\dot{V}(t) = V(t)\left[\dot{\varepsilon}(t) + 3\,\varepsilon(t)\frac{\dot{a}(t)}{a(t)}\right].$$

Consequently, equation (6.15) becomes: $V(t)\left[\dot{\varepsilon}(t) + 3\varepsilon(t)\frac{\dot{a}(t)}{a(t)} + 3P\frac{\dot{a}(t)}{a(t)}\right] = 0.$

We finally obtain:

$$\dot{\varepsilon}(t) + 3\varepsilon(t)\frac{\dot{a}(t)}{a(t)} + 3P\frac{\dot{a}(t)}{a(t)} = 0 \tag{6.17}$$

which is the important "fluid equation." It can also be written:

$$\dot{\varepsilon}\,\frac{a}{\dot{a}} = -\,3\,(\varepsilon + P) \tag{6.18}$$ ■

6.2.4 The Acceleration Equation

We will use the Friedmann equation (6.14) in conjunction with the fluid equation:

Let's multiply equation 6.14 by a^2: $\dot{a}^2 = \frac{8}{3c^2}a^2 G\pi\varepsilon(t) - \frac{kc^2}{R_0^2}$, then: let's make a time derivative: $2\,\dot{a}\ddot{a} = \frac{8G\pi}{3c^2}\left(a^2\,\dot{\varepsilon} + 2\,a\dot{a}\varepsilon\right)$.

Dividing by 2 $a\dot{a}$ yields: $\frac{\ddot{a}}{a} = \frac{4G\pi}{3c^2}\left[\frac{a}{\dot{a}}\dot{\varepsilon} + 2\varepsilon\right]$.

Then with equation 6.18, we finally obtain the "Acceleration Equation":

$$\frac{\ddot{a}}{a} = -\frac{4G\pi}{3c^2}\left[\varepsilon + 3P\right] \blacksquare \tag{6.19}$$

We notice that we have an important problem: this equation implies that the universe expansion must be decelerating, since the matter or radiation that we know have a positive energy and a positive pressure (due to the spatial Momentum of the particles, massive or not). However, this expansion deceleration is in contradiction with observations.

The solution to this important problem is still subject of research. The current direction is to imagine the existence of a matter-energy, called the "dark energy," such that $P < -\frac{1}{3}\varepsilon$. A possible explanation for this new component lies in the cosmological constant, denoted by Λ (lambda), that was first issued by Einstein in 1917 with the purpose of making his field equation compatible with a static universe.

6.3 Cosmological Models

6.3.1 The Cosmological Constant

The underlying principle behind the cosmological constant also applies with Newtonian gravitation, where calculations are simpler: if we add a constant Λ in the Poisson equation (cf. Section 3.6.6.1), we obtain: $\Delta^2\Phi + \Lambda = 4\pi G\rho$, then if $\Lambda = 4\pi G\rho$, we have: $\Delta^2\Phi = 0$, meaning that the universe is static.

Indeed, $0 = \Delta^2\Phi = \mathrm{div}(\overrightarrow{\mathrm{grad}\Phi}) = -\mathrm{div}(\vec{g})$; then the Ostrogratski theorem shows that the flow of $\vec{g}$ over the surface of the sphere is null, meaning $g=0$ on the sphere surface (assuming spherical symmetry).

Einstein noticed that a constant term could be added in his field equation, and hence suggested to do so. His equation gave for the cases of very weak fields:

$$\Delta^2\Phi = 4\pi G\rho - \Lambda. \tag{6.20}$$

Hence, in order for the universe to be static, he suggested: $\Lambda = 4\pi G\rho$.

Equation 6.20 then yields a gravitational field $\vec{g} = -\overrightarrow{\mathrm{grad}}(\Phi)$; its intensity at the distance r of the sphere center can again be calculated with Ostrogratski's theorem, the result being:

$$g = \frac{MG}{r^2} - \frac{\Lambda r}{3}.$$

Thus, this constant has a repulsive effect that increases proportionally with r.

Further, the Friedmann equation could be adapted to Einstein's new field equations with the Λ constant, and became:

$$\left(\frac{\dot{a}}{a}\right)^2 = \frac{8}{3c^2} G\pi\varepsilon(t) - \frac{kc^2}{a^2 R_0^2} + \frac{\Lambda}{3}. \tag{6.21}$$

This is equivalent to adding in the universe a new component whose energy density is:

$$\varepsilon_\lambda(t) = \frac{c^2}{8\pi G}\Lambda. \tag{6.22}$$

If we want this new component to have an energy density that is constant in time, then the fluid equation 3.18, which remains unchanged, implies:

$$P_\lambda = -\varepsilon_\lambda = -\frac{c^2}{8\pi G}\Lambda. \tag{6.23}$$

The pressure is thus negative, which is a novelty but consistent with the repulsive effect of this component. Finally, the Friedmann equation (6.21) yields with a static universe ($\dot{a} = 0$):

$$0 = \frac{8}{3} G\pi\rho - \frac{kc^2}{R_0^2}\frac{1}{a^2} + \frac{4}{3} G\pi\rho = 4G\pi\rho - \frac{kc^2}{a^2 R_0^2}.$$

This shows that the space is positively curved ($k=+1$) with the curvature radius being: $R_0 = \frac{c}{2\sqrt{G\pi\rho}} = \frac{c}{\sqrt{\Lambda}}$

The likely assumption today for such a component is the vacuum energy, a possibility that is also[1] supported by Quantum Mechanics. This theory indeed showed that virtual particle-antiparticle pairs are created in the vacuum during very short time-lapses, while respecting the Heisenberg uncertainty principle, $\Delta E \cdot \Delta T < h$. The lambda constant thus indicates the energy of the vacuum (cf. equation 6.22), also named the "dark energy."

If the lambda constant is greater than Einstein's value corresponding to the static model, the universe expands and its expansion accelerates. Its repulsive effect remains constant, while the energy density of the known matter-energy decreases as the volume expands.

However, the current indications based on Quantum Mechanics are very far from agreeing with this model.

Remark 6.3A: The vacuum existence was denied by some important philosophers, notably Aristotle and Pascal.

6.3.2 State Equations and Universe Evolution Models

We saw the universe expansion follows two main equations:

- the Friedmann equation (6.14): $\left(\frac{\dot{a}}{a}\right)^2 = \frac{8}{3c^2}G\pi\varepsilon(t) - \frac{kc^2}{a^2R_0^2}$,
- the fluid equation (6.18): $\varepsilon = -3\frac{a}{a}(\varepsilon + P)$.

These equations can be solved if we can prescribe the relation between the energy density ε(t) and the pressure P(t). Such a relation, called the equation of state, is not simple in the general case. However, it is simpler in ideal cases, where the universe is composed of only one type of component. We will indeed see that in these cases, there is a linear relation between P and ε, denoted by: $P = w\,\varepsilon$.

6.3.2.1 The State Equations

Case where the component is ***non-relativistic*** *matter only:*

We will show that $P = \mathbf{w}_m\,\varepsilon$, with $\mathbf{w}_m \approx 0$:

Consider an ideal gas composed of massive particles having very low velocities compared to c. The ideal gas law yields: $P = \frac{\rho}{\mu}k'T$, where k′ is the Boltzmann constant[2], μ the average particle mass, and ρ the mass density. The latter is equal to ε/c^2 since $v \ll c$. Hence: $P = \frac{\varepsilon}{c^2\mu}k'T$.

From equipartition, we also have: $k'T = \frac{1}{3}\mu\langle v^2\rangle$, with $\langle v^2\rangle$ being the average velocity square of the particles. So finally: $\frac{P}{\mu} = w_m = \frac{v^2}{3c^2} \approx 0$ ■

Case where the component is ***radiation only****:* We have: $P = \frac{1}{3}\varepsilon$, so: $\mathbf{w}_r = \frac{1}{3}$.

Remark 6.4: This 1/3 factor is explained by i) the equality between the spatial momentum and the energy of the photon; ii) the pressure exerted on a surface is only due to photons whose momentum have a nonzero component along the perpendicular direction to that surface; iii) photons move in all directions, so that the pressure they exert is the same along 3 perpendicular directions, so that the direction perpendicular to the considered surface receives 1/3 of the total energy.

Case where the component is the ***dark energy only*** *(Lambda constant):* $P = -\varepsilon$, *so:* $\mathbf{w}_\Lambda = -1$.

6.3.2.2 Universe Evolution

Once the value of w is known, the Fluid Relation tells us how the density evolves as a function of the scale factor: this equation can indeed be written: $\dot{\varepsilon} = -3\frac{\dot{a}}{a}\varepsilon(1 + w)$, then: $\frac{\dot{\varepsilon}}{\varepsilon} = -3\frac{\dot{a}}{a}(1 + w)$.

This yields, with the current energy density being denoted by ε_0:

$$\varepsilon(a) = \varepsilon_0 \, a^{-3(1+w)}. \tag{6.24}$$

Consequently, we have for the matter:

$$\varepsilon_m(a) = \varepsilon_{m,0}(a) \, a^{-3}. \tag{6.24a}$$

For radiation:

$$\varepsilon_r(a) = \varepsilon_{r,0} \, a^{-4}. \tag{6.24b}$$

For the dark energy: ε_Λ is constant.
We can thus see that as a(t) increases, the radiation density decreases more rapidly than the matter. In both cases, the number of particles per unit of volume reduces, but one additional effect concerns radiation: all wavelengths get longer with the universal expansion; $\lambda(t) = a(t)\lambda_0$, implying a decrease of each photon's energy.

Remark 6.5: The radiation density evolution is related with the temperature by the Stefan-Boltzmann law regarding black body radiation: $\varepsilon_r = \sigma T^4$. We notice that the exponent of T is also 4; hence, equation 6.24b implies that the temperature linearly decreases with the scale factor: Let T_e be the temperature at the emission time, we have: $\sigma T_e^4 = \sigma T_0^4 a(t_e)^{-4}$, so: $T_e = T_0 / a(t_e)$. Thus, with the cosmic microwave background, we have: T_e=2.970 K×1,090=**3,237 K.**

The universe actually contains a mix of these components; moreover, there are neutrinos that are relativistic particles, whose w is close to 1/3.

We may assume that the resulting energy density is the sum of the energy density of the different components, whereby matter accounts for **32%**, dark energy (Λ constant) for **68%** and only **0.009%** for radiation.

Under the current estimates, 90% of the current radiation energy in the universe is due to the cosmic microwave background (CMB), representing $\varepsilon_{CMB} = 0.2606$ MeV/m³, the rest being star light from the galaxies.

Besides, physical considerations tell us that there should also exist a Neutrino cosmic background.

Dark energy is thus dominant today, but from the relations (6.24) we can deduce that it was the opposite in the past, and that the time when matter and lambda densities were equal corresponds to the scale factor $\mathbf{a} \approx \mathbf{0.778}$. Similarly in the very far past, radiation was dominant: at the time when radiation and matter represented an equal energy density, the scale factor was: $\mathbf{a} \approx \mathbf{2.8 \times 10^{-4}}$ (cf. Problem 6.4).

6.3.3 The 4 Ideal Scenarios

Looking at the future, the dark energy will be even more dominant. This invites us to consider first the type of universe expansion that the Friedmann equation predicts in the ideal case of dark energy only and with a flat universe.

6.3.3.1 Ideal Scenario of a Flat Universe with Lambda Only

In a flat universe (k=0), the Friedmann equation yields: $\left(\frac{\dot{a}}{a}\right)^2 = \frac{8G\pi\Lambda}{3c^2}$.

We notice that the right-hand term is constant. Let's use equation 6.9:

$H_0 = \frac{a(\dot{t}_0)}{a(t_0)}$; hence: $\frac{\dot{a}(t)}{a(t)} = \frac{\dot{a}(t_0)}{a(t_0)} = H_0 = \left(\frac{8G\pi\Lambda}{3c^2}\right)^{1/2}$, which yields:

$$a(t) = e^{H_0(t-t_0)}. \tag{6.24}$$

Thus, a(t) increases exponentially with time: in this scenario, the universe expands exponentially.

Imagine that you observe a galaxy at the time t_0 and you notice its redshift z. Its photons were emitted at the time t_e. Helpfully, relation (6.12) indicates the relation between t_e and z, so:

$$z = \frac{1}{a(t_c)} - 1 = e^{-H_c(t_c - t_c)} - 1.$$

At the time t_0, the relations (6.12) and (6.24) enable us to know the distance to this galaxy:

$$d(t_0) = \int_{t_e}^{t_0} \frac{c}{a(t)} dt = c\int_{t_e}^{t_0} e^{H_0(t_0-t)} dt = -\frac{c}{H_0}\left(e^{H_0(t_0-t_0)} - e^{H_0(t_0-t_e)}\right)$$

$$= \frac{c}{H_0}\left(e^{H_0(t_0-t_e)} - 1\right) = \frac{c.z}{H_0} \tag{6.25}$$ ■

Remark 6.6: The ratio $\frac{1}{H_0}$ is called the Hubble time, its value being 14.3 billion years.

We notice that the relation between $d(t_0)$ and z is linear whatever the redshift, which is specific to this ideal case.

Combining, the photons emitted at t_e came from the distance:

$$d(t_e) = a(t_e)d(t_0) = \frac{c}{H_0}\frac{z}{1+z}.$$

If z is much greater than 1, then $d(t_e) \approx \frac{c}{H_0}$. We saw that beyond this value, the recession speed of the galaxy is greater than c.

6.3.3.2 Ideal Scenario of an Empty Universe

If the dark energy assumption were not confirmed, then as the universe continues to expand, its energy density will become very small. Hence, it could be approximated by the extreme case of an empty universe: $\varepsilon=0$. The Friedmann equation becomes:

$$\dot{a}^2 = -\frac{kc^2}{R_0^2}. \tag{6.26}$$

If we assume the universe is flat, then $k=0$ implies $\dot{a}=0$, meaning that there is no expansion. Thus, as the energy density becomes extremely small, the expansion rate also becomes very small (which is not what is observed).

However, relation (6.26) also permits the case $k=-1$, meaning a negatively curved surface (the case $k=1$ is rejected since it leads to an imaginary value of $\dot{a}$). In such case, we have: $\dot{a}(t) = \pm\frac{c}{R_0}$, meaning that the universe either expands or reduces linearly with time. We will consider the expansion case, which better fits observations. We then have: $a(t) = \frac{c}{R_0}t$.

Then, as $a(t_0)=1$, we have: $t_0 = \frac{R_0}{c}$.

Besides, $\dot{a}(t_0) = H_0 a(t_0)$, then: $H_0 = \frac{c}{R_0} = \frac{1}{t_0}$, so finally:

$$a(t) = \frac{c}{R_0}t = \frac{t}{t_0} = H_0\, t. \tag{6.27}$$

Imagine you observe a galaxy at the time t_0 and you notice its redshift z. Its photons were emitted at the time t_e, which is given by equations 6.12 and 6.27: $a(t_e) = \frac{1}{Z+1}$, then $t_e = \frac{1}{H_0(Z+1)}$. At the time t_0, the distance to this galaxy is:

$$d(t_0) = \int_{t_e}^{t_0} \frac{c}{a(t)}dt = \int_{t_e}^{t_0} \frac{c}{H_0 t}dt = \frac{c}{H_0}\ln\left(\frac{t_0}{t_e}\right) = \frac{c}{H_0}\ln\left(\frac{H_0(1+z)}{H_0}\right) = \frac{c}{H_0}\ln(1+z).$$

If z is close to 0, this relation is linear: $d(t_0) \approx \frac{c}{H_0}z$.

If z is much greater than 1: $d(t_0) \approx \frac{c}{H_0}\ln(z)$.

It may seem amazing that we can see galaxies at a distance greater than c/H_0 while the universe age is $1/H_0$, but this is explained by the universe expansion.

At the time t_e, the proper distance to this galaxy was:

$$d(t_e) = d(t_0)\frac{a(t_e)}{a(t_0)} = d(t_0)H_0 t_e = \frac{c}{H_0}\ln(1+z)H_0\frac{1}{H_0(z+1)} = \frac{c\ln(1+z)}{H_0(z+1)}.$$

This distance has a maximum for galaxies with a redshit of $z = e - 1 \approx 1.72$; the corresponding distance is: $0.37\frac{c}{H_0}$.

6.3.3.3 Ideal Scenario of a Flat Universe with Radiation Only

The case of radiation only is interesting as it approximates the initial phase of the universe.

From equations 6.14 and 6.24b we have: $\left(\frac{\dot{a}(t)}{a(t)}\right)^2 = \frac{8G\pi}{3c^2}\varepsilon_{r,0}a(t)^{-4}$, then: $\dot{a}(t)^2 a(t)^2 = \frac{8G\pi\varepsilon_0}{3c^2}$.

By definition, we have: $\dot{a}(t_0) = H_0 = \sqrt{\frac{8G\pi\varepsilon_{r,0}}{3c^2}}$; and so: $2\,\dot{a}(t)\,a(t) = 2H_0$.

Let's integrate: $a(t)^2 = 2H_0\,t$, then: $a(t) = \sqrt{\frac{t}{t_0}}$ (since $a(t_0)^2 = 1 = 2H_0\,t_0$) ■

The age of the universe in this scenario is: $t_0 = \frac{1}{2H_0} \approx 7.1$ billion years.

The energy density follows the relation:

$$\varepsilon_r(t) = \varepsilon_{r,0}\,a^{-4} = \varepsilon_{r,0}\left(\frac{t}{t_0}\right)^{(1/2)(-4)} = \varepsilon_{r,0}\left(\frac{t}{t_0}\right)^{-2}.$$

Thus, the energy density decreases with the square of the time. At $t=0$, the density was infinite, but the laws of physics are only valid after the Plank Time. Hence, let's denote respectively by t_P and ε_P the Planck time and the energy density at that time, and we obtain: $\varepsilon_t = \varepsilon_P\left(\frac{t}{t_P}\right)^{-2}$ ■

A galaxy seen at the time t_0 with a redshift z has its proper distance: $d(t_0) = \frac{c}{H_0}\frac{z}{(1+z)}$ (cf. Problem 6.5).

6.3.3.4 Ideal Scenario of a Flat Universe with Matter Only

From equations 6.14 and 6.24a we obtain: $\left(\frac{\dot{a}(t)}{a(t)}\right)^2 = \frac{8G\pi}{3c_2}\varepsilon_{m,0}a(t)^{-3}$.

Currently, we have: $\dot{a}(t_0) = H_0 = \sqrt{\frac{8G\pi\varepsilon_{m,0}}{3c_2}}$, and so: $\left(\frac{\dot{a}(t)}{a(t)}\right)^2 = H_0\,a(t)^{-3}$.

Let's set: $a(t) = \left(\frac{t}{t_0}\right)^x$, and we will find x: $\left(\frac{x\,t^{(x-1)}}{t^x}\right)^2 = \frac{x^2\,t^{2(x-1)}}{t^{2x}} = H_0\left(\frac{t}{t_0}\right)^{-3x}$;

hence, $2x-2-2x=-3x$, so: $x=2/3$. So finally:

$$a(t)=\left(\frac{t}{t_0}\right)^{2/3} \qquad (6.28) ■$$

We notice that the expansion rate is greater than with the radiation case.

Let's calculate t_0, *which is the universe's age according to this scenario:*

From $\dot{a}(t_0)=H_0$, and $\dot{a}(t_0)=\frac{2t_0^{-1/3}}{3t_0^{2/3}}=\frac{2}{3t_0}$, we obtain:

$$t_0=\frac{2}{3H_0}\approx 9.5 \text{ billion years} \qquad (6.29) ■$$

A galaxy seen at the time t_0 with a redshift z is at the proper distance: $d_0(t_0)=\int_{t_e}^{t_0}\frac{c}{a(t)}dt=c\int_{t_e}^{t_0}\left(\frac{t}{t_0}\right)^{-2/3}dt$. Let's set $u=\frac{t}{t_0}$, and we have:

$$d_0=c\int_{t_e}^{t_0}t_0u^{-2/3}du=3ct_0\left[u(t_0)^{1/3}-u(t_e)^{1/3}\right]=3ct_0\left[1-\left(\frac{t_e}{t_0}\right)^{1/3}\right]. \qquad (6.30)$$

Combining equations 6.12 and 6.28 we have: $a(t_e)=\frac{1}{z+1}=\left(\frac{t_e}{t_0}\right)^{2/3}$, so: $\left(\frac{t_e}{t_0}\right)=\left(\frac{1}{z+1}\right)^{3/2}$.

Then with equations 6.29 and 6.30, we finally obtain:

$$d_0=3ct_0\left[1-\left(\frac{1}{z+1}\right)^{(1/3)(3/2)}\right]=\frac{2c}{H_0}\left(1-\frac{1}{\sqrt{1+z}}\right) \qquad (6.31) ■$$

Let's calculate the "horizon distance," which is the distance to the furthest object that can be seen at t_0: it is the distance to the first light that was emitted at $t_e=0$. From equation 6.31, it is: $d_{0\max}=\frac{2c}{H_0}\approx 28$ billion light-years.

**

The table hereafter gives a comparison of these ideal cases, showing the different expansion rates and distances at t_0 to a galaxy having z for redshift.

Ideal Model Type	Expansion Rate	Distance to a Galaxy as a Function of its Redshift
Empty Universe – Flat	0	–
Empty Universe Curved	$a(t)=e^{H_0(t-t_0)}$	$d(t_0)=\frac{c}{H_0}\ln(1+z)$
Radiation only – Flat	$a(t)=\left(\frac{t}{t_0}\right)^{1/2}$	$d(t_0)=\frac{c}{H_0}\frac{z}{(1+z)}$
Matter only – Flat	$a(t)=\left(\frac{t}{t_0}\right)^{2/3}$	$d(t_0)=\frac{2c}{H_0}\left(1-\frac{1}{\sqrt{1+z}}\right)$
Lambda only – Flat	$a(t)=e^{H_0(t-t_0)}$	$d(t_0)=\frac{c}{H_0}z$

6.3.3.5 Concluding Comments

More precision can be obtained with models having a mix of radiation, matter, dark energy, and introducing a possible universe curvature, even if the universe appears to be amazingly flat.

Besides, we should bear in mind that important aspects are still the subject of research and the difficulties are numerous: we mentioned the dark energy which requires a better understanding at the quantum level. There is also evidence that the mysterious dark matter, probably non-baryonic for an important part, accounts for an important part (27%). In addition, about 85% of the baryonic matter is in the very tenuous intergalactic gas that cannot be detected today; the black holes are not visible directly. Other important open questions concern the great inflation phase. Thus, cosmology is a domain where several breakthroughs are expected.

6.4 Questions and Problems

Problem 1) Relations between redshift, distance and scale factor.

You observe at the current time t_0 two galaxies A and B, and you mark their redshits: $z_a=2$ and $z_b=4$.

Q1A) Is the distance at t_0 to the galaxy B twice that to the galaxy A?

Q1B) Is there an ideal model where this is true?

Q2) The photons received at t_0 were respectively emitted at t_{eA} and t_{eB}. Is the scale factor at t_{eB} twice that of t_{eA}?

Problem 2) Relations between redshift, distance and scale factor – Part 2.

In this problem, we assume the universe is flat with matter only.

Q1A) What are the distances of these galaxies to you at t_0?

Q1B) Imagine that you also see at t_0 two other galaxies C and D, with their redshits: $z_c=500$ and $z_d=1{,}000$. Same question as Q1A.

Q2) At what time were the photons of these four galaxies emitted? Give the general relation.

Q3) The photons received from the galaxy D have an energy which corresponds to 3.50°K. What was their temperature when they were emitted?

Problem 3) Same questions as Problem 2, but assuming a flat universe with lambda only, but for Q2), give the values for $t_0 - t_e$.

Problem 4) The scale factor at the main transition phases

Q1) What is the scale factor at the time when dark energy density was equal to matter density?

Q2) What is the scale factor at the time when radiation and matter represented an equal energy density?

Problem 5) Ideal Scenario of a Flat Universe with Radiation Only

At the time t_0 you observe a galaxy having a redshift z.

Q1) What is its proper distance at t_0?

Q2) What is its proper distance at the time t_e when it was emitted?

Problem 6) Impact of the Earth curvature

You are on Earth and you see an object that is at the distance of 50 km from you. We make the assumption that the photons emitted by this object toward your eye follow the curved Earth surface. What is its width, knowing that the angle with which you see the object is 0.01 radian, and that the Earth radius: 6,371 km? What is the difference if the Earth were flat?

Notes

1 The vacuum energy may not be quantum; the potential energy for scalar fields, such as the Higgs, has a constant term that is not determined by the particle properties, and was used to offset the quantum vacuum energy before the cosmological constant was determined to be nonzero.

2 The notation k′ is used in order to avoid confusion with the parameter k in the Friedmann equation.

7

Further Axiomatic Considerations

Introduction

Relativity has considerably changed our vision of the fundamental notions of time, distance, energy and matter. The impacts of Special and General Relativity on the time will be first precisely recapitulated, including a discussion on the causality principle. The very concepts of time, distance and event will then be analyzed from an axiomatic perspective, and as these notions are also linked to matter-energy, the Momentum will also be subject to this analysis. We will show that the Momentum is an essential concept expressing the 4D notion of energy, and that its fundamental system conservation is linked to the universal homogeneity–isotropy, which thus appears to be the main postulate on which Relativity is built. Moreover, we will see that the light speed invariance is actually a consequence of the universal homogeneity–isotropy. Finally, the hypothetical tachyons' world, where an object can surpass c will be explored, leading to the conclusion that if it exists, it cannot interact with us.

7.1 Further Considerations on Time, Causality and Event

Soon after Relativity was accepted, it was frequently said (but not by Einstein) that "time does not exist," which generated tense discussions with some important philosophers, in particular Bergson. At the end of his life, Einstein said: "the distinction between past, present and future is only a stubbornly persistent illusion."

7.1.1 Time Revisited by Relativity

Preliminary remark: We will often use the terms "intrinsic" and "primitive"; hence, the reader is invited to see their definitions in Section 1.3.

DOI: 10.1201/9781003201359-7

7.1.1.1 Restrictions Brought by Special Relativity on the Notion of Time

Special Relativity shows that time is not absolute, but relative to an inertial frame. Moreover, there is no privileged frame that would give the reference time or the "true time." The simultaneity between distant events being relative to an inertial frame, the notions of past, present and future are also relative to an inertial frame: some events take place before some other events in one inertial frame, but it can be the opposite in other inertial frames.

Nevertheless, these notions of past, present and future could survive, but with the restriction that they are relative to an event: an event E* is in the past of an event F* if in any inertial frame K, E* is prior to F*. This is indeed possible if the *causality condition* is met, meaning that the time separation between E* and F* in K is longer or equal to the duration that a photon (i.e., the fastest object) would take to go from the point E to the point F in K. This condition will then be met in all inertial frames due to light speed invariance. The space–time interval E*F* is then said "time-like" or "light–light" in the case of equality. Using a Minkowski diagram, E* is inside the lower part of the light cone whose summit is F* (cf. Volume I Section 3.2.2).

Another surprising novelty is the possibility of travels in the future (cf. accelerated trajectories and Langevin's paradox).

7.1.1.2 Further Restrictions Brought by General Relativity

The above notions of past–present–future rely on the concept of inertial frame, but General Relativity shows that perfect inertial frames don't exist: these are only local approximations, our universe being not Euclidean.

Additionally, General Relativity brings another problem that makes it impossible to have a synchronized time in the general case: we saw that time synchronization requires the possibility to exchange signals (cf. Volume I Section 1.3.2.2), but if two events E* and F* are not close to each other, the (Lorentzian) distance between them is not only more complex to compute than in Special Relativity because of the chrono-geometrical deformations, but this distance depends on the trajectory chosen between E* and F*, as there can indeed exist different trajectories between them, with different distances. Even light can take different paths (null geodesics) between two events. This is a significant difference from Special Relativity, where there always exists one straight line between E* and F*.

However, General Relativity did not abolish all properties regarding the time: the proper time universality is confirmed. Besides, the cosmic time commonly used in cosmology is the proper time of a "fundamental" observer (i.e., free-falling in the cosmic fluid), and it is assumed to be the same for all such observers (universal) and even synchronized, which makes an exception to the above restriction in General Relativity. Moreover, it is remarkable that events that occurred 10^{-12} seconds after the Big Bang could be dated. Besides, time has become the physical quantity, which is known with the highest precision: 10^{-18} seconds! Hence, for a notion that was said not to exist, it is still quite meaningful…

General Relativity confirms the possibility of travels in the future, and even offers another means: gravitational time dilatation. Time indeed has another important specificity: its orientation, which we encountered with the Causality principle. *Hence, let's further examine this principle and its relation with Relativity:*

7.1.2 The Causality Principle and the Impossibility of Greater Speeds than C

The Causality principle states that "a consequence cannot precede its cause." It is worth noting that Relativity does not require the Causality principle and that the only addition, extremely important though, that this principle brings to Relativity is the impossibility for objects to move at a greater speed than c (cf. Volume I Section 3.2.1). We are then induced to further examine the causality principle, and it first appears that its wording is more adapted to classical physics than Relativity for the following reasons:

- The word "precede" refers to the past, but we saw that Relativity brought important restrictions on the notion of the past.
- This principle does not mention the "event," although it is the elementary concept of our universe according to Relativity.

A consistent statement with Relativity would be: "For an event E* to be the cause of an event F*, E* must be in the past of F*." Even though the notion of past refers to the inertial frame, but General Relativity shows that it is only valid locally.

Besides, it is worth noting that the orientation of time, implied by the causality principle, is a breach in the time-distance equivalence revealed by Relativity. Note that this time orientation also appears in the second principle of thermodynamics, which states that the entropy of a system always increases, meaning that the level of disorder always increases (a more ordered situation has a lower probability of occurring, and thus a lower entropy). In simple words, this second principle says that time averages, randomizes and damages everything. Like for Causality, this second principle is exterior to Relativity, but Relativity is compatible with it.

Informally, the notion of *causality* is often associated with the notion of *responsibility* and raises the issue of *determinism*. This should be clarified as it may be a source of confusion: it is because someone decides to take an action that he (or she) is responsible for the consequences *caused* by his or her action. We implicitly assume that this person has *free will*. However, the word *cause* can also be used in a context that does not involve free will, nor even the hazard: for instance, the fact that it is daytime is *caused by* the sun being above the horizon. In Relativity, the Causality principle applies to all types of causes: with free will, hazard or without (meaning determinism). Before Heisenberg's uncertainty principle of Quantum Mechanics (1927), all physical laws were considered deterministic; hence, some important scientists, in particular Laplace, stated that the universe is fully deterministic, but it is too complex to determine the initial conditions. Today, this vision is abandoned by the vast majority.

Besides, in a deterministic world or system, the common notion of causality raises some issues, as pointed out by some philosophers, particularly A. Schopenhauer. The following example illustrates this difficulty: consider a billiard with balls moving and having perfectly elastic collisions, and we assume that there are no frictions. We observe the positions and the velocities of all balls at two different times, t1 and t2, with t1 being prior to t2 in the billiard frame K. From the knowledge of the situation at t1, we can deduce the situation at t2; hence, we commonly say that the situation at t1 is the *cause* of the one at t2. However, the knowledge of the situation at t2 also enables us to deduce and explain the situation at t1, but we wouldn't say that the situation at t2 is the *cause* of the one at t1, the only reason being that the latter is prior to the former. This shows that the common notion of cause includes the causality principle, and that in a deterministic system, the situation at any time is equivalent regarding its capacity to explain the situation at any other time. In this example, one can say that the real cause explaining the evolution of the system is the system's Momentum conservation law.

To overcome this difficulty, the causality principle can be replaced by a postulate stating that the chronological order between any couple of events belonging to the trajectory of a real object is the same in all inertial frames (i.e., intrinsic). First, this postulate implies the impossibility of greater speeds than c (cf. Volume I Section 3.2.1), and then it also implies the causality principle since Quantum Mechanics shows that any interaction between objects requires the motion of objects between them, even massless ones such as photons or gravitons.

However, we mentioned that the chronological order is only a local notion (since it relies on the concept of inertial frame); hence, W. van Stockum showed in 1937 that General Relativity permits the existence of Closed Time-like Curves (CTC), whereby it is possible for an object to travel from an event A* and then to reach the same event A*. This opens the possibility of interventions in the past, breaking a fundamental axiom well expressed by Aristotle: "Even Zeus cannot do so that what existed didn't exist." This theoretical possibility of CTC was confirmed by K. Gödel in 1949, but never observed in reality. Stephen Hawking then issued the "chronology protection conjecture" that the laws of physics prevent time travel on all but microscopic scales.

The event being the elementary concept, one could alternatively postulate the "existing event inalterability," whereby it is impossible to delete or modify an existing event.

Besides, we show in Section 7.3.2 that hypothetical objects moving at greater speeds than c, called "tachyons," cannot have any interaction with our world; hence, we can ignore them.

7.1.3 Axiomatic Considerations on the Concepts of Time and Event

First, the time of an inertial frame cannot be primitive since it is constructed from the proper time in one point, and then extended to the whole inertial frame thanks to a synchronization method (cf. Volume I Section 1.3.2). Moreover, General Relativity shows that the inertial frame is an ideal construction that does not really exist, but can be approximated locally or when we are very far from masses and energies.

We are then induced to check if the proper time can be the primitive notion from which time is derived. We will see that this possibility raises some objections; hence, other options will be explored.

7.1.3.1 The Proper Time and the Proper Distance as Primitive Notions?

The proper time is based on a first postulate: the existence of periodic phenomena at one point with an absolutely stable period, and this constitutes a perfect clock. Then, the universal homogeneity postulate implies that the same perfect clock beats at the same pace anywhere in an inertial frame. Moreover, the inertial frame equivalence postulate implies that the same clock also beats at the same pace in all inertial frames (cf. Section 1.2.2.1.1). An additional postulate, the Clock Postulate, states that the pace of a perfect clock is not affected by its acceleration; and finally, the principle of equivalence implies that the proper time of a perfect clock is not affected by gravity: the proper time is thus universal. The term "universal" is preferred to "absolute" because there still exist limitations: the impossibility to synchronize time in the general case; and of course the proper time of a perfect clock is perceived differently by observers moving relative to this clock or located in a different gravitational potential.

The proper distance is linked to the proper time by the relation $d = ct$. Hence, the proper time universality also means the proper distance universality, but locally (since in General Relativity, this relation is valid only locally).

The proper time and the proper distance can yield the notion of event, as being the set (time, position in the 3D space) with which an observer considers an event.

Alternatively, we are induced to examine the axiomatic possibility whereby the event is the primitive notion, since the proper time and the proper distance can be defined as being the Lorentzian distance between events (with natural units): $\tau_{MN} = \sqrt{ds^2(MN)}$.

7.1.3.2 The Event as a Primitive Notion?

Setting the event as the primitive notion has the interest of offering a common root for the proper time and the proper distance, both being the Lorentzian distance between a couple of events. This induced the time-distance equivalence.

The event definition requires postulates, as for any primitive notion, and we already saw these:

- The event is an intrinsic notion: cf. Section 1.3.2.1.
- The "event coincidence invariance": cf. Section 1.3.1.1.
- As mentioned in Section 7.1.2, an additional postulate should be added: the "Existing Event Inalterability."

These postulates are often used without mentioning because they seem quite natural.

7.1.3.2.1 The Distance between Events

Placing the event at the root from which time and distance are derived requires to postulate that our space–time universe can be represented by a 4D pseudo-Riemannian manifold, as described in Section 4.2. This paradigm assumes that in the vicinity of any event, the distances can be approximated as if the universe were Euclidean, enabling us locally to define Minkowski frames.

However, for the distance to be Lorentzian, it must discriminate between time-like and space-like intervals; hence, it is necessary that the values in the diagonal of the metric tensor are: (+1, −1, −1, −1). Mathematically, the difference between the number of positive and negative values in the diagonal is called the "signature" of matrix, and it has the property of being intrinsic. A postulate must then be taken, stating that the signature of the metric is always: −2, but this postulate seems artificial.

Besides, defining the proper time as $\tau_{MN} = \sqrt{ds^2(MN)}$ assumes that the proper time between M and N is only a function of the events M and N, and not of those in between. This actually corresponds to the clock postulate, which also seems artificial in this context.

Then, when passing from a point (event) M to neighboring points and then distant points N, the length of the trajectory from M to N requires to know the affine connection coefficients (cf. Section 4.2.2). These involve the different g_{ij} parameters, whose values are linked to the space–time deformations due to the presence of mass-energy. Hence, the notion of event is linked to the notion of matter-energy.

When placing the event as the primitive notion, we wished to solve the issue regarding the link between time and distance, but we find that the event is also always linked to another concept: mass-energy. Moreover, we saw that the metric requires postulates that seem artificial.

Furthermore, placing the event at the root has other limitations: the event scheme is not able to describe all physical realities: in particular, the photon cannot be characterized by the concept of event: we cannot tell where a photon is at any point of time, but only at its emission and detection. Additionally, quantum mechanics shows that this is also the case of particles since they also have wave aspects, as stated by the French scientist L. de Broglie. This is well shown by Young's famous double-slit experiment, where an electron interferes with itself. The problem with the event scheme is that particles don't have trajectories that can be defined as a succession of events, but they have a probability of presence, which follows the Schrodinger equation, until some external conditions make them materialize at some point, whose location is random within the huge number of possibilities offered by the Schrodinger equation.

Another aspect is not explained by the event scheme: the specificity of the time dimension having one direction.

Besides, from a pedagogical perspective, the event scheme is quite abstract and has the drawback of masking the universal character of the proper time, which is physically important and significantly helps with understanding (and avoiding common errors).

There can be different sets of primitive concepts for the same theory. At this point, both options (proper time and proper distance) or (event) have their interests and their limitations; both can be considered quasi-primitive notions. We will see that another possibility exists: the Momentum as the primitive notion.

7.2 Axiomatic Analysis of the Laws of Dynamics

We will examine the axiomatic aspects of the Relativistic laws, meaning that we will identify the logical links between the different relativistic concepts and laws, and therefrom we will derive the primitive concepts and postulates.

7.2.1 The Root: The Law of Inertia

We will show that the fundamental law of Inertia is a consequence of the Universal Homogeneity and Isotropy:

The trajectory of a point object is characterized by a succession of events in the 4D space–time universe. If an object-point is inertial, meaning not subject to any force,[1] the evolution of the events characterizing its trajectory only depends on the properties of the universe in its vicinity. In this respect, the universal homogeneity and isotropy postulate state that the universe's properties are the same everywhere, in all directions and at all times.

In Special Relativity, if the point object is considered in an inertial frame K with its time being denoted by t, this translates by the 4D relation:

$$\frac{\overrightarrow{dM(t)}}{dt} = \overrightarrow{cst}. \tag{7.1}$$

The spatial part of this relation shows that an isolated point object evolves along a straight line and at a constant velocity, meaning that the classical law of inertia is a consequence of the universal homogeneity–isotropy postulate.

General Relativity brought important changes, but the principle of equivalence states that the laws of Special Relativity still apply locally, including the law of inertia, which expresses the constancy of the time rate of change of the events characterizing the object's trajectory. However, this time rate of change cannot be expressed using the time t since we saw in Section 4.2.5 that the only parameter u which ensures the constancy of $\frac{\overrightarrow{dM(u)}}{du} = \overrightarrow{cst}$ is the object's proper time τ. The law of inertia then reads: an inertial object (i.e., free-falling) has its proper velocity constant:

$$\frac{\overrightarrow{dM(\tau)}}{d\tau} = \overrightarrow{cst}. \tag{7.2}$$

Significantly, we saw in Section 4.2.6 that this implies that the trajectory of a free-falling object is a geodesic.

7.2.2 From the Law of Inertia to the Momentum

For the sake of simplicity, we will skip the arrow on 4D vectorial quantities such as the Momentum, P, when there is no ambiguity as to their vectorial character.

Relation (7.2) shows that there is an invariant and constant element characterizing the evolution of an inertial point object: $\frac{dM}{d\tau}$. Hence, it is logical to wonder whether reciprocity holds: does the existence of a force on an object necessarily imply a change of its proper velocity?

In classical physics, the answer was positive, but Relativity shows that forces can also affect the object's mass. We are then induced to postulate the existence of a scalar quantity, m, which is intrinsic to the object and such that the 4D quantity $P = m\frac{dM}{d\tau}$ changes if, and only if, a Force is acting on the object. We have chosen the letters m and P by anticipation, as for the time being we don't know their meaning. However, we will show that the properties of m match with the mass and those of P with the relativistic Momentum:

We first notice that P is intrinsic (m being intrinsic by definition): P is a contravariant -vector (rank 1 tensor).

7.2.2.1 P Matches with the Relativistic Momentum

In the context of forces that don't impact the mass, m is an approximation of the classical inertial mass for low speeds compared to c, since the variation of the spatial part of P, $mv\gamma$, matches with the classical Newtonian force. Consequently, the spatial part of P is an approximation of the classical momentum for low speeds.

Similarly, the variation of the time part of P, $\Delta(m\,c\,\gamma)$, matches with the classical kinetic energy variation, $\Delta(1/2\ mv^2)$ for speeds low compared to c.

Then, for an isolated system composed of several objects, the fact that P is a contravariant rank 1 tensor ensures that if the sum of the spatial parts of all the P_i is constant in an inertial frame K, this sum is also constant in all inertial frames. The same applies to the time part of P, consequently the law stating the constancy of the relativistic system Momentum for an isolated system complies with the principle of Relativity. Moreover, this shows that both the classical system conservation laws (momentum and energy) are derived from this single relativistic conservation law.[2]

7.2.2.2 The Scalar m Is the Object's Intrinsic Energy

When expressing P with fully natural units, which better expresses the physical meaning, its time part is $m\gamma$. It is even m in the object's own frame, meaning that the object has an energy equal to m in its own frame, which is a major novelty brought by Relativity. We can say that m is the object's intrinsic energy.

This intrinsic energy can be explained by the first principle of thermodynamics, which acknowledges that there are various forms of energy, with possible exchanges between them, but the overall quantity of energy remains constant within an isolated system. Thus, m is the energy lost by the system to create the object, considered in its Center of Momentum frame.

7.2.2.3 The Inertia-Energy Equivalence

The inertia-energy equivalence expressed with fully natural units is actually an identity, both being equal to $m\gamma$. It is remarkable that the concepts of inertia and energy are opposite: the capability to change (energy) versus the capability to oppose any change (inertia). This coincidence should be explained: in classical physics, the inertial mass could be defined by the relation $m = f/a$, but this is not true in Relativity (cf. relation 3.14). In Relativity the inertia $m\gamma$ still expresses the capability to oppose a change in speed, and it increases with speed in such a way that no massive object can reach or surpass the speed of light. Thus, the energy-inertia equivalence preserves the fundamental Causality principle from violation.

7.2.2.4 The Momenergy and the System Mass

Having seen that the scalar m represents the object's intrinsic energy, the relativistic Momentum $P = m\frac{dM}{d\tau}$ appears to be the 4D (space–time) notion of energy. Let's specify the definition of energy: energy is the resource necessary to make any change regarding the velocity or the state of an object (its constitution, structure). Using the same word energy for both the kinetic and the intrinsic energy is justified by experiments showing that kinetic energy can be changed into mass (intrinsic energy), and vice versa.

However, the word *energy* is so commonly used as being a scalar number that we cannot change this general practice. Hence, other names were invented that better convey the physical meaning of the relativistic Momentum: in particular, the "Momenergy" and the "Energy-Impulse tensor." Hence, for the sake of clarity, we will use the word Momenergy instead of Momentum in the rest of this chapter, as it better conveys its meaning: space–time energy.

Momenergy is a system concept since we have the fundamental system Momenergy conservation law (for isolated systems). This system concept yields an important new concept: the system-mass, which is the Lorentzian norm of the system Momenergy, with the important properties of being intrinsic and constant. The system mass is the system energy in its Centre of Momentum frame; it is its intrinsic energy, the minimal energy that was required for its creation.

This new system-mass concept is all the more important as most real objects actually are systems: in particular, an atom is composed of electrons and nucleons (protons, neutrons) that themselves are made of more elementary particles (quarks…). Note that the intrinsic energy (i.e., mass) of a part of a system can vary, especially during inelastic collisions (infinitesimally in usual cases).

This axiom scheme is based on the event and the existence of a scalar m representing the object's intrinsic energy. *However, we saw that the event scheme has some limitations; hence, we are induced to consider another possibility, whereby the Momenergy is the primitive concept:*

7.2.3 Alternative Possibility: The Momenergy as a Primitive Concept

Let's then make precise this alternative axiom scheme placing the Momenergy as the primitive concept:

Momenergy is a resource, intrinsic to an object or a system, which characterizes its capacity to change the speed or the structure of other object(s) or system(s). Everything that exists has a Momenergy: paraphrasing Minkowski, no one has ever observed an object else way than through its Momenergy since the very detection of an object requires an exchange of Momenergy.

This resource has to be 4D: Any object has a 3D spatial orientation, given by its velocity; even the photon has one. A fourth dimension is required to reference the different states: the time. If we assume the, universal homogeneity–isotropy, but not the absolute character of the time, then there is necessarily an invariant speed as shown in Section 7.3.1. This has important consequences: time and space are linked together, we are in a space-time universe where physical laws are 4D by essence, and these laws are those of Relativity.

The Momenergy is then a 4D quantity, whose properties match with the relativistic Momentum, $P = m\frac{dM}{d\tau}$, as seen in Sections 7.2.1 and 7.2.2.

The Lorentzian norm of the Momenergy is the mass (i.e., intrinsic energy), for an object as for a system. Thus, the mass is no longer a primitive notion. Regarding massless objects (photons), they also have a 4D Momenergy which can be seen by their interactions with massive objects, showing that they have both a spatial momentum and an energy.

The system Momenergy is the sum of the Momenergies of all its parts. The constancy of the system Momenergy for isolated systems is a consequence of the universal homogeneity and isotropy postulate: Indeed, the overall capacity for change of an isolated system can be considered as being a universe property concerned by the universal homogeneity and isotropy postulate, and hence must remain constant.

Remark: The same conclusion was obtained by the Noether theorem, also based on the constancy of the physical laws.

7.2.3.1 The Intimate Links between Proper Time, Proper Distance and Momenergy

The proper time and distance can be derived from the Momenergy of the photon: its energy being the time part, $E = h\nu$, we have a relation between a period of time and energy: $\Delta t = 1/\nu = h/E$. Thus, the second can be defined as the duration of ν consecutive periods of a photon whose energy is $E = h\,\nu$.

The same reasoning applies to massive objects: L. de Broglie showed that a particle also has a wave aspect, with the relation $E = mc^2 = h\,\nu$ (in the particle's frame). This provides the particle's proper time; moreover, the proper time universality can be explained by the fact that this relation is valid independently of the location of the particle, its gravity potential and its state of motion.

Similarly, the distance unit can also be defined with the photon Momenergy: its wavelength λ is observable, and we have the relation: $\lambda = c / \nu = ch / E$.

Thus, the proper time and proper distance can be defined using the Momenergy. Moreover, time, distance, energy and mass can be expressed with the same unit, and this is indeed commonly used in the physics of particles, where the electron volt is used for the time, the distance, the energy and the mass:

$$[\text{distance}] = [\text{time}] = \left[\text{mass}^{-1}\right] = \left[\text{energy}\right]^{-1}.$$

For example, 10^{-15} m $= [200 \text{ MeV}]^{-1}$.

The fact that both time and distance represent the inverse of the energy provides another explanation for the time–distance equivalence.

Time and energy are called dual notions, and the same for distance and the spatial part of the Momenergy. Indeed, when changing inertial frames, it is the same Lorentz transformation that applies for the Momenergy and for the event (assuming the origin-events of K and K′ are the same).

Compared to the approach of placing the event at the axiomatic root, this approach has several interests:

- It encompasses the cases of particles such as the photon, which do not fit in the event scheme.
- It explains the artificial postulates of the Riemannian manifold: the metric signature (–2), the g_{ij} coefficients and the proper time universality.
- It is consistent with Einstein's statement that "the mass (energy) generates the space." Hence, it is logical to place the Momenergy at the root.

Another aspect of the Momenergy is also important: it is a system concept, and there is another case where the energy appears in a fundamental law concerning systems: the second principle of thermodynamics.

If we further imagine extreme situations, such as a world where all that exists is composed of radiation, we cannot define the notions of time and distance. R. Penrose pointed out that these notions cannot exist since there is no observer, no one who matters about these notions. For R. Penrose, such a state of the universe is possible in its very beginning, and also at the end of the current accelerated expansion phase.

Thus, the event scheme is the state of the Momenergy where time and distance simultaneously appear, and it is characterized by the existence of localized massive particles. The current hypothesis relates this apparition to the occurrence of the Higgs field and its associated bosons (but this is out of the scope of this book).

Still one may wonder: Why is time not absolute? We indeed mentioned that this statement is necessary to demonstrate the existence of an invariant speed (cf. Section 7.3.1). Relativity shows that this invariant speed is also the maximum one. This prohibits infinite speeds, which has important consequences: With infinite speeds, objects have an infinite energy; instantaneous distant interactions are possible; a consequence can occur at the same time as its cause; computers can run extremely complex programs instantaneously; there is no chicken and egg problem any longer, etc. The reject of the absolute time and space definitely prohibits these possibilities. However, before the apparition of time and space, these possibilities were open; hence, some scientists could have said: "At the Big Bang there is no time, everything happened at once."

**

As far as fundamental notions in life are concerned, it is also interesting to know the views of some important philosophers:

7.2.3.2 Philosopher's Views

In the 4th century BC, the Greek philosopher Anaxagoras perceived the existence of a fundamental constancy in nature, and stated: « Nothing arises, nothing dies, but existing things together combine, and then separate. » This phrase well applies to the Momenergy.

Heraclitus stressed the omnipresence of change: "The only thing that is constant is change -" His vision of the constancy of the capability of change was quite prophetic.

In the 18th century, the French scientist, A.-L. Lavoisier famously said about the mass: « Nothing is lost, nothing is created, everything is transformed.» These words perfectly apply to the system Momenergy, and also to the system-mass or even to the system-energy, but not the mass of an individual object.

Besides, several other important philosophers had a luminous vision about the link between space and matter:

- Aristotle (–360 before JC) stated « In a world where there is no movement, how can we define time?» He even added: « Time is the number of the movement according to a before and an after.
- Leibnitz stated that space, time and matter are linked together.
- Spinoza and Descartes stated that matter and space–time constitute the same entity: all properties and relations of matter are those of space and time.
- Spinoza went even further, stating that not only matter and time are the same entity, but the spirit also.
- Mach showed the existence of a link between matter and space (cf. Mach principle). Besides, he stressed that as science progresses, concepts should evolve.
- Nietzsche confirms: There is no ultimate truth.

Newton was wrong about the absolute time and space, but still there is an important element of truth in his vision of unity in the universe: had he stated that the *proper* time and the *proper* distance were absolute, he would have been correct.

Finally, we should acknowledge that matter/energy/time still contain a great deal of mystery: we indeed ignore 95% of the matter/energy that compose our universe (27% dark matter; 68% dark energy). With the notions of time and space being linked to energy/matter, we should admit that the final word on these notions has not been said.

7.3 Axiomatic Considerations Concerning the Speed of Light

Light-speed invariance was the compelling finding that led to Relativity, and this was a consequence of Maxwell's equations. Hence, one may question if before Maxwell scientists could have made a reasoning leading to the light-speed invariance. The answer is positive: after the theory of Relativity was built, it was shown that light-speed invariance is implied by the other two fundamental postulates: universal homogeneity–isotropy and the principle of Relativity, but provided that the Newtonian postulate of the absolute time and space is rejected.

A formal demonstration was published by J.M. Lévy-Leblond in 1976 in the American Journal of Physics: "One more derivation of the Lorentz transformation." A more accessible one is presented hereafter.

In the second part, we will question the possibility of greater speeds than c, which will lead us to explore the hypothetical tachyon's world.

7.3.1 The Invariance of Light Speed Is Not a Necessary Postulate

We will first show that the velocity composition law must still be a homographic function. We will then apply the physical constraints on the possible homographic functions, which will enable us to deduce the result.

7.3.1.1 The Velocity Composition Law Must Still Be a Homographic Function

We cannot use the Lorentz transformation in our demonstration because its demonstration required the light-speed invariance.[3] However, we will first show that the event coordinates transformation function must still be linear.

7.3.1.1.1 The Event Coordinate Transformation Function Must Still Be Linear

Consider two inertial frames, K and K′, with K′ moving at the speed v along the X-axis of K and the X′-axis of K′. We saw that the universal homogeneity postulate implies that the transformation by which the coordinates in K of any space–time separation vector are transformed in K′ must be linear. We also saw that the distances are invariant along the transverse directions. Then, if (x, y, z, t) are the coordinates of a random space–time separation vector in K, the transformation giving the coordinates of the same vector in K′, must still have the form:

$$x' = a\,x + b\,t \tag{7.3}$$

$$t' = e\,x + f\,t \tag{7.4}$$

$$y' = y \ \text{ and } \ z' = z. \tag{7.5}$$

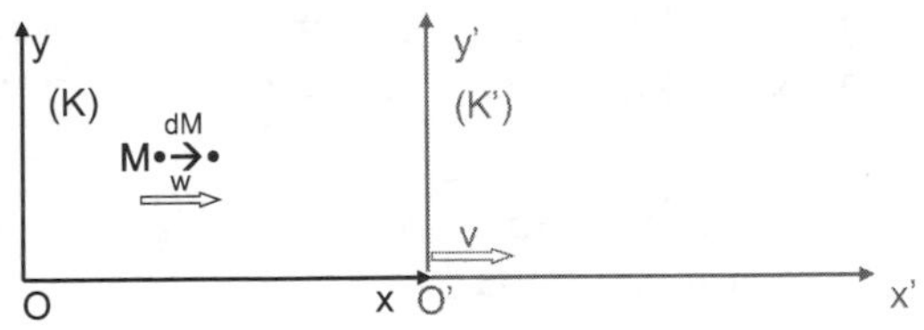

FIGURE 7.1
A moving object M seen in two inertial frames.

These four parameters a, b, e and f are only functions of the speed v of the frame K′ relative to K, and not of the coordinates of the considered event (due to the universal homogeneity–isotropy).

7.3.1.1.2 The Velocity Composition Law

Consider an object moving at the speed w in K along a parallel direction to the OX-axis. At the time t in K, its abscissa is x; then at the time t+dt, it is x+dx. These two steps define two events, denoted by M* and M*+dM*, their coordinates being in K: M(x, t) and M+dM(x+dx, t+dt). The space–time separation vector between these two events, dM*, is seen in K as dM (dx, dt). The speed w of this object in K is: w=dx/dt (Figure 7.1).

In K′, the same vector dM* has its components denoted by dM′(dx′, dt′). The universal homogeneity implies that dM′(dx′, dt′) is obtained from dM(dx, dt) by a linear transformation (cf. Section 1.4.1). Besides, the object speed in K′, denoted by w′, is: w′=dx′/dt′. We then have from equations 7.3 and 7.4:

$$w' = dx'/dt' = (a\,dx + b\,dt)/(e\,dx + f\,dt).$$

Let's divide all by dt, and then replace dx/dt with w, and we obtain: w′=(a w+b)/(e w+f). The inverse transformation has the same form; hence:

$$w = (a'w' + b')/(e'\,w' + f'). \tag{7.6}$$

Besides, we know that: a′=a(–v) and the same for the other coefficients b′, e′ and f′.

Relation (7.6) is a homographic function, except for the case where e′=0. However, relation (7.4) shows that in this case, t′ is not a function of x, meaning that time and space are not related, as it is in the Newtonian absolute frame. However, we place ourselves in the general case and don't take the absolute frame assumption, consequently the velocity composition function is a real homographic function. These functions are well known mathematically: they all have the same general shape with two perpendicular asymptotes.

7.3.1.2 Characteristics of Homographic Functions Matching with the Velocity Composition Law

Homographic functions are of two kinds: their slopes are either always positive, or always negative. We will demonstrate that the only physically acceptable form has positive slopes. Moreover, we will show that their slopes must always be decreasing, which will imply the existence of an invariant speed.

7.3.1.2.1 First Conditions

First, we must have f(0) = v. Then, for low speeds compared to c, the velocity composition function must be very close to the classical Galilean additive law, implying that the slope of the curve point of abscissa w′ = 0 must be very close to 1 as shown in Figure 7.2.

7.3.1.2.2 The Acceptable Homographic Functions Must Always Have Positive Slopes

We mentioned that the slope at the point of abscissa w′ = 0 is close to 1; hence, the slope at any point must be positive, meaning that the function is always increasing (except at its point of discontinuity).

There are only two possible forms for the homographic function matching the previous constraints: one with the slopes always decreasing (as in the left figure below), and the other one with the slopes always increasing (as in the right figure). *We will now show that the latter is unacceptable:*

7.3.1.2.3 The Acceptable Homographic Functions Must Have Their Slopes Always Decreasing

Homographic functions with increasing slopes are not acceptable because we could obtain an infinite speed by composing one finite speed, W′2, with the speed V, as shown in the right hand of the above Figure 7.2. This is not

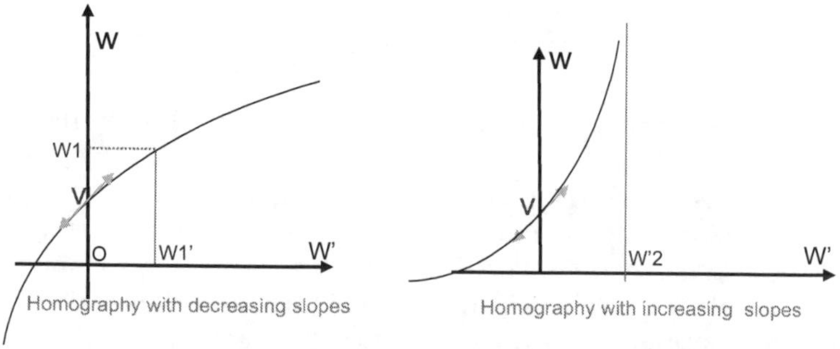

FIGURE 7.2
Homographic functions with positive slopes.

acceptable since it would be possible to generate an infinite amount of energy from objects with finite energy. Furthermore, when composing V with a slightly greater speed than W′2, we would obtain a negative speed with a very high value, which is absurd.

Hence, the only acceptable homographic functions are those with positive and always decreasing slopes. We will now show that these functions have one, and only one, invariant speed.

7.3.1.3 There Is One, and Only One, Invariant Speed; and It Is the Same in All Inertial Frames

The asymptote of a homographic function with decreasing slopes is necessarily horizontal. Besides, the slope at the point (0, v) being close to 1, there must exist an intersection between the homographic function and the bisector line w=w′. This intersection is denoted by L, and we have: L=f(L) (Figure 7.3).

We will show that the speed L is the same in all inertial frames by applying a reasoning based on the existence of a permanent center of symmetry, O″, between two inertial frames (as in Section 1.2.2):

Consider the following scenario: The referee observer in O″ simultaneously sends a pulse toward O of K and another one toward O′ of K′. When the pulse arrives at O′, an observer in O′ sends an object A toward O of K at the speed L, which is the invariant speed in K. The observer in O sees this object coming at the speed L, since L is the invariant speed in K.

When the pulse arrives in O, the observer in O sends an object B toward O′ of K′ at the speed L. The observer in O″ can say that the observer in K′ sees the object B moving at the same speed as the observer in O of K sees the object A moving, due to the symmetry of the scenario. Consequently, L is also the invariant speed in K′.

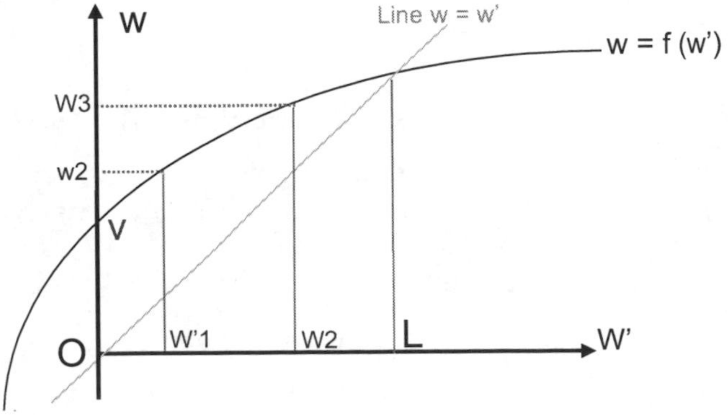

FIGURE 7.3
The only invariant speed.

7.3.1.4 Conclusion

We have shown that the existence of an invariant speed results from the universal homogeneity–isotropy postulate, in conjunction with the rejection of the absolute character of time and space. Since this invariant speed is reserved for massless objects (according to $E=mc^2\gamma$), it is the speed of light c.

However, are we really sure that the photon mass is absolutely null? It is extremely difficult to prove it experimentally: Many experiments were conducted to assess the photon mass, and until now they concluded that $m \leq 4 \times 10^{-51}$ kg. A similar investigation arose about the mass of the neutrino, which was assumed to be null until experiments showed that it was not. Besides, the photon is not the only particle assumed to be massless: e.g., the Gluon and the Graviton.

Then if one day we discover that the photon mass is not exactly null, it will mean that c is not invariant but another speed "L" is, with $L>c$. Consequently, the laws of Relativity will need to be changed by replacing c with L.

Thus, there were initially three main postulates on which the theory of Relativity was grounded: the universal homogeneity–isotropy, the principle of Relativity and the light-speed invariance. It now appears that the whole theory of Relativity can be built upon the universal homogeneity–isotropy, and a reduced principle of Relativity stating that the universal homogeneity–isotropy is valid in any inertial frame.

7.3.2 Visit in the Hypothetical Tachyon's World

We saw that objects cannot reach c, nor any greater speed than c, by composing velocities below c, nor even by applying a huge force on an object. Still, there could a priori exist objects called "Tachyons" that would be born moving at a greater speed than c. However, we saw that the causality principle forbids their existence: a tachyon going from a point A to a point B in an inertial frame K would be seen in other inertial frames arriving in B before departing from A.[4] *However, we also saw in Section 7.1.2 that the causality principle raises some issues; hence, it is worth investigating if other considerations forbid the existence of Tachyons.*

7.3.2.1 The Coefficients of the Lorentz Transformation Are Purely Imaginary

We are in the context where the absolute frame is rejected, hence the theory of Relativity applies. The Lorentz transformation is valid whatever the speed of K′ relative to K. For example, our demonstration of Section 2.2 is valid whatever this speed. We notice that the gamma factor is purely imaginary since $\gamma = \sqrt[2]{\frac{1}{1-v^2/c^2}}$, and when $v>c$ the term under the square root is negative. All coefficients of the Lorentz transformation are then purely complex numbers.

Consequently, a real event in K is seen as purely imaginary in a tachyon's frame (its coordinates being imaginary).

The reverse is also true: the reverse Lorentz matrix is also purely imaginary; hence, a real event in the tachyon's world is seen as purely imaginary in ours, and indeed, no one has ever observed such an event.

7.3.2.2 ***There Is No Correspondence between the Tachyon's Proper Time and Our Time***

Let's consider two consecutive beats of a clock in an inertial frame K. In the tachyon's frame, the duration between these beats is dilated by the gamma factor, but the latter is purely imaginary. Hence, 1 second in K is seen as an imaginary duration in the Tachyon's frame. Similarly, 1 second in the Tachyon's frame is seen as an imaginary duration in our inertial frame.

Tachyons thus don't have real rest frames. Photons also don't have rest frames, but they have real Momentum, which is not the case with tachyons, as we will see:

7.3.2.3 ***The Energy of the Tachyon Is Purely Imaginary, and Consequence of the Causality Principle***

The relativistic laws of dynamics still apply for higher speeds than c since the postulates, which were used did not include the obligation for objects to move at a lower speed than c. Consequently, the Momentum follows the Lorentz transformation when changing frames; and the famous energy relation, $E=mc^2\gamma$, shows that the energy of the tachyon would be purely imaginary. This implies that the tachyon cannot exchange any energy with real objects; hence, it cannot be detected (any detection mechanism requires an exchange of energy, even a very small one, such as a photon).

Consequently, we don't need the causality principle to state the impossibility of greater speeds than c: tachyons, if they exist, would be in a different world than ours, and without any interaction between the two. We can then ignore the hypothetical tachyon world.

Notes

1. The force has not been defined yet, but the state of no force can be defined as being incurred by any object which is extremely far from any other object.
2. This has also been seen in Section 3.1.
3. In particular, the time dilatation scenario.
4. Cf. Volume I Section 3.2.1.1.

8

Answers to the Questions and Problems

8.1 Answers to Problems of Chapter 1

Answers to Questions

Q1.1) Michelson–Morley. No, because the two beams come from a common beam (cf. Remark 1.1). This experiment shows that the ether has no impact on light propagation. Maxwell's equations imply in this context that light speed is independent of the speed of its source.

Q1.2) Principle of Relativity. No, because you incur a centrifugal inertial force that you can feel, which implies that your frame is not inertial. Moreover, this centrifugal force shows that your frame has a centripetal acceleration, which implies that your trajectory is circular relative to an inertial frame.

Q1.3) The Observer in Relativity. Yes, but provided that he considers the information from his senses and instruments as raw material, from which he or she makes reasoning, which correctly applies the laws of physics including Relativity. When saying that two events are simultaneous, the observer forgets that time is not intrinsic, and so he should have said that in his inertial frame, these events are simultaneous.

Q1.4) Intrinsic notion-1. No, it is the principle of relativity and the universal homogeneity and isotropy, which imply that the proper time flows at the same pace in all inertial frames (cf. Section 1.2.2).

Q1.5) Intrinsic notion-2. Yes, the event being a good example. Each coordinate of an event represents a non-intrinsic notion, which is meaningful in a given frame (e.g., time, distance). However, the event is intrinsic, being a phenomenon that exists independently of any frame. When the whole set characterizes an intrinsic notion, there is a one-to-one relationship between the set of numbers in K and in K'.

Q1.6) Intrinsic and primitive notion. No, because a non-intrinsic notion refers to a frame, and the frame is an artificial construction of men,

DOI: 10.1201/9781003201359-8

which relies of quasi-primitive notions. Hence, non-intrinsic notions rely on quasi-primitive notions, meaning that they cannot be quasi-primitive (cf. Section 1.3.3.3).

Q1.7) The Tennis ball scenario: The event coincidence postulate is violated: Indeed, consider the point A which is the center of the ball, and the point B which is the center of the hole. The observer who saw the ball traversing saw one event where A and B coincide. Hence, the event coincidence invariance implies that this coincidence is also seen by any observer in any frame.

Answers to Problems

Problem 1.1) Calculation of the Stellar Aberration Angle

In K, the photon velocity along OX (horizontal) is: $W_h = -c.\cos\theta$; and along OZ (transverse): $W_t = -c.\sin\theta$

Q1) *Calculation with the classical velocity addition law:* $\overline{W'} = \overline{W} + \overline{V}$. In our convention, the speeds V and W are negative; hence, to have the absolute value of the angle θ, let's consider the absolute values of these speeds. Thus: $W'_h = c.\cos\theta - v$, and $W'_t = c.\sin\theta$, and then: $tg\,\theta' = \frac{W'_t}{W'_h} = \frac{\sin\theta}{\cos\theta - \beta}$.

For θ=75°, the calculation yields: $\beta = 30/300{,}000 = 10^{-4}$.

Having sin(75°)=0.96592583 and (cos75°) = 0.25881905, tg θ′=3.73349332.

Then, artg θ′=75.0055345 and so we obtain the stellar aberration angle of: 0.005°.

1 second of arc=0.000277778°, and then 0.005° represents 0.005/0.000277778=19.9 seconds of arc, which is extremely small.

Q2) *Calculation with the Relativistic velocity composition law* (cf. Volume I Section 2.4): using the same notations as previously part, we obtain: $W'_h = (v - c.\cos\theta)/(1 - v.c.\cos\theta/c^2)$ and $W'_t = -c.\sin\theta/\gamma(1 - v.c.\cos\theta/c^2)$.

We then have: $tg\,\theta' = \frac{W'_t}{W'_h} = \frac{\sin\theta}{\gamma(\cos\theta - \beta)}$ *with* β=v/c. The value of γ is 1.000000005, meaning extremely close to 1. Thus, the relativistic calculation gives an extremely close result to the classical one, the difference being 0.00026 seconds of arc, which is extremely small and represents a difference of $5 \times 10^{-7}\%$.

Problem 1.2) Michelson and Morley Experiment

Q1) The interference pattern will be like if there is no ether, because both beams are identically affected by the ether wind; hence, their phases are identical when they recombine on the half-silvered mirror.

Q2) We obtain the same result as with the initial layout, it is just that mirror 2 plays the role of mirror 1 and vice versa. Comment: when repeating the same experiment all along the year with the layout being fixed, the motion of the Earth makes the layout progressively turn relative to the ether (assuming it exists). Hence, we should observe these variations on the interference pattern during the year.

Q3) No, when stating that the beam speed is c+v, and then c−v, we did not apply the relativistic velocity composition rule.

Q4) *We will apply the relativistic velocity composition law, denoted by* ⊕: The speed relative to the layout (or to the Earth) of the beam going to the mirror M_1 is: c ⊕ v=c, and when going back to the half-silvered mirror: c ⊕ (−v)=c. Hence: $T_1 = 2d/c$.

Regarding the speed of the beam going to the mirror M_2: in the ether frame, it is perpendicular to the initial beam trajectory; hence in the Earth frame, it is the composition of the speed in the ether and the ether speed relative to the Earth. It then takes an oblique direction, with the transverse part of the speed being given by the relativistic law: $W_y = \dfrac{W'_y}{\gamma_v\left(1+vW'_x/c^2\right)}$. In our case, we have $W'_x = 0$, so that: $W_y = \dfrac{W'_y}{\gamma_v} = \dfrac{c}{\gamma_v}$.

The duration for the beam to reach the mirror M_2 is then: $\dfrac{d}{W_y}$; hence, $T_2 = 2d.\gamma_v/c$. So finally: $T_2 - T_1 = (\gamma_v - 1)2d/c$ ■

With v=30 km/h, we have $\gamma_v = 1.000000005$; hence: $\gamma_v - 1 = 5\times10^{-9}$; then: $T_2 - T_1 = 20\times5\times10^{-9}/3\times10^{8} = 3.3\times10^{-16}$, which is smaller than the calculation made with classical physics.

Problem 1.3) Calculation of the speed S of the Center of Symmetry between K and K′.

Q1) *We will apply the relativistic velocity composition law, denoted by* ⊕. The speed of O′ relative to O″ is equal to S, and it also is the result of the composition of the speed of O′ relative to O (which is V) with the speed of O relative to O″ (which is –S). We thus have: S=V ⊕ (–S).

Then applying the relativistic speeds composition law, we have: $S=(V - S)/(1 - VS/c^2)$, which is a second-degree equation. To solve this equation, we will first simplify its expression by introducing the velocities: $\beta=V/c$ and $x=S/c$. This equation thus becomes: $x=(x-\beta)/(1-\beta x)$, then $x(1-\beta x)=x-\beta$, and then: $x^2\beta-2x+\beta=0$.

This equation has two solutions: $\frac{2+\sqrt{4-4\beta^2}}{2\beta}$ and: $\frac{2-\sqrt{4-4\beta^2}}{2\beta}$, which simplifies: $\frac{1+\sqrt{1-\beta^2}}{\beta}$ and $\frac{1-\sqrt{1-\beta^2}}{\beta}$. Only $x=\frac{1-\sqrt{1-\beta^2}}{\beta}$ is acceptable because x must be smaller than 1 since S must be smaller than c.

Q2) For usual speeds that are far from c, $x=\frac{1-\sqrt{1-\beta^2}}{\beta}$ can be approximated by:

$$x \approx \frac{1-\left(1-\frac{1}{2}\beta^2\right)}{\beta} = 1/2\beta.$$

With β=c / 100, the exact value of x is: 0.005000125, which is very close to: ½ β=0.005, the difference representing 0.0025%.

Problem 1.4) Resolving Contradictory Statements

Q1) This observer must not deduce that the length of the train TX has reduced, but that he or she sees it as shorter than its proper length (i.e., the length of the train when it is at rest relative to the observer).

Q2) Figure 1.16 is not an evidence for passengers in the train TX because this picture is not "neutral" since it has been taken from the platform, and this has an impact: the four points A, B, C and D are simultaneous in the platform frame, but not in the train TX's frame[1] (relativity of simultaneity). Hence, passengers in TX see a different picture, and indeed if they take a photo, it will be different.

Q3) Yes, the frame of the permanent center of the symmetry, which always exists between two inertial frames (cf. Section 1.2.1).

Problem 1.5) Time Dilatation and Distance Contraction

Q1) It is later, because my travel duration of 20 years was of my proper time, and it is seen dilated in a moving frame relative to me, such as the frame K.

Q2) My travel duration has been dilated in K by the gamma factor of 2. My speed must have been very close to 935,000,000 km/h.

Q3) My speed of 935,000,000 km/h corresponds to 0.866 c. Hence, I can say that I have covered the distance of: 20×0.8663=17.3 light-years.

However, this is not a proper distance since I was moving relative to the segment AB. Hence, I see this segment contracted compared to its proper length, which is then: 34.6 light-years. It is also the length seen by observers in K since they are fixed relative to this segment.

Problem 1.6) Gamma Factor Approximation Using Gamma = 1 + β²/2

Q1) From equation 1.17, the gamma factor approximation with $\gamma \approx 1+\beta^2/2$ is better than β^3. Expressed in percentage, we have: $d\gamma/\gamma \approx \beta^3 \times 100\%$.

Hence: 1,000 km/h $\rightarrow \beta^3 = \left(9.3\times10^{-7}\right)^3 = 8\times 10^{-1} \rightarrow 8\times 10^{-17}\%$.

For 10,000 km/s $\rightarrow 3.7\times 10^{-3}\%$

For $c/10 \rightarrow 0.1\%$. For $c/3 \rightarrow 3.7\%$.

Q2) To obtain: 0.1%, we have: $\beta^3 = 0.001$, and then $\beta = 0.1 \Rightarrow$ $v \approx 30{,}000$ km/s. For 0.5%, $\beta^3 = 0.005$, then $v \approx 51{,}000$ km/s. For 1% $\rightarrow$ 66,000 km/s.

Q3) $\gamma \geq 1+\beta^2/2$, because the next term in the Taylor polynomial is positive: $+3/8\,\beta^4$.

Problem 1.7) The Gamma Factor Approximation with Gamma $\gamma \approx 1+1/2\ \beta^2+3/8\,\beta^4$

Q1) From equation 1.16, the gamma factor approximation with $\gamma \approx 1+1/2\ \beta^2+3/8\,\beta^4$ is better than β^5. Expressed in percentage, we have: $\frac{d\gamma}{\gamma} < \beta^5 \times 100\%$ Hence: 10,000 km/h $\rightarrow \frac{d\gamma}{\gamma} < (0.033)^5 \times 100 = 4\times 10^{-6}\%$.

For $\beta = c/10 \rightarrow \frac{d\gamma}{\gamma} < 10^{-3}\%$. For $\beta = c/3 \rightarrow \frac{d\gamma}{\gamma} < 0.4\%$.

For $\beta = 0.7c \rightarrow \frac{d\gamma}{\gamma} < 17\%$.

Q2) To obtain 0.5%, we have: $\beta^5 = 0.005$, and then $\beta = 0.35 \Rightarrow v \approx 105{,}000$ km/s.

Then, 1% $\rightarrow$ 120,000 km/s.

Problem 1.8) The Fizeau Experiment

The speed relative to the ground of the beam, which propagates against the water flow is:

$$s' = c/n \oplus (-v). \text{ So: } s' = \frac{c/n - v}{1 - \frac{vc/n}{c^2}} \approx (c/n - v)(1+\beta/n) \approx c/n - v + v/n^2.$$

$$\text{Then, } dt = \frac{L}{s} - \frac{L}{s'} = \frac{10}{c/n - v + v/n^2} - \frac{10}{c/n + v - v/n^2}.$$

8.2 Answers to Problems of Chapter 2

Problem 2.1) The Train and Tunnel Scenario

Q1) Seen from the ground, the moving train is shorter than the tunnel due to the length contraction law.

Q2) Seen from the train, the moving tunnel is shorter than the train, due to the same law.

Q3A) An observer on the ground will answer that the bomb will explode inside the tunnel, since the train is shorter than the tunnel.

Q3B) An observer inside the train will answer that the bomb will explode outside the train, since the tunnel is shorter than the train.

Q4A) The proper length L of the train is equal to the proper length of the tunnel since the train at rest measures the same as the tunnel. The observer in O′ of the train says that the explosion will occur at the time $t'=L/c$, and that the length of the tunnel measures L/γ_v. The end of the tunnel will have moved toward him at the speed v during L/c, so its distance to him at the time of the explosion is: $L/\gamma_v - vL/c$, which is smaller than L. Hence, he will conclude that the explosion will take place while the front of the train is outside the tunnel.

Q4B) The observer in O (entrance of the tunnel) will see the signal also going at the speed c due to light speed invariance, and besides, he sees the train with length L/γ_v. Let's denote by t the time at which the explosion occurs in K. At the time t, the train will have covered the distance vt. Hence, the signal will have covered the distance: $L/\gamma_v + vt$. This signal also goes at the speed c relative to the ground; hence: $t=(L/\gamma_v + vt)/c$. He will deduce that the explosion will occur at the time $t = \dfrac{L}{c\gamma_v(1-\beta)}$ with $\beta=v/c$. During this time, the train will have covered the distance $d = \dfrac{Lv}{c\gamma_v(1-\beta)}$, so that the distance of the front of the train will be: $d = \dfrac{L}{\gamma_v} + \dfrac{Lv}{c\gamma_v(1-\beta)}$.

Let's simplify: $d = \dfrac{L}{\gamma_v}\left[1+\dfrac{\beta}{(1-\beta)}\right] = \dfrac{L}{\gamma_v(1-\beta)}$; multiplying the numerator and the denominator by $(1+\beta)$, we obtain:

$d = \dfrac{L(1+\beta)}{\gamma_v(1-\beta^2)}$ Then knowing that: $\gamma_v^2 = \dfrac{1}{(1-\beta^2)}$, we finally obtain:
$d = L\,\gamma_v\,(1+\beta)$, which is greater than L.

Q5) We saw that the time at which the explosion occurs in the train is $t' = L/c$. Besides, it occurs at the distance L in the train frame. Therefore, we can apply the Lorentz transformation to find the distance in K at which the explosion occurs, knowing that the speed of the ground relative to the train is –v. Hence, we have: $d = \gamma_v(L + vL/c) = L\gamma_v(1+\beta)$. We see that the Lorentz transformation greatly simplifies the calculations.

Q6) Let's denote by A the point of K inside the tunnel where the explosion is seen occurring by observers in K. Due to the event coincidence invariance postulate, this explosion is seen occurring in A by observers in all frames. Besides, the proper length OA is shorter than the proper tunnel length, hence observers in moving frames will see these two lengths contracted by the same ratio, hence they will also see A inside the tunnel.

Problem 2.2) Previous Train and Tunnel Scenario with a Minkowski–Loedl Diagram

See Figure 8.1.

Q1) The train is fixed in its frame K′, and corresponds to the segment O′F′. Seen from the ground at the time t = 0, the front of the train is at the intersection between the OX-axis (since t=0) and the worldline of F′, which is the parallel to OcT′ passing by F′. This intersection is the point A, and we can see that the proper

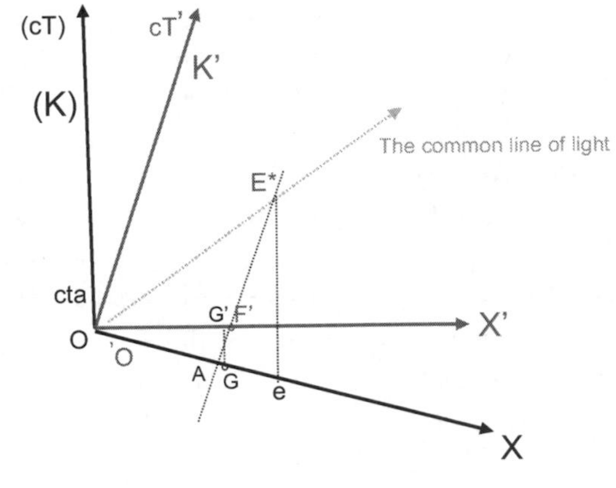

FIGURE 8.1
The Train and Tunnel scenario. Minkowski–Loedel diagram.

length of the train O′F′ is seen contracted from the ground: OA<OF′; and with the Minkowski–Loedl diagram, the scales of the lengths and the times are the same in K and K′.

Q2) The tunnel is fixed in K and corresponds to the segment OG, such that OG = O′F′. In K′ at t′=0, the two ends of the tunnel respectively are at the points O′ and G′, which is at the intersection between OX′ and the worldline of G (the end of the tunnel) which is parallel to OcT. In conclusion, the proper length of the tunnel, OG, is seen contracted by observers in K′ since O′G′<OA.

Q3) The signal emitted in O′ follows the common line of light. Let's call E* the event: "the signal reaches the front of the train." The abscissa of E* in the train frame K′ is then equal to O′F′, which implies that F′E* is parallel to the axis O′cT′. The event E* also corresponds to the bomb explosion, and we can see that in K its abscissa, e, is necessarily greater than OG, since E*A is perpendicular to OX.

Problem 2.3) Bob and Alice, Part 2: Amazing Effects When Traveling in the Cosmos

Q1) The 30-year duration of Alice is measured with her proper time; hence, it is seen dilated by Bob by the γ factor of 3, and Bob's clock will show $3 \times 30 = 90$ years! He will thus be 120 years old (hopefully). Remark: It will take even longer for the call to reach Bob due to the transmission delay (the rocket's speed being close to c). We can thus see that there are serious limitations in the communication possibilities between the Earth and extremely fast travelers in the cosmos!

Q2) Photons from Alice's emitter takes time to reach Proxima: the distance Earth-Proxima considered by Alice being: $30 \times 0.943 = 28.29$ light-years; thus, they take 28.29 years to reach Proxima. Then if a photon leaves Alice's emitter at Earth time $t = 30 - 28.29 = 1.71$ years, it will reach Proxima at year 30 (Earth time), exactly when Bob's rocket reaches Proxima.

Q3) Light coming from the rocket's clock takes time to reach Alice's telescope, let's call T this time duration measured in the Earth frame. T also represents the distance D covered by this ray expressed in light-years and measured in the Earth fame.

a. The time duration necessary for the rocket to cover this distance D is: $t = T/0.943$ (in Earth frame).

b. Hence, this ray arrives on Earth at the time $t + T = T(1 + 1/0.943) = 2.06T$.

c. This ray is seen on Earth at the year 30, thus: $t + T = 30 = 2.06T$, so $T = 14.56$ years.

T also is the distance D covered by the rocket when the ray was emitted; hence D=14.56 light-years (measured in Earth frame).

d. This distance is seen contracted in the rocket's frame, so: $D'=D/\gamma=D/3=4.853$.

e. The time measured in the rocket to cover the distance D′ is then: $t'=D/v=4.853/0.943=5.1466$.

Problem 2.4) Minkowski–Loedel Diagram of Bob and Alice

See Figure 8.2.

K′: Bob's Rocket frame / K: Alice's Earth frame

The angle α is such that sin $\alpha=0.943$; hence, $\alpha=70.6°$. Remark: For the sake of clarity, the above figure does not exactly reflects this value.

O and O′ correspond to the event E: "Bob leaves Alice and starts his cosmos trip."

Event F: "Bob arrives at the distance of Proxima."

Q1-1) Event G: "Alice initiates a call to congratulate Bob for his arrival at Proxima": FG is parallel to OX since F and G are simultaneous in K. Bob's time at the event G is: OG′, with OG′ parallel to O′X′.

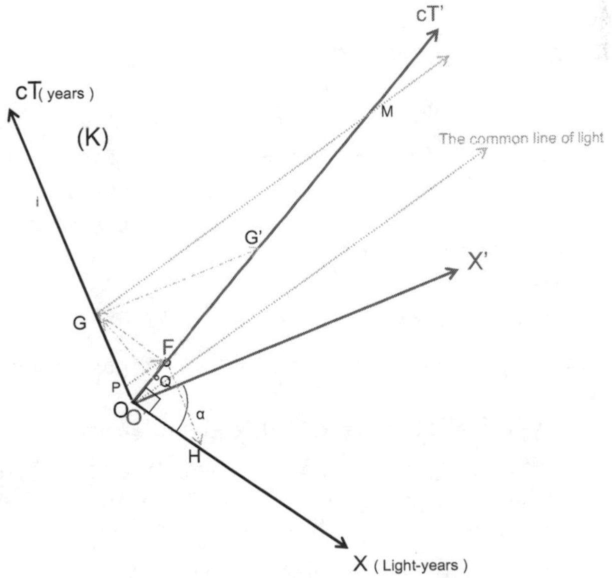

FIGURE 8.2
Minkowski Loedel Diagram of Bob and Alice.

Q1-2) The Photon emitted by Alice, and which reaches Bob at the event F, leaves Alice at the point **P** which is on the OcT-axis, and such that PF is parallel to the line of light.

Q1-3) The photon emitted by Bob's clock, which reaches Alice at the event G, has followed a line perpendicular to the line of light. Hence, it has been emitted at the point **Q**.

Q2) The Distance Earth-Proxima measured from the Earth is given by the point **H** which is the intersection between the OX-axis and the parallel to OcT passing by F.

Problem 2.5) Time Dilatation and De-synchronization Effect

In both diagrams, the clock A is at the point O=O′; the clock B is on the O′X′-axis at the time t′=0 in the train, and the same for the clock A; besides, the clock B is in the middle of O′C (Figure 8.3).

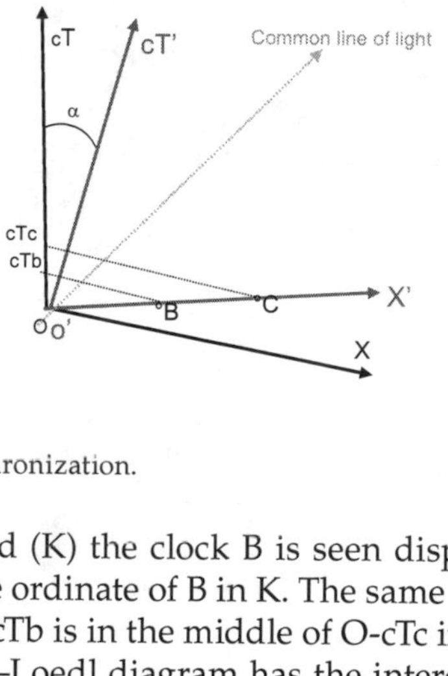

FIGURE 8.3
Time dilatation and de-synchronization.

From the ground (K) the clock B is seen displaying the time Tb, with cTb being the ordinate of B in K. The same applies for the clock C. Geometrically, cTb is in the middle of O-cTc in both diagrams.

The Minkowski–Loedl diagram has the interest of enabling us to calculate O-Tb: Indeed: O-cTb=OBtgα. Besides, in Minkowski–Loedl diagrams, we have: sinα=β.

Subsequently, $c.Tb = OB.tg\alpha = AB.\beta\sqrt[2]{1/\left(1-\beta^2\right)} = AB.\beta\gamma$, which matches the result of question 1 of Problem 2.13) of Volume I.

Problem 2.6) Hyperbolic Rotation Angle Calculation

Q1) We saw that cosh β= $\gamma_v = 1.512$. Hence, the angle θ which gives: 1.512 on the curve is approximately: 1.0. With a calculator, the value is given by arcosh(1.512)=0.973.

Q2) We calculate γ_v from the relation cosh, and we see on the cosh curve that cosh(1.76) = 3. Then, with the table giving the gamma factor, we find that: v = 0.94 c.

Problem 2.7) Velocities Composition

We will use the following result (cf. Section 2.5.2.1): artanh (s/c) = artanh (v/c) + artanh (w/c), with: $s = v \oplus w$. Case 1) We can see that artanh (0.4) ≈ 0.45 and artanh (0.6) ≈ 0.70. Hence, artanh(W/c) = 0.45 + 0.7 = 1.15. Then, the abscissa whose ordinate is: 1.15 is ≈ 0.80. So: W ≈ 0.80c.

Case 2) We can see that: artanh (0.6) ≈ 0.70 and: artanh (0.8) ≈ 1.15.

Hence, artanh(W/c) = 0.70 + 1.15 = 1.85. Then, the abscissa whose ordinate is 1.85 is: ≈0.95. So: W = 0.95c.

Problem 2.8) An Accelerated Rocket with Constant Power

Q1A) The rocket moves along the OX-axis, so its trajectory is: $\frac{d^2x}{d\tau^2} = k/m$, with k/m being constant.

Q1B) In a mixed-mode graph, the rocket trajectory is a horizontal parabola: We indeed have: $x = k/2mc^2.c^2\tau^2$. We set: $k' = k/2mc^2$, so: $x = k'c^2\tau^2$ or $c\tau = \sqrt[2]{x/k'}$ (Figure 8.4).

Q2A) The time Ta is given by the curvilinear length of the parabola curve from O to A*. We can see that this curvilinear length is greater than $c\tau_a$, which means that the time of A* seen from the ground is greater than in the rocket.

Q2B) This curvilinear length is given by the mathematical expression:

$$Ta = \int_o^A \sqrt[2]{dx^2 + c^2 d\tau^2} = \int_o^A \sqrt[2]{1 + \frac{c^2 d\tau^2}{dx^2}}dx = \int_o^A \sqrt[2]{1 + \frac{1}{4k'x}}dx = \int_o^A \sqrt[2]{1 + \frac{mc^2}{2kx}}dx.$$

Problem 2.9) The Event Horizon and the Rocket Trajectory

Q1A) $\lambda = 1$ light-year $= 3 \times 10^8$ m/s $\times 365 \times 24 \times 60 \times 60$ seconds $= 9.46 \times 10^{15}$ m.

Q1B) The curve equation is given by using the formula: $\cosh^2(\theta) - \sinh^2(\theta) = 1$.

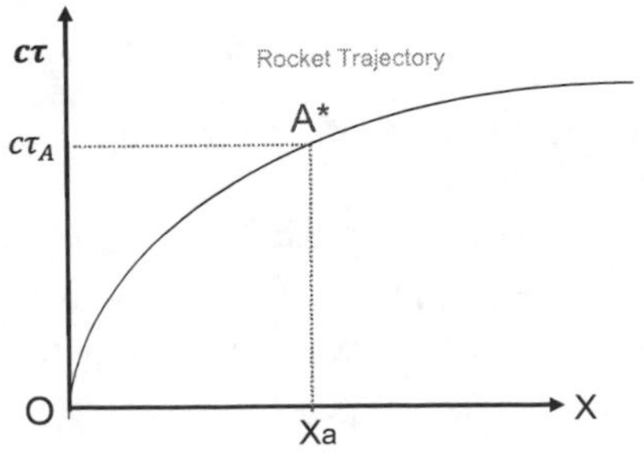

FIGURE 8.4
Rocket trajectory.

Thus, $(x/\lambda+1)^2 - c^2t^2/\lambda^2 = 1$; then: $x^2 - c^2t^2 + 2x\lambda = 0$.

Q1C) The asymptote slope is given by the terms of higher order: ct=x, (the line of light).

Q2A) $ds^2(dM) = c^2dt^2 - dx^2 = \lambda^2\left[\cosh^2(\theta) - \sinh^2(\theta)\right]d\theta^2 = \lambda^2 d\theta^2$. Here, we know that: $ds^2(dM) = c^2\,d\tau^2$; hence: $\tau = \lambda\,\theta/c + cst$; this cst=0 because at $\tau=0$, we have $\theta = x = 0$.

We then have: $\mathbf{x/\lambda = \cosh(c\tau/\lambda) - 1}$ and: $\mathbf{ct/\lambda = \sinh}(c\tau/\lambda)$.

Q2B) The asymptote crosses the OcT-axis at the point H given by limit(ct−x) when x tends to the infinite. Hence: $OH = \lim\left[\lambda\sinh(c\tau/\lambda) - \lambda(\cosh(c\tau/\lambda)) - \cosh(c\tau/\lambda) - 1\right] = \lambda$.

We indeed have: $\lim\left[\sinh(c\tau/\lambda) - \cosh(c\tau/\lambda)\right] = 0$, since: $\sinh(c\tau/\lambda) - \cosh(c\tau/\lambda) = -e^{-c\tau/\lambda}$, which tends to zero when τ tends to the infinite. The asymptote equation is: $ct = x + \lambda$.

Q3) The rocket speed is: $\beta(t) = \dfrac{dx}{d(ct)} = \dfrac{\sinh(c\tau/\lambda)d\tau}{\cosh(c\tau/\lambda)d\tau} = \tanh(c\tau/\lambda)$.

The limit of $\tanh(c\tau/\lambda)$ when τ tends to the infinite is 1. Hence, the rocket speed limit is $\beta = 1$ so: V=c.

Q4) The rocket acceleration is:

$$\frac{dV(t)}{dt} = \frac{cd\beta(t)}{dt} = \frac{cd\beta(t)}{cd\tau} = \frac{cd\tau}{dt} = c\frac{d\left(\tanh\left(\frac{c\tau}{\lambda}\right)\right)}{}\frac{c}{\gamma_v} = \frac{1}{\lambda\cosh^2(c\tau/\lambda)}\frac{c^2}{\gamma_v}.$$

Also, $\gamma_v = \left(1 - \dfrac{V^2}{c^2}\right)^{-1/2} = \left[1 - \tanh^2(c\tau/\lambda)\right]^{-1/2} = \cosh(c\tau/\lambda)$.

So finally: $\dfrac{dV(ct)}{dt} = \dfrac{d\beta(ct)}{dt} = \dfrac{c^2}{\lambda}[\cosh(c\tau/\lambda)]^{-3}$.

Q5) The relation (2.11) gives:

$$A'(c^2) = A(ct) = A(ct)\lambda_v^3 = \frac{c^2}{\lambda}[\cosh(c\tau/\lambda)]^{-3}[\cosh(c\tau/\lambda)]^3 = \frac{c^2}{\lambda}.$$

Bob thus incurs a constant acceleration of $c^2/\lambda = \dfrac{9\times10^{16}}{9.46\times10^{15}} = 9.51\,m/s^2$. This value can be read on his accelerometer, and it also reflects the engine thrust.

Problem 2.10) Event Horizon and Rocket Trajectory-2

Q1) We saw in Section 2.6.3.1.1 that the trajectory of an object having a constant acceleration as seen in its inertial tangent frame is a hyperbola, and that its parametric equations are: $t = \dfrac{\mathbf{c}}{\mathbf{a'}}\sinh\left(\dfrac{a'\tau}{c}\right)$ and $x = \dfrac{\mathbf{c^2}}{\mathbf{a'}}\left(\cosh(a'\tau/c) - 1\right)$.

We have: $c = 3 \times 10^8$ m/s, and then: $a'/c = 3.27 \times 10^{-8}\ s^{-1}$.

Q2) With the distances in light-years and the times in years, the value of c is 1. The value of the acceleration $a' = 9.8\,m/s^2$ expressed in light-year/year² is denoted by a″. Then: $t = \frac{1}{a''}\sinh(a''\tau)$ and $x = \frac{1}{a''}(\cosh(a''\tau) - 1)$. 1 year=31,356,000 seconds; hence:

$$a'' = a'\frac{31{,}536{,}000^2}{31{,}536{,}000 \times c} = a'\frac{31{,}356{,}000}{c} = 9.8 \times 0.1052 = 1.031\ \text{Year}^{-1}.$$

With $\tau = 1$ **year**, we obtain: t = 1/1.031 × sinh (1.031.x1) = 0.970 × 1224 = 1.187 **years**.

Then $x = 1/1.031 \times (\cosh((1.031 \times 1) - 1)) = 0.970 \times 0.508 =$ **0.563 light-years**.

Q3) $\tau = 10$ years yields: $t = (1/1.031) \times \sinh(1.031 \times 10) = 0.970 \times 18{,}511 = 15{,}016$ years.

And $x = (1/1.031) \times (\cosh(1.031 \times 10) - 1) = 0.970 \times 15{,}015 =$ **14,564 light-years.**

The difference with Q2) is huge: the cosh and sinh functions are indeed very close to exponential functions.

Q4) The ordinate of the point where the asymptote crosses the OcT-axis it is given by: lim (t−x) when τ tends to the infinite. Remaining with natural units, we have:

$$\lim(t - x) = \lim\left[\frac{1}{a''}\sinh(a''\tau) - \frac{1}{a''}(\cosh(a''\tau) - 1)\right]$$

$$= \lim\frac{1}{a''}[\sinh(a''\tau) - \cosh(a''\tau) + 1] = \frac{1}{a''}$$

Indeed, $\sinh(a''\tau) - \cosh(a''\tau) = -e^{-a''\tau}$, which tends to zero when τ tends to the infinite. Hence, the year of the event horizon is: 1/1.031 = **0.97** years. Then, Bob won't receive any news from Alice.

Problem 2.11) The Line of Simultaneity

This problem is a continuation of the previous one.

Q1) Alice can establish the parametric equation of Bob's trajectory (worldline), as seen in Section 2.6.3.1.1. We thus have for the time part: $t = \frac{c}{a'}\sinh\frac{a'\tau}{c}$ with t being Alice's time (frame K) and τ Bob's proper time, and a′ being Bob's acceleration relative to his inertial tangent frame (ITF).

Let's express all values in natural units: c = 1; a′ = 1 year^{-1}, then: $\mathbf{t = \sinh(\tau)}$. We then have at the event A*: Ta = sinh(1) = **1.175** year. Alice is thus 31.175 years at the event A* whereas Bob is 31 years.

Q2A) The abscissa of the event A* seen by Alice is also given by the parametric equation of Bob's trajectory: $x = \frac{c^2}{a'}(\cosh(a'\tau / c) - 1)$. With natural units, a′ = 1, so we have: $\mathbf{x} = (\cosh(\tau) - 1)$. Then with $\tau = 1$, we have: $\mathbf{Xa = cosh(1) - 1 = 0.543}$ year.

Q2B) The Minkowski diagram is shown in Figure 8.5.

Q3A) The set of events which occur simultaneously with A* for Bob is on the "simultaneity line," which is the AX-axis of Bob's frame K'_A at the event A*. This AX-axis is obtained as follows: We first draw the time axis of K'_A, knowing that it is tangent to Bob's trajectory at the event A*. Then we draw the line of light at the event A*, which makes a 45° angle with the parallel to OX passing by A*. Then, the AX-axis is the line such that the line of light is the bisector between this AX-axis and the AcT′-axis.

Q3B) The slope of the tangent line to Bob's trajectory in A* is obtained from the parameterized equation. With natural units, and having seen that a′ = 1, we have: $c\frac{dt}{dx} = \frac{d(\sinh(\tau))}{d(\cosh(\tau))} = \frac{\cosh(\tau)d\tau}{\sinh(\tau)d\tau} = \coth(\tau)$.

Consequently, the slope of the simultaneity line in A* is the inverse (being symmetrical relative to the line ct = x + const); hence this slope is: $\tanh(\tau)$. For the event A*, $\tau = 1$, so the slope of the simultaneity line is: $\tanh(1) = 0.762$. We then have: Ta − Tab = Xatanh(1). So: Tab = Ta − Xa. tanh(1). Then, Tab = sinh (1) − [cosh(1) − 1]tanh(1) = tanh(1) = **0.762** years.

Finally, Alice's age at the time Tab is: **30.762** years, which means that Bob considers that Alice is younger than him at the event A*. This is the opposite of the final outcome when they meet again.

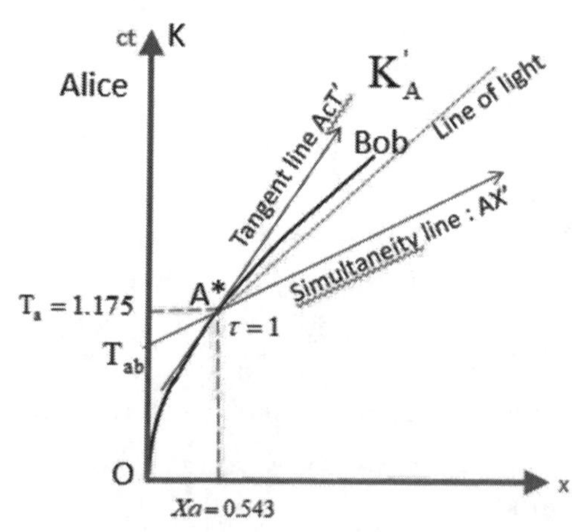

FIGURE 8.5
Line of simultaneity.

Problem 2.12) The Components of the Proper Velocity: S

We have: $\frac{1}{\gamma^2} = 1 - \frac{\vec{v}^2}{c^2}$. Let's differentiate with respect to τ:

$$\frac{-2}{\gamma^3}\frac{d\gamma}{d\tau} = -\frac{2}{c^2}\vec{v}\cdot\frac{d\vec{v}}{d\tau} = -\frac{2}{c^2}\vec{v}\cdot\frac{d\vec{v}}{d\tau}\frac{dt}{d\tau} = -\frac{2\gamma}{c}\vec{\beta}\cdot\vec{a}.$$

Then $\frac{d\gamma}{d\tau} = \frac{\gamma^4}{c}\vec{\beta}\cdot\vec{a}$; so for the time part of **S**: $\frac{dc\gamma}{d\tau} = \gamma^4\vec{\beta}\cdot\vec{a}$.

Regarding the spatial part:

$$\frac{d(\vec{v}\,\gamma)}{d\tau} = \gamma\frac{d\vec{v}}{d\tau} + \vec{v}\frac{d\gamma}{d\tau} = \gamma^2\vec{a} + \vec{\beta}\,\gamma^4\left(\vec{a}\cdot\vec{\beta}\right).$$

Problem 2.13) Spaceship Appearance

Q1) When you are in front of the point M, the distances between your eye and the different points of the station are not the same. The shrinking of the distances that are parallel to the velocity V does not change the fact from that the distance of the point M to your eye is shorter than that of the edge C, since the distance MC remains unchanged, and so C moves along the line CD which is the same line in K and K′. Consequently, the photon emitted from the edge C and received by your eye was emitted before that of the point M. Your rocket being moving to the right relative of the station, your see the station moving to the left; hence, you see the point M on the left relative to the points A and C, and so the general shape of the station is like in Figure 8.6.

Q2) The photons received from the point C were sent at the time $T'_0 - T'_C$. We have: $T'_C = CE/c$, and we also have: $CE^2 = ME^2 + CM^2 = 20^2 + 5^2 + x^2$. Besides, the photons received from the point M were sent at the time $T'_0 - T'_M$ and we have: $T'_M = ME/c = 20/c$. Between T'_C and T'_M, the station has moved by the distance:

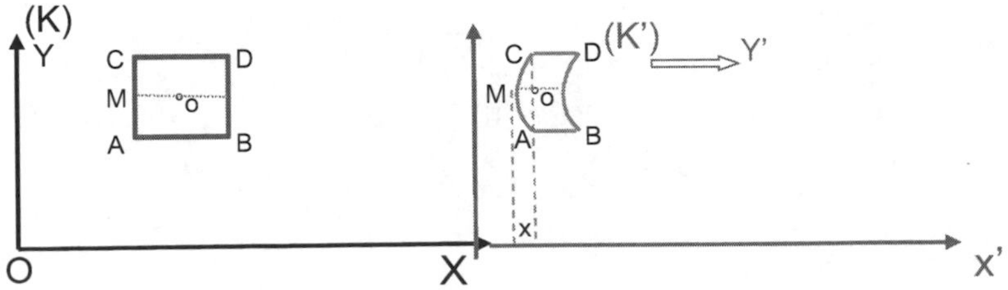

FIGURE 8.6
The cubic station seen from the Spaceship.

$x = (T'_C - T'_M)V = (T'_C - T'_M)0.9c$. Thus, $x = 0.9(\sqrt{425 + x^2} - 20)$. The calculation yields x = 0.561 m. We can thus see that the difference of the distances to the eye of the different points of the space station plays an important role in distorting the vision of an extremely fast traveler.

Q3) Let's consider a point N on AC which is at the proper distance MN=y. The previous reasoning leads to: $T'_N = NE/c$, and $NE^2 = ME^2 + NM^2 = 20^2 + y^2 + x^2$. Besides, $x = (T'_N - T_M)0.9c = 0.9\left(\sqrt{400 + x^2 + y^2} - 20\right)$. The coordinates (x, y) of N satisfy the relation: $x + 18 = 0.9\sqrt{400 + x^2 + y^2}$; then: $(x+18)^2 = 0.81\,(400 + x^2 + y^2)$.

So finally: $0.19\,x^2 - 0.81\,y^2 + 36\,x - 324 = 0$, which is a hyperbola.

Q4) The photon received from edge C was sent at the time $T'_0 - T'_C$. At that time, some points of the front side of the station had the same distance to your eye as the point C, and the front end of the station was on your right. Consequently, the photons they emit at $T'_0 - T'_C$ are received by you at T'_0. We thus have this strange situation where at the time when you see the station in front of you, you also see part of its front end, which gives you the impression of a rotation.

Problem 2.14) The determinant of the Lorentz matrix

The Lorentz transformation is a hyperbolic rotation (cf. section 2.5); its determinant is then: $\cosh^2(\theta) - \text{sinsh}^2(\theta) = 1$■

8.3 Answers to Problems of Chapter 3

Problem 3.1) The speed of a Neutrino with an energy of 10 MeV

Q1) In natural units, E=mγ; hence: 1.0 MeV=25 eV γ, and then γ=40,000. Applying the ultra-relativistic approximation, $\gamma \approx \frac{1}{\sqrt{2\varepsilon}}$; hence: $\varepsilon = \frac{1}{2\gamma^2} = \frac{1}{2 \times 16 \times 10^8} = 3.1 \times 10^{-10}$.

Then, $1 - \beta = \frac{c - v}{c} = 3.12 \times 10^{-10}$; hence:

$c - v \approx 3.12 \times 10^{-10} \times 3 \times 10^8 \approx 9.3$ cm/s.

Q2) $T = d/v = \frac{c10^5}{c(1-\varepsilon)} \approx 10^5(1+\varepsilon) \approx 10^5 + 3.1 \times 10^{-5}$ years. The additional time taken by the neutrino is:

$3.1 \times 10^{-5} \times 3.1 \times 10^7 \approx 16$ minutes.

Problem 3.2) The Muon energy

We have: $\gamma \approx \frac{1}{\sqrt{2\varepsilon}} = \frac{1}{\sqrt{2(1-0.992)}} = 7.92$. Then, E = $mc^2\gamma$ = $106 \times 7.92 = 838.5$ MeV, whereas the energy at rest is: 106 MeV; and the kinetic energy: 733 MeV, which is 87% of the total energy.

Problem 3.3) Cosmic Rays: Extremely Energetic Protons

Q1) From $E = mc^2\gamma$, $\gamma = \frac{5\times10^{19}}{10^9} = 5\times10^9$.

Then: $\varepsilon = \frac{1}{2\gamma^2} = \frac{1}{50\times10^{18}} = 2\times10^{-20}$ with $\varepsilon = (1-\beta)$.

Then: $2\times10^{-20} = (c-v)/c$; hence: $c - v = 10^{-19}\times 2\times10^{-20}\times 3\times10^8 = 6\times10^{-12}$ m/s. We can see that its speed is extremely close to c. The distance difference with c in 1 year: 1 year is 3.15×10^7 seconds; hence, it is: 4.7×10^{-4}m or 0.47 mm.

Q2) The proton mass is $\approx 10^9\ \text{ev}/c^2 = \frac{10^9\times1.6\times10^{-19}}{9\times10^{16}} = 1.7\times10^{-27}$ kg.

The spatial part of the system momentum conservation yields: $1.7\times10^{-27}.\,3\times10^8.\,5\times10^9.v\gamma$.

Thus: $v\gamma = 25.5\times10^{-27+8+9+6} = 2.55\times10^{-3}$m/s.

For such low speed: $\gamma \approx 1$, $v = 2.55\times10^{-3}$ m/s.

With the finest particle $v\gamma = 25.5\times10^{-27+8+9+15} = 2.55\times10^6$ m/s=2,550 km/s. For such speed, $\gamma < 1+10^{-5}$; hence: v = 2,550 km/s. We can thus see that such very fine particles can be pushed away very fast by cosmic rays. Remark: On the other hand, these collisions begin at the atomic level, and it takes many meters of lead to stop an energetic particle. In outer space, light pressure has a bigger effect than cosmic rays.

Problem 3.4) Accelerated Electrons

Q1) The final inertia is: $m\gamma = m\times1.400$. The 3D work of the Newtonian-like Force is equal to the relativistic energy variation (cf. Section 3.3.1) which is:

Q2) The work of the Force is: $\Delta E = W = qE\times10 = 204$ keV; hence: E=20.4 kV.

Q3) The relation (3.15) applies: $F_s = m\gamma a\left(1+\frac{\gamma^2v^2}{c^2}\right)$ and then

$$a = \frac{F_s}{m\gamma\left(1+\frac{\gamma^2v^2}{c^2}\right)}.$$

At the start, $a = \frac{20.4\ \text{KeV/m}}{511\ \text{KeV}/c^2} = 0.0399\times c^2 = 3.59\times10^{15}$ m/s^2.

At the end, $a = \dfrac{20.4\ \text{KeV/m}}{511 \times 1.4\left(1 + \dfrac{0.7^2 \times 1.4^2}{1^2}\right)\text{KeV/c}^2} = 1.31 \times 10^{15}$ m/s².

We can see that the acceleration is less at the end than at the start, despite the constancy of the Force, and this is due to the electron inertia increase with the speed.

Problem 3.5) Electrons in a Magnetic Field

The 3D relation (3.22) applies, $\vec{F_s} = q\left(\vec{E} + \vec{v} \times \vec{B}\right)$, with F_s being the spatial part of the Newtonian-like Force. Hence, $\vec{F_s} = q\left(\vec{v} \times \vec{B}\right)$, implying that $\vec{F_s}$ is perpendicular to both $\vec{v}$ and $\vec{B}$. Consequently, the electron remains in K, and its acceleration is given by the relation (3.14): $\vec{a} = \dfrac{1}{m\gamma} \cdot \vec{F_s} - \dfrac{\vec{F_s} \cdot \vec{v}}{m\gamma c^2}\vec{v}$. We can see that: $\vec{a} = \dfrac{1}{m\gamma} \cdot \vec{F_s}$ since $\vec{F_s} \cdot \vec{v} = 0$. The acceleration is then parallel to the Force; hence, it is perpendicular to the speed. The trajectory is then circular, its radius being given by the relation: $a = v^2/r$, and then: $r = v^2/a = v^2 m\gamma/(evB) =$ **$v m\gamma/(eB)$**, which gives:

$$r = 0.7 \cdot 3 \times 10^8 \times 9.093 \times 10^{-31}\ \text{kg} \times 1.51 / e \times 0.2$$

$$= 1.602 \times 10^{-19} \times 0.2 = 9.0 \times 10^{-3}\ \text{m} = 0.9\text{cm}.$$

With v=0.9999c, we find: 6.2 m.

Problem 3.6) Cosmic Protons Deflected by the Terrestrial Magnetic Field

The same reasoning as in Problem 3.5 applies, showing that the proton trajectory will bend with a curvature radius given by: **$r = v m\gamma/(eB)$**. To calculate the γ and v, we have: $E = mc^2\gamma$, and then $\gamma = \dfrac{12}{0.938} = 12.79$; hence, the speed is very close to c.

Then $r = c \times 12.79 \times 1.673 \times 10^{-27}/1.602 \times 10^{-19} \times 5.7 \times 10^{-5} = 699$ km.

Then, with the less energetic proton: $\gamma = \dfrac{5}{0.938} = 5.33$. The speed is again very close to c; so finally: r = 291 km.

Problem 3.7) Electrons Entering a Transverse Electrical Field

From relation (3.14), we have: $a_y = \dfrac{qE}{m\gamma} - \dfrac{qEV_y^2}{m\gamma c^2}$ and $a_x = -\dfrac{qEV_y}{m\gamma c^2}V_x$.

The first relation shows that a_y is always positive and will decrease as V_y increases; consequently V_y will always be positive, implying that a_x is negative, meaning that the electrons speed along OX will decrease. The value of a_x is small because of γc^2 in the denominator, so that when V_y approaches c, a_x will be close to 0.

Problem 3.8) The Fusion Helium-3 Neutron

The sum of the masses/energies must remain constant:
$3.01393u + 1.00867u = 4.00151u + x.$

So: $x = 0.00211u = 0.00211u \times 0.9315$ GeV/c^2 per u = 20.58 MeV.

Problem 3.9) Binding Energies in Uranium

Q1) ${}^{235}_{92}U$: mass defect = 1.856 u = 0.8% of the total mass, represents a binding energy of 1.74 GeV. Barium: 1.233 u; 0.875%; 1.149 GeV; Krypton: 0.822 u; 0.894%; 0.766 GeV.

Comment: the uranium 235 has a very low binding energy. When transforming into smaller atoms with higher binding energy, some of its mass will be converted into kinetic energy.

Q2) From equation 3.3, we have

$$\sum_i Ki = (1.149 + 0.766) - 1.74 = 0.175 \text{ GeV} = 175 \text{ MeV}.$$

Problem 3.10) Creation of an electron-positron pair

Let's call x the minimal energy of the gamma ray. We denote by E the energy of each resulting particle, and p their spatial momentum. We will use fully natural units. The system Momentum conservation gives: (m, 0) + (x, x) = (3E, 3p); hence: m + x = 3E (1), and x = 3p. (2)

For each electron or positron, we have: $E^2 = m^2 + p^2$ (cf. Vol 1 Sections 5.1.3.3 and 5.2.4). Then with (2): $E^2 = m^2 + \frac{x^2}{9}$. Then with (1): $E^2 = m^2 + \frac{x^2}{9} = \frac{(m+x)^2}{9}$ so: $9m^2 = m^2 + 2mx$; finally: x = 4m = 2.044 MeV.

Problem 3.11) Decay of a Lambda Particle

Q1) Relation (3.5A) concerning disintegration into two bodies yields:

$$E_p = \frac{m_\lambda^2 + m_p^2 - m_\pi^2}{2m_\lambda} \cdot c^2 = 0.9436 \text{ GeV},$$

$$\text{and } E_\pi = \frac{m_\lambda^2 + m_\pi^2 - m_p^2}{2m_\lambda} \cdot c^2 = 0.1721 \text{ GeV}.$$

Q2) The pion speed can be obtained by: $E = mc^2\gamma$, so: $\gamma = \frac{E}{mc^2} = 1.232$; then using $\beta^2 = 1 - \frac{1}{\gamma^2}$, β = 0.585, which is 175,378 km/s.

**

Topics related to the Potential Energy

Problem 3.12) The Speed of a Falling Object on the Earth

The constancy of the sum of the potential and the classical kinetic energy yields: $1/2mv^2 = \frac{mMG}{Z_A} - \frac{mMG}{Z_B}$. Hence:

$$v^2 = 2\times 4\times 10^{14}\times\left(\frac{1}{6\times 10^6} - \frac{1}{106\times 10^6}\right) = 8\times 10^{14}\left(\frac{100\times 10^6}{636\times 10^{12}}\right) = 1.23\times 10^8,$$

so: v = 11.2 km/s.

With Relativity: $c^2(\gamma - 1) = 0.63\times 10^8$ m²/s². So: $\beta = v/c = 3.7\times 10^{-5}$, we can approximate: $\gamma(\beta) \sim 1 + \beta^2/2$, so that the relativistic calculation gives a result which is extremely close to the classical one, the approximation of the gamma factor being in the range of $\beta^3 = 10^{-13}$.

Problem 3.13) The Speed of a falling object on the Sun

Q1) We now have: $MG = 2\times 10^{30}\times 6.67408\times 10^{-11} \approx 1.33\times 10^{20}$.

The constancy of the sum of the potential and classical kinetic energy gives: $mc^2(\gamma - 1) = \frac{mMG}{Z_B} - \frac{mMG}{Z_A}$. Then,

$$c^2(\gamma - 1) = 1.33\times 10^{20}\times\left(\frac{1}{10^5} - \frac{1}{10^{10}}\right) = 1.33\times 10^{20}\left(\frac{10^{10}}{10^{15}}\right) = 1.33\times 10^{15},$$

$$\text{so } \gamma = 1 + \frac{1.33\times 10^{15}}{9\times 10^{16}} = 1 + 1.48\times 10^{-2} = 1.015.$$

We have $\beta^2 = 1 - \frac{1}{\gamma^2} = 0.0293$ so $\beta = 0.171$, so v = 51,385 km/s.

Q2) In the ITF, we have: $F \approx ma = \frac{mMG}{Z_A^2}$. Hence,

$$a = \frac{1.33\times 10^{20}}{10^{2\times 5}} = 1.33\times 10^{10} \text{ m/s}^2 \text{ (huge!)}.$$

Q3) From equation 2.9, we have: $A' = \frac{a}{\gamma^3(1-\beta^2)^3} = \frac{1.33\times 10^{10}}{1.0456\times 0.914} =$ 1.39×10^{10} m/s², which represents ≈ 5% difference with the one in the IFT.

Problem 3.14) Calculation of the Escape Velocity from Earth

Let's denote by Vo the escape speed, and R the Earth radius. The object leaves the Earth at the speed Vo and goes upward virtually until the infinite where its speed is zero. Let's call B the "arrival" point in the infinite. From equations 3.16 and 3.20, we have using the classical kinetic energy: $\frac{1}{2} m Vo^2 - mMG/R = 0 - mMG/\infty = 0$.

Hence: ½ mVo^2 = mMG/R, so: $Vo = \sqrt[2]{2GM/R}$, which gives: 11.2 km/s=40,320 km/h.

The difference between the relativistic kinetic energy and the classical one is extremely small with this speed, the approximation being in the range of $\beta^3 \approx 5\times10^{-14}$.

Problem 3.15) Theory – The Work of the Four-Force

The Lorentzian scalar product being invariant, the work of the 4-Force is the same in any frame K and K° (object frame). Hence in the context of constant mass: $(cdt, dx, dy, dz)\bullet\left(\frac{dE}{cd\tau}, \Phi_x, \Phi_y, \Phi_z\right) = c^2 dm = 0$ (since cst mass). Hence in K: $dE.dt/d\tau = dE.\gamma = dx\Phi_x + dy\Phi_y + dz\Phi_z$ ■

Alternative method) From equation 3.9, we have: $\Phi = \gamma\, \mathbf{F}$. Besides, we have: $dE = dx\, F_x + dy\, F_y + dz\, F_z$. Hence: $dE.\ \gamma = dx\Phi_x + dy\Phi_y + dz\Phi_z$ ■

Problem 3.16) Theory – The Newtonian Force

Q1) In K and in K′, the laws represented by the Newtonian-like Force are the same:

$F = \frac{dP}{dt}$ and $F' = \frac{dP'}{dt'}$; hence, the Newtonian-like Force respects the principle of Relativity.

In classical physics, the acceleration was identical in all inertial frames; hence, so was the Newtonian force. But the principle of Relativity by itself does not oblige a quantity to be equal in all frames (and this is indeed the case of non-intrinsic quantities).

Q2) The quantity dt has no meaning in K′; hence, we cannot convert $\frac{dP}{dt}$ into K′, which stems from the non-intrinsic character of the Newtonian-like Force. The information required for F′ is the time rate of change of the object Momentum, the time used for this rate being t′ of K′ (cf. more in the next question).

Q3) We have $\Phi = F\gamma_v$. Hence, if we know the speed v of the object experiencing the Force relative to K, we can calculate the Force of Minkowski. Then, if we know the speed v′ of the same object relative to K′, we have: $\Phi' = F'\gamma_{v'}$. Besides, we have: $\Phi' = (\Lambda)\Phi$ with Λ being the Lorentz transformation from K to K′. So:

$$F' = \frac{\Phi'}{\gamma_{v'}} = \frac{(\Lambda)\Phi}{\gamma_{v'}} = \frac{\gamma_v}{\gamma_{v'}}(\Lambda)F \ ■$$

This shows that in Q2, we needed the information as to the speeds of the object relative to K and to K′, and also the speed of K′ relative to K.

8.4 Answers to Problems of Chapter 4

Answer to Questions

Questions 1) **S** is a contravariant tensor of rank 1 (cf. Volume 1 Section 4.2.2.2). **A** also is a contravariant tensor of rank 1 for the same reason, i.e., in K′ we have $\mathbf{A}' = \frac{d\mathbf{S}'}{d\tau} = \frac{\Lambda d\mathbf{S}}{d\tau} = \Lambda\mathbf{A}$ where Λ is the Lorentz transformation.

Question 2) **S** being a tensor, its norm is invariant, and we indeed have: $ds^2(\mathbf{S}) = \mathbf{S} \bullet \mathbf{S} = \mathbf{S}^2 = c^2$.

Being constant, its derivative relative to the proper time τ is null; hence: $(\mathbf{S} \bullet \mathbf{S})' = \mathbf{0} = \mathbf{2} \cdot \mathbf{S} \bullet \frac{d\mathbf{S}}{d\tau}$; so: $\mathbf{S} \cdot \mathbf{A} = 0$.

Question 3) Yes, because $\vec{\Phi}' = (\Lambda)\vec{\Phi}$ as seen in Volume I, relation (5.15). Contravariant.

Question 4) No, because V is not an intrinsic notion: its definition indeed refers to an inertial frame (due to the time t).

Question 5) No because the Newtonian-like Force is not an intrinsic notion: its definition indeed refers to an inertial frame (due to the time t).

Question 6) $\mathbf{S}^* = \left(c, -\frac{dx}{d\tau}, -\frac{dy}{d\tau}, -\frac{dz}{d\tau}\right)$.

Question 7) We saw in Section 4.2.3.1 that the resulting vector depends on the trajectory on which the parallel transport is performed between A and B. However, its norm remains the same since dV is constantly null.

Question 8) We saw in the previous question that the norm is conserved. Hence, the scalar product is also conserved, considering: $2\,\mathbf{U} \bullet \mathbf{V} = ds^2(\mathbf{U} + \mathbf{V}) - ds^2(\mathbf{U}) - ds^2(\mathbf{V})$.

Answer to Problems

Problem 4.1) Tensors of rank 1: Difference between $\mathbf{e}^1$ and $\mathbf{e}_1^*$

Q1) Both are linear forms, and hence they are covariant rank 1 tensors. However, they are different: $\mathbf{e}^1$ is such that for any vector $\mathbf{e}_j$, we have: $\mathbf{e}^1(\mathbf{e}_j) = \delta_j^1$.

Also, $\mathbf{e}_1^*$ is such that $\mathbf{e}^*_1(\mathbf{e}_j) = \mathbf{e}_1 \bullet \mathbf{e}_j = g_{1j}$; hence, if the space is curved, and then $\delta_j^1 \neq g_{1j}$, meaning that these tensors are different; otherwise, they are identical.

Q2) From $\mathbf{e}^*_1(\mathbf{e}_j) = \mathbf{e}_1 \bullet \mathbf{e}_j = g_{1j}$, we deduce $\mathbf{e}_1^* = g_{1j}\mathbf{e}^j$ and more generally: $\mathbf{e}_i^* = g_{ij}\mathbf{e}^j$.

Problem 4.2) Tensors of rank 1: Difference between the scalar products $\mathbf{e}^i \odot \mathbf{e}^j$ and $e_i \bullet \mathbf{e}_j$.

First answer: $\mathbf{e}^i \odot \mathbf{e}^j = g^{ij}$ and $\mathbf{e}_i \bullet e_j = g_{ij}$. Hence, they are different in case of curved space because the matrix $G'(g^{ij})$ is different from its inverse which is the matrix $G(g_{ij})$. Fundamentally, the reason is that by definition $\mathbf{e}_i \bullet e_j = \mathbf{e}_i^* \odot \mathbf{e}_j^*$, but $\mathbf{e}_j^* \neq \mathbf{e}^j$ as seen in the previous problem, except if there is no chrono-geometrical deformations.

Problem 4.3) Show that all metrics in the space-time satisfy the relation: $g^{ij}g_{ji} = 4$

We saw in Section 4.1.3 that g^{ij} is the generic element of the inverse matrix of the Metric tensor $G\left(g_{1j}\right)$. Let's consider the case i = 3: the expression $g^{3j}g_{j3}$ means the multiplication of line 3 of G^{-1} with column 3 of G: it is equal to 1. Then, as i successively takes the values 0, 1, 2 and 3, the same reasoning applies, leading to $\mathbf{g}^{ij}\mathbf{g}_{ji} = 4$.

Problem 4.4) The Trace operator

For the sake of clarity, consider the second term of the diagonal of the matrix (A)(B): it is: $\sum_{i=1}^{4} a^{2i}\, b^{i2}$, with Einstein's compact notation: $a^{2i}b^{i?}$. The sum up of all terms in the diagonal is: $\sum_{j=1}^{4}\sum_{i=1}^{4} a^{ji}b^{ij}$ or: $a^{ji}b^{ij}$.

We can see that this sum is equal to $b^{ji}a^{ij}$, which is the Trace of (B)(A) ■

Problem 4.5) Mixed covariant-contravariant tensor

In K, we have: $(w^i) = (C_j^i)(v^j)$ where V and W are contravariant rank 1 tensors. Then in K', we have: $\left(w^i\right)' = \left(C_j^i\right)'\left(v^j\right)'$, then: $(\Lambda)\left(w^i\right) = \left(C_j^i\right)'(\Lambda)\left(v^j\right)$, then: $\left(w^i\right) = (\Lambda)^{-1}\left(C_j^i\right)'(\Lambda)\left(v^j\right)$, and so: $\left(C_j^i\right) = (\Lambda)^{-1}\left(C_j^i\right)'(\Lambda)$ or, $\left(C_j^i\right)' = (\Lambda)\left(C_j^i\right)(\Lambda)^{-1}$ ■

Problem 4.6) Twice covariant rank 2 tensor

Being twice contravariant, there exists two contravariant rank 1 tensors, **A** and **B,** such that: $F = \mathbf{A} \otimes \mathbf{B}$, meaning: $F^{\alpha\beta} = a^{\alpha}b^{\beta}$. The covariant tensor associated to **A** is: $\mathbf{A}^* = (\mathbf{A})^T(G)$, thus: $a_i = g_{\mu i}a^{\mu}$; similarly: $b_j = g_{\nu j}a^{\nu}$. We then have: $F^* = \mathbf{A}^* \otimes \mathbf{B}^*$, thus: $F_{ij} = a_i b_j = g_{\mu i}\, a^{\mu}\, g_{\nu j}\, a^{\nu} = g_{\mu i}\, g_{\nu j}\, F^{ij}$

Problem 4.7) The Christoffel Coefficients of the Surface of a Sphere

Q1) $MN^2 = R^2\, d\alpha^2 + R^2 \sin^2\alpha\, d\varphi$.

Q2) $MN^2 = g_{11}d\alpha^2 + g_{22}\,d\varphi^2 + 2g_{12}\,d\alpha d\varphi$.

Hence, $g^{11} = R^2$; $g^{22} = R^2 \sin^2\alpha$; $g_{12} = g_{21} = 0$.

Q3) We saw that the g^{ij} constitute the inverse matrix of the g_{ji}; hence:

$$g_{ji} = R^2\begin{pmatrix} 1 & 0 \\ 0 & \sin^2\alpha \end{pmatrix}$$

Q4A) To calculate $\Gamma^{\alpha}_{\varphi\varphi}$ using the relation (4.31), we must set: $\mu=1$ and $c=b=2$; then:

$\Gamma^{\alpha}_{\varphi\varphi} = \Gamma^{1}_{22} = \frac{1}{2}\left(\partial_{\varphi} g_{a2} + \partial_{\varphi} g_{2a} - \partial_{a} g_{22}\right) g^{21}$. It is null except if $a=2$;

hence: $\Gamma^{\alpha}_{\varphi\varphi} = \frac{1}{2}\left(\partial_{\varphi} g_{11} + \partial_{\varphi} g_{22} - \partial_{\alpha} g_{22}\right) g^{11}$.

This simplifies:

$$\Gamma^{\alpha}_{\varphi\varphi} = -\frac{1}{2}\left(\partial_{\alpha} g_{22}\right) g^{11} = -1/2\left(2R^2 \sin\alpha \cos\alpha\right)/R^2 = -\sin\alpha \cos\alpha.$$

Q4B) For calculating $\Gamma^{\varphi}_{\alpha\varphi}$, we set: $\mu = 2$, $c = 1$ and $b = 2$; then:

$$\Gamma^{\varphi}_{\alpha\varphi} = \Gamma^{2}_{12} = \frac{1}{2}\left(\partial_{\alpha} g_{a2} + \partial_{\varphi} g_{1a} - \partial_{a} g_{22}\right) g^{a2}.$$

It is null except if $a=2$; hence: $\Gamma^{\varphi}_{\alpha\varphi} = \Gamma^{2}_{12} = \frac{1}{2}\left(\partial_{\alpha} g_{22} + \partial_{\varphi} g_{12} - \partial_{\varphi} g_{22}\right)$

$$= \frac{1}{2}\left(2\sin\alpha\cos\alpha\right) / R^2 \sin^2\alpha = \cot\alpha \;\blacksquare$$

8.5 Answers to Problems of Chapter 5

Problem 5.1) $\overrightarrow{\text{curl}}\,(\overrightarrow{\text{grad}}\,(\varphi)) = 0$

The first component of $\overrightarrow{\text{curl}}\,(\overrightarrow{\text{grad}}\,(\varphi))$ is:

$$\frac{\partial B_z}{\partial y} - \frac{\partial B_y}{\partial z} = \frac{\partial\left(\frac{\partial\varphi}{\partial z}\right)}{\partial y} - \frac{\partial\left(\frac{\partial\varphi}{\partial y}\right)}{\partial z} = \frac{\partial^2\varphi}{\partial y\,\partial z} - \frac{\partial^2\varphi}{\partial z\,\partial y} = 0$$ thanks to Clairaut's theorem. The same applies to the other components.

Problem 5.2) $\text{Div}(\overrightarrow{\text{curl}}) = 0$

The curl of (u, v, w) is: $\left(\frac{\partial w}{\partial y} - \frac{\partial v}{\partial z}, \frac{\partial u}{\partial z} - \frac{\partial w}{\partial x}, \frac{\partial v}{\partial x} - \frac{\partial u}{\partial y}\right)$

Then,

$$\text{div}\left(\overrightarrow{\text{curl}}(\vec{B})\right) = \frac{\partial}{\partial x}\left(\frac{\partial B_z}{\partial y} - \frac{\partial B_y}{\partial z}\right) + \frac{\partial}{\partial y}\left(\frac{\partial B_x}{\partial z} - \frac{\partial B_z}{\partial x}\right) + \frac{\partial}{\partial z}\left(\frac{\partial B_y}{\partial x} - \frac{\partial B_x}{\partial y}\right)$$

$$= \frac{\partial^2 B_z}{\partial x \partial y} - \frac{\partial^2 B_z}{\partial y \partial x} - \frac{\partial^2 B_y}{\partial x \partial z} + \frac{\partial^2 B_y}{\partial z \partial x} + \frac{\partial^2 B_x}{\partial y \partial z} - \frac{\partial^2 B_x}{\partial z \partial y} = 0$$ thanks to Clairaut's theorem.

Problem 5.3) First Maxwell Equation: $\mathrm{div}(\vec{E}) = \frac{\rho_0}{\varepsilon_0}$.

The relation (5.13) is $\nabla F = \mu_0 J$ meaning: $\frac{\partial F^{\mu\nu}}{\partial x^\mu} = \mu_0 J^\nu$. The 4 components of ∇F, respectively, are the 4 Divergences of the four 4-D vectors formed with the 4 columns of F. Hence, the first component of $\mu_0 J^\nu$, which is $\mu_0 \rho_0 c$, is equal to the divergence of the first column of F, which we denote by F°. We thus have: $\mathrm{Div}(F^\circ) = \mu_0 \rho_0 c$.

Let's expand:

$$\mathrm{Div}(F^\circ) = \mathrm{Div}(0, E_x/c, E_y/c, E_z/c) = \frac{\partial E_x}{c\partial x} + \frac{\partial E_y}{c\partial y} + \frac{\partial E_z}{c\partial z} = \mathrm{div}(\vec{E})/c.$$

So finally: $\mathrm{div}(\vec{E}) = \mu_0 \rho_0 c^2 = \frac{\rho_0}{\varepsilon_0}$ ■

Problem 5.4) Second Maxwell Equation: $\overrightarrow{\mathbf{curl}}\ \vec{\mathbf{E}} = -\frac{d\vec{\mathbf{B}}}{dt}$

With the relativistic Maxwell Equations, we still have the relation (5.14): $\vec{B} = \overrightarrow{\mathrm{Curl}}\ \vec{A}$. The divergence of a curl is always null, and hence $\mathrm{div}(\vec{B}) = 0$.

Problem 5.5) Third Maxwell Equation: $\overrightarrow{\mathbf{Curl}}\ \vec{\mathbf{B}} = \mu_0 \vec{j} + \varepsilon_0 \mu_0 \cdot \frac{d\vec{\mathbf{E}}}{dt}$

Let's again use relation (5.13), and apply it to columns 2, 3 and 4, denoted by F^1, F^2, F^3.

$$[F] = \begin{pmatrix} 0 & -E_x/c & -E_y/c & -E_z/c \\ E_x/c & 0 & -B_z & B_y \\ E_y/c & B_z & 0 & -B_x \\ E_z/c & -B_y & B_x & 0 \end{pmatrix}$$

Thus, from $\frac{\partial F^{\mu\nu}}{\partial x^\mu} = \mu_0\ J^\nu$, we have: $\mathrm{Div}\ (F^1) = \mu_0 j_x$. Let's expand:

$\mathrm{Div}(F^1) = -\frac{\partial E_x}{c^2 \partial t} + \frac{\partial B_z}{\partial y} - \frac{\partial B_y}{\partial z} = \mu_0 j_x$. The first term matches with the first component of $-\mu_0 \varepsilon_0 \frac{d\vec{E}}{dt}$. The next two terms match with $\overrightarrow{\mathrm{Curl}}\ \vec{B}$ since we have: $\overrightarrow{\mathrm{Curl}}\ \vec{B} = \frac{\partial B_z}{\partial y} - \frac{\partial B_y}{\partial z}$. Moreover, we can see that this is also the case for F^2 and F^3; hence, we have: $-\varepsilon_0\mu_0\frac{d\vec{E}}{dt}+\overrightarrow{Curl}\,\vec{B}=\mu_0\vec{j}$, so finally: $\overrightarrow{Curl}\,\vec{B}=\mu_0\vec{j}+\varepsilon_0\mu_0\frac{d\vec{E}}{dt}$ ■

Problem 5.6) Fourth Maxwell Equation: $\overrightarrow{curl}\,\vec{E}=-\frac{d\vec{B}}{dt}$.

We will use equation 5.15 and then equation 5.14: $\vec{E}=-\overrightarrow{grad}(\Phi)-\frac{d\vec{A}}{dt}$ and $\vec{B}=\overrightarrow{curl}\,\vec{A}$.

Let's apply the $\overrightarrow{curl}$ operator to equation 5.15: $\overrightarrow{curl}(\vec{E})=-\overrightarrow{curl}\left(\overrightarrow{grad}(\Phi)\right)-\overrightarrow{curl}\left(\frac{d\vec{A}}{dt}\right)$.

Besides, the Schwarz theorem enables us to write: $\overrightarrow{curl}\left(\frac{d\vec{A}}{dt}\right)=\frac{d(\overrightarrow{Curl}\vec{A})}{dt}=\frac{d\vec{B}}{dt}$.

Hence, we finally obtain: $\overrightarrow{curl}\left(\vec{E}\right)=-\frac{d\vec{B}}{dt}$ ■

Note that this demonstration is the same as in classical physics. Relativity brings a confirmation since the relations (5.14) and (5.15) are consequences of the fundamental relativistic relation $A^i=\mu_0 J^i$ with F being the "Jacobian's Dissymmetry" of **A**.

Problem 5.7) Relativistic potential when changing frame

Q1) The 4-Vector **A** is a contravariant tensor, and hence it is transformed in K′ using the Lorentzian transformation; hence: $\frac{\Phi'}{c}=\gamma_s\frac{\Phi}{c}-\gamma_s\beta A^x, A'^x=-\gamma_s\beta\frac{\Phi}{c}+\gamma_s A^x, A'^y=A^y, A'^z=A^z$.

Q2A) The 4-Vector **A** being a contravariant tensor, its Lorentzian norm is constant, which is in the context of Minkowski's frames:

$$c^2(\Phi/c)^2-(A^x)^2-(A^y)^2-(A^z)^2$$ ■

Q2B) The d'Alembertian of each component of **A** being equal to the corresponding component of **J**, if the latter is not constant, so is the former.

Problem 5.8) Schwarzschild Solution and Black Hole

Q1) The event horizon is:

$$r_s=\frac{2GM}{c_2}=\frac{2\times 6.674\times10^{-11}\times10\times1.989\times10^{30}}{2.989^2\times10^{16}}=29.538\text{ km}.$$

Q2) The norm² of the Schwarzschild time basis vector at A is: $1-\frac{r_s}{r}=0.09$. Its norm is then: 0.30. Consequently, 1 second of A's proper time is: $\frac{1}{0.3}=3.33$ units of Schwarzschild time at A.

Q3) At infinity, observers see the time in A running slower by 3.33. Hence, a photon of frequency ν_A is seen at the infinite with the frequency $V_A/3.333$.

Problem 5.9) Black Hole – Approaching the Danger – Part 1

Q1) The Newtonian gravity is:

$$-2\frac{GM}{r^3} = \frac{-2 \times 1.52 \times 10^{12}}{29{,}538} = 1.03 \times 10^8 (m/s^2)/m, \text{or } s^{-2}.$$

The result of General Relativity is even greater.

Q2) The gradient is $= -2\frac{GM}{r^3} = \frac{-2 \times 1.52 \times 10^{12}}{29{,}538} = 1.03 \times 10^8 \, s^{-2}$

This represents a gravity difference for 1 m, and the result of General Relativity is again greater. It is an enormous tide effect, absolutely unbearable by humans and which destroys virtually all objects.

Q3A) The ISCO radius is: $3 \times 29{,}538 = 88{,}614$ m.

Q3B) The Earth angular potential momentum is:

$$L = r^2\omega = \left(1.50 \times 10^{11}\right)2\frac{2\pi}{365 \times 24 \times 3{,}600} = \frac{14.13 \times 10^{22}}{3.15 \times 10^7} = 4.48 \times 10^{15} m^2/s.$$

Its constancy implies that at the ISCO radius: $\omega_{isco} = \frac{L}{r^2} = \frac{4.48 \times 10^{15}}{(88{,}614)^2} = 5.7 \times 10^5$ rad/s.

One rotation takes: $2 \times 3.14 / (5.7 \times 10^5) = 1.1 \times 10^{-5}$ second.

Q3C) The planet speed is: $2 \times 3.14 \times 88{,}614 / 1.1 \times 10^{-5} = 5.06 \times 10^{10}$ m/s. Warning: it is greater than c!

Q3D) Hence, the relativistic angular potential must be used; from the relativistic momentum, it is: $L = r\, v\, \gamma_v$. Thus, $L = 4.48 \times 10^{15}$ $m^2/s = r\, v\, \gamma = 88{,}614\, v\, \gamma$. So: $v\,\gamma = 5.05\ 10^{10}$ m/s. We set: $v = c\,(1-\varepsilon)$ and using the ultra-relativistic approximation (cf. Section 3.2.5), we obtain: $\varepsilon \approx 1.8 \times 10^{-5}$. So $v = 0.999982c$. Note that the assumption that such a planet can orbit at the ISCO radius is not realistic, as shown in the next problem.

Problem 5.10) Black Hole – Approaching the Danger – Part 2

Q1) From relation (5.40): $E_{eff} = \frac{-r_s c^2}{2r} - \frac{r_s L^2}{2r^3} + \frac{L^2}{2r^2}$. At the ISCO:

$$E_{eff} = \frac{-r_s c^2}{2 \times 3r_s} - \frac{r_s L^2}{2 \times 27\, r_s^3} + \frac{L^2}{2 \times 9\, r_s^2} = \frac{-c^2}{6} + \frac{L^2}{27\, r_s^2}.$$

Q2A) For an object to have a circular orbit, the derivative of E_{eff} with respect to r must be null for $r = 3\, r_s$. Thus:

$$0=\frac{r_s c^2}{2(3\,r_s)^2}+\frac{3r_s L^2}{2(3\,r_s)^4}-\frac{2L^2}{2(3\,r_s)^3}=\frac{c^2}{2\times 9r_s}+\frac{L^2}{2\times\,27r_s^3}-\frac{2L^2}{27r_s^3}$$
$$=\frac{3r_s^2c^2+L^2(1-2)}{2\times\,27r_s^3}. \qquad (8.1)$$

So: $L_{isco}=\sqrt{3}\ \mathbf{c}\ \mathbf{r_s}=1.732\times 3\times 10^8\times 88{,}614=4.6\times 10^{13}$ m²/s.

Having $L_{isco}=Vr\gamma=\sqrt{3}cr_s$, and $r = 3\ r_s$, we obtain: $\gamma\beta=\sqrt{3}/3$, or $\gamma^2\beta^2=\frac{1}{3}$.

Also, $\gamma^2\beta^2=\frac{\beta^2}{1-\beta^2}=\frac{1}{1/\beta^2-1}$. Hence, $1/\beta^2=4$, so finally: $\beta = ½$ ■

Q2B) The additional term compared to classical physics, $\frac{r_s L^2}{2r^3}$, generates an acceleration at the ISCO radius which is equal to the Newtonian one, as seen by the relation (8.1).

Q3) This speed is lower than the one calculated in Problem 5.9 (Q3C), and hence an object having the same angular momentum as the Earth cannot reach the ISCO, nor fall in the black hole: its centrifugal force will always be greater than gravitation.

8.6 Answers to Problems of Chapter 6

Problem 6.1) Relations between redshift, distance and scale factor.

Q1A) No, because it would mean that the expansion is linear with time.

Q1B) Such linearity is true with the ideal model of a flat universe comprising lambda only.

Q2) No, because: $a(t_{ea})\ =\frac{1}{z_a+1}=1/3$ and $a(t_{eb})=1/5$.

Problem 6.2) Relations between redshift, distance and scale factor – Part 2.

Q1A) From equation 6.31, we have: $d_{0,a}=(2\times 14.5)\times 0.42=12.18$ Billion light-years (Bly); $d_{0,b}=15.95$ Bly.

Q1B) $d_{0,c}=27.7$ Bly; $d_{0,d}=28.1$ Bly.

Q2) From equation 6.28, we have: $\left[a(t_e)\right]^{3/2} = \frac{t_e}{t_0}$, and then from equation 6.12: $t_e = t_0\left(\frac{1}{z+1}\right)^{3/2} = \frac{2}{3H_0}\times\left(\frac{1}{z+1}\right)^{3/2}$.

Hence: $t_{e,a} = 1.94$ Billion years (By); $t_{e,b} = 0.90$ By;

$t_{e,c} \approx 900{,}000$ years; $t_{e,d} \approx 319{,}000$ years.

Q3) From Remark 6.5, we have: $T_e = T_0 / a(t_e)$; so: $T_e = 3{,}503\text{K}$.

Problem 6.3) The same as Problem 6.2, but assuming a flat universe with lambda only, but for Q2), give the values for $t_0 - t_e$.

Q1A) We have: $d_{0,a} = 14.5\times2 = 29$ billion light year (Bly); $d_{0,b} = 58$ Bly.

Q1B) $d_{0,c} = 5{,}250$ Bly; $d_{0,d} = 14{,}500$ Bly.

Q2) We have: $a(t_e) = e^{H_0(t_e - t_0)} = \frac{1}{z+1}$, then $H_0(t_e - t_0) = \ln(a(t_e))$, and

then $t_0 - t_e = -\frac{1}{H_0}\left(\ln\left(\frac{1}{z+1}\right)\right)$.

Hence: $t_{e,a} = -14.5\times(-1.09) = 15.9$ Bly; $t_{e,b} = 23.3$ Bly;

$t_{e,c} \approx 90.14$ Bly; $t_{e,d} \approx 100.17$ Bly.

Q3) Same as previous problem because from the redshift, we have $a(t_e)$ independently of the model universe.

Problem 6.4) Scale factor at the main transition phases

Q1) The scale factor at the time when Dark energy density was equal to matter density

We denote the desired time by x. We have $\varepsilon_m(a) = \varepsilon_{m.0}a^{-3}$ and $\varepsilon_\Lambda(a) = \varepsilon_{\Lambda,0}$. So $\varepsilon_m(a) = \varepsilon_{m.0}a^{-3}$, then $a^3 = \frac{\varepsilon_{m,0}}{\varepsilon_{\Lambda,0}} = \frac{0.32}{0.68} = 0.470$, and

so: $a = 0.778$.

Q2) We have $\varepsilon_r(a) = \varepsilon_{r.0}a^{-4}$ and $\varepsilon_m(a) = \varepsilon_{m.0}a^{-3}$, so $\varepsilon_m(a) = \varepsilon_{m.0}a^{-3} = \varepsilon_{r.0}a^{-4}$, so finally: $a = \frac{\varepsilon_{r,0}}{\varepsilon_{m,0}} = \frac{0.009}{32} = 2.8\times10^{-4}$.

Problem 6.5) Ideal Scenario of a Flat Universe with Radiation Only

Q1) We have $d(t_0) = \int_{t_e}^{t_0} \frac{c}{a(t)}dt = c\int_{t_e}^{t_0}\left(\frac{t}{t_0}\right)^{-1/2} dt$. Let's set $u = \frac{t}{t_0}$, then $dt = du.t_0$ and $d(t_0) = ct_0\int_{t_e/t_0}^{1} u^{-1/2}du = 2ct_0\left[1-\left(\frac{t_e}{t_0}\right)^{1/2}\right]$. Besides,

$$\mathbf{a}(\mathbf{t_e}) = \frac{1}{\mathbf{Z}+1}, \text{ and } \mathbf{a}(\mathbf{t_e}) = \left(\frac{t_e}{t_o}\right)^{1/2}.$$

Hence: $d(t_0) = 2ct_0\left[1 - a(t_e)\right] = 2ct_0\left[1 - \frac{1}{1+z}\right] = \frac{c}{H_0}\frac{z}{1+z}.$

Q2) We have: $d(t_e) = d(t_0)a(t_e) = \frac{c}{H_0}\frac{z}{1+z}\frac{1}{1+z} = \frac{c}{H_0}\frac{z}{(1+z)^2}.$

Problem 6.6) Impact of the Earth Curvature

If the Earth were flat, the object width would be:
$50{,}000\ \text{m} \times 0.01 = 500\,\text{m}.$
With the Earth curvature, equation 6.3 yields: $D = d\alpha.R\sin\left(\frac{r}{R}\right).$

We will use the approximation for sin(x) when $x \to 0$: $\sin x \approx x - \frac{x^3}{6}.$

With $x = \frac{50}{6731} = 0.00742832$; and: $\frac{x^3}{6} = 6.8 \times 10^{-8}.$

Hence $D = 500\ \text{m} - 67310 \times 6.8 \times 10^{-8}\ \text{m} = 500\ \text{m} - 4.60 \times 10^{-3}\text{m}$. Thus, the object is seen larger than it is by 4.6 mm, which represents: 0.0009%.

Note

1 In our reasoning, we assume that a photo displays events that are simultaneous in K. Actually, we should take into account the travel time of the photons from these events to the camera. This bias can be overcome if the camera is at equal distance of the concerned events.

Bibliography

Abbott, B.P., et al. (2016) "Observation of Gravitational Waves from a Binary Black Hole Merger", *Phy. Rev. Lett.* 116: 061102. doi: 10.1103/PhysRevLett.116.061102

Berche, B. (2006) « *Licence de Physique et un peu après...Relativité* », Nancy, France: Université Poincaré Nancy 1.

Bergmann, P.G (1942) "*Introduction to the Theory of Relativity*", Garden City, NY: Dover Publications.

Clark, R.W. (1971) "*Einstein the Life and Times*", Cleveland, OH: The World Publishing Company.

Damour, T. (2006) "*Once upon Einstein*", or: "*Si Einstein m'était conté...* ", Flammarion, Boca Raton, FL: CRC Press.

Durandeau, J.P. and Decamps, E.A. (2000) « *Mécanique Relativiste* », Malakoff, France: Dunod.

Einstein's (1912) manuscript on "*The Special Theory of Relativity, a Fac Simile*", New York, NY: G. Braziller Publishers.

Feynman, R.P., Leighton, R.B., and Sands, M. (2011) « *The Feymann, Lectures on Physics* », Pasadena, CA: Caltech.

French, A.P (1968) "*Special Relativity*", New York, NY: WW Norton & Company.

Galison, P. (2003) "*Einstein's Clocks, Poincaré's Maps - Empires of Time*", W.W. Norton & Company.

Gourgoulhon E . (2012) "*3+1 Formalism in General Relativity: Bases of numerical Relativity.*" Heidelberg, Germany: Springer.

Halpern, P. (2016) « *Le dé d'Einstein et le Chat de Schrödinger* », Malakoff, France: Dunod.

Hawking, S. (1988) "*A Brief History of Time, from Big Bang to Black Holes*", New York, NY: Bantam Press.

Hobson, M.P., Efstathiou, G.P., and Lasenby, A.N. (2006) « *General Relativity. An Introduction for Physicists* », Cambridge: Cambridge University Press.

Iliopoulos, J. (2014) « *Une Introduction aux Origines de la Masse* », Les Ulis, France A: EDP Sciences.

Kogut, J.B. (2001) "*Introduction to Relativity*", Amsterdam, Netherlands: HAP.

Landau and Lifshitz, L.D. and E.M.. (1980) « *Theory of fields* », Amsterdam: Pergamon - Elsevier

Le Bellac, M. (2015) « *Les Relativités: Espace, Temps, Gravitation* », Les Ulis, France A: EDP Sciences.

Lévy-Leblond, J.M. (1976) "One More Derivation of the Lorentz Transformation", *Am. J. Phys.* 44, 271.

Ougarov, V. (1974) « *Théorie de la Relativité Restreinte* », Moscow: Editions de Moscou.

Penrose, R. (2004) "*The Road to Reality*", New York, NY: Vintage Books.

Resnick, R. and Halliday, D. (1992) "*Basic Concepts in Relativity and Early Quantum Theory*", Basingstoke: Mac Millan.

Rougé, A. (2002) « *Introduction à La Relativité* », Palaiseau, France: les éditions de L'école Polytechnique.

Rovelli, C. (2014) « *Et si le Temps n'existait pas?* », Malakoff, France: Dunod.
Ryden, B. (2017) "*Introduction to Cosmology*", Cambridge: Cambridge University Press.
Taylor, E.F. and Wheeler, J.A. (1992) « *Spacetime Physics* », New York, NY: W.H. Freeman & Company.
Thorne, K.S. (1994) « *Blacks Holes and Times Warps* », New York, NY: W.W. Norton & Company.
Villain, L. (2015) « *Relativité Restreinte* », Paris, France: De Boeck.

Index

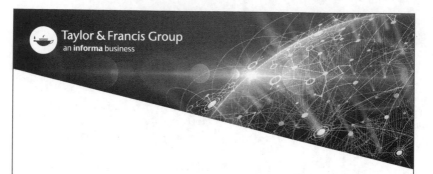

9781032056760